Methods for Studying Nucleic Acid/Drug Interactions

Methods for Studying Nucleic Acid/Drug Interactions

Meni Wanunu • Yitzhak Tor

CRC Press
Taylor & Francis Group
Boca Raton London New York

CRC Press is an imprint of the
Taylor & Francis Group, an **informa** business

CRC Press
Taylor & Francis Group
6000 Broken Sound Parkway NW, Suite 300
Boca Raton, FL 33487-2742

First issued in paperback 2018

© 2012 by Taylor and Francis Group, LLC
CRC Press is an imprint of Taylor & Francis Group, an Informa business

No claim to original U.S. Government works

ISBN-13: 978-1-4398-3973-7 (hbk)
ISBN-13: 978-1-138-38203-9 (pbk)

Library of Congress Cataloging-in-Publication Data

Methods for studying nucleic acid/drug interactions / [edited by] Meni Wanunu, Yitzhak Tor.
   p. ; cm.
   Includes bibliographical references and index.
   ISBN 978-1-4398-3973-7 (hardback : alk. paper)
   I. Wanunu, Meni. II. Tor, Yitzhak.
   [DNLM: 1. Nucleic Acids--drug effects. 2. Binding Sites. 3. Drug Discovery--methods. 4. Ligands. 5. Molecular Probe Techniques. 6. Nucleic Acids--metabolism. QU 58]

615.7'04--dc23   2011041491

Visit the Taylor & Francis Web site at
http://www.taylorandfrancis.com

and the CRC Press Web site at
http://www.crcpress.com

Contents

SECTION 1   Classical Techniques

SECTION 2   Emerging Techniques

Foreword

Although at first viewed only as molecules of heredity, our contemporary understanding of nucleic acids and the flow of information from DNA to RNA to proteins suggests much more complex genetic and regulatory roles for these biomolecules. The relatively recent discovery of RNA interference, where noncoding RNA sequences engage in gene silencing, serves as a case in point. Adding to the complexity of cellular-based processes involving nucleic acids is their intimate relationship with endogenous and exogenous small molecules. Shedding light on structural and biochemical features of nucleic acids and their ligand-binding characteristics is therefore more important than ever.

By focusing on a selection of novel and emerging techniques, Wanunu and Tor provide a remarkable overview of biophysical and computational advances in structure-based investigations of the interactions between nucleic acids and small-molecule ligands. An impressive collection of approaches is described: analytical biophysical techniques alongside novel exploitation of chemical and computational tools. Indeed, the smart combination of results obtained by classical methods with state-of-the-art biophysical approaches reveals astonishing insights, thus paving the new research avenues in this central area of research.

The biophysical methods presented in this book include capillary electrophoresis, Fourier transform infrared spectrometry (FTIR), ultraviolet (UV)–visible and circular dichroism (CD) spectroscopy, real-time surface plasmon resonance (SPR), optical tweezers, mass spectrometry (including electrospray ionization), fluorescence correlation spectroscopy (FCS), atomic force microscopy (AFM), and electron paramagnetic resonance spectroscopy (EPR). The addition of chemical approaches yields hybrid procedures, such as microarray-based two-dimensional combinatorial screening, the use of nanopore ion microscopes for single-molecule analysis, electrochemistry, relaxation kinetics analysis, and design of novel fluorescent nucleoside analogs. The inclusion of theoretical perspective and molecular modeling provides unique means for monitoring RNA–drug interactions and for multidisciplinary insights into one of the most fundamental issues in life sciences.

The vast amount of information documented in this book illustrates the changing landscape of modern nucleic acids research, where instrumental and computational advances drive the evolution of new technology. It provides younger researchers with a glimpse into nascent tools and, in many respects, inspires them to look into the future. By focusing on small-molecule binders, this book uncovers new routes that can facilitate the improvement of existing therapeutic agents and discovery of new drugs. As such, it raises the expectations for the rational design and synthesis of innovative nucleic acid–targeting drugs. Although only a handful of the techniques described in this book are likely to find their way into mainstream biophysics and drug-screening technology, this compilation is nevertheless a celebration of human creativity!

Ada Yonath
Weizmann Institute of Science
2009 Nobel Laureate in Chemistry

Introduction

Nucleic acids contain a linear sequence of nucleobases, which encode, to a first approximation, a living being's unique features. A well-regulated information transfer process dictates the function of living cells from DNA to RNA and ultimately to proteins. Defects, or interruptions in this delicate flow process, have been implicated in a wide array of disorders. From pathogenic infections to genetic disorders, nucleic acids have been identified as drug targets. Small molecules that target these biopolymers in key regions of interest can mediate cellular processes and determine a cell's fate, thereby providing therapeutic avenues. The discovery and analysis of drugs based on sequence-specific nucleic acid binders has therefore become the goal of both academia and the pharmaceutical industry.

Since most therapeutic efforts have been predominantly focused on pharmaceuticals that target proteins, there is an unmet need to develop drugs that intercept cellular pathways that critically involve nucleic acids. Progress in the discovery of nucleic acid-binding drugs naturally relies on the availability of analytical methods that assess the efficacy and nature of interactions between nucleic acids and their putative ligands. This can tremendously benefit from new methods that probe nucleic acid/ligand interactions both rapidly and quantitatively. Since a variety of novel methods for these studies have emerged in recent years, this book is intended to highlight new and nonconventional methods for exploring nucleic acid/ligand interactions. It is partly designed to present drug-developing companies with a survey of possible future techniques, as well as to highlight their drawbacks and advantages with respect to commonly used tools. Perhaps more importantly, however, this book is designed to inspire young scientists to continue and advance these methods into fruition, especially in light of current capabilities for assay miniaturization and enhanced sensitivity using microfluidics and nanomaterials.

To put new and emerging methods in perspective and provide the appropriate background, this book commences with a survey of established techniques commonly used for the study of nucleic acid/ligand interactions. Studying metal complexes as prototypical nucleic acid binders, Chapter 1 introduces several classical techniques, including crystallography, nuclear magnetic resonance (NMR) and mass spectroscopy, and optical (absorption, emission, circular dichroism (CD), and linear dichroism (LD)) and calorimetry-based techniques (ITC). Over the years, such techniques, both individually and in combination, have provided profound insight into the covalent- and noncovalent-binding modes of nucleic acid binders. Further insight into the utility of established techniques is provided in Chapter 2, which focuses on the interactions of biogenic polyamines, a family of well-studied naturally occurring ligands, with nucleic acids. In addition to spectroscopy-based techniques, the reader is exposed to electrophoresis-based techniques and computational modeling. Accessibility of many of these more classical tools to researchers in the field makes them extremely attractive for the investigation of nucleic acid/ligand interactions, and a bulk of scientific literature is available. Until new techniques as the ones described in this book slowly transition into mainstream biophysics, classical techniques will continue to serve this community.

Chapter 3 describes in detail advancements in electrospray mass spectrometry (ESI-MS) techniques for discovering and studying complexes formed between small molecules and nucleic acids. In addition to high-throughput screening, the chapter, focusing on RNA/ligand interactions, illustrates the utility of such sensitive techniques for the determination of binding constants and identification of ligand-binding sites. These techniques have proven particularly useful for identifying and analyzing the binding of low-molecular-weight ligands to bacterial and viral RNA targets. Specifically, the chapter looks at the binding of naturally occurring and synthetic ligand to the bacterial-decoding site and the hepatitis C virus (HCV) internal ribosomal entry site (IRES) element, two important drug targets of contemporary interest.

Researchers have long relied on fluorescence-based techniques to shed light on the ligand recognition features of biomolecules. While many proteins contain fluorescent aromatic amino acids, nucleic acids present a challenge, as they are practically nonemissive. This requires either labeling with established fluorophores or the development of new nucleic acid-specific probes. Not surprisingly, this book includes several chapters that detail diverse aspects of powerful fluorescence-based techniques. Chapter 5 describes the theory behind fluorescence correlation spectroscopy, a tool that can be used to study nucleic acid/ligand interactions in solution. The chapter focuses on measuring diffusion times of fluorescently labeled nucleotides at the single-molecule level, which provides insight into complex formation in extremely small volumes. Bright and photostable fluorophores are critical for the success of such experiments. As case studies the authors investigate aptamers, single-stranded oligonucleotides selected as high affinity and selective binders of either small ligands or high-molecular-weight biopolymers.

Other single-molecule techniques for investigating nucleic acid binders have been recently developed. Chapter 6 reports on the use of optical tweezers to pull on individual DNA molecules and measure their force trajectories, yielding rich and unique information on the effect of ligands on the thermodynamics and kinetics of ligand binding. Chapter 8 reports on investigations of the structure of individual DNA molecules using atomic force microscopy (AFM), a relatively young technique that has over the past three decades found its way into biology. In Chapter 12, the use of nanopores as ion microscopes that scan individual nucleic acids and detect ligand binding is described, a technique that enables label-free detection of single molecules at high throughput. The primary advantage of probing single molecules is that structural and topological features that are masked by ensembles can be observed. In addition, the ability to probe a small sample is attractive for future drug-screening applications.

End labeling of oligonucleotides is frequently employed for the analysis of nucleic acid systems. However, such approaches do not typically provide information at the nucleotide level. Minimally perturbing fluorescent nucleoside analogs that judiciously replace selected native nucleosides can provide significant insight into otherwise spectroscopically silent events such as nucleic acid dynamics, recognition, and damage. Chapter 7 discusses the development of such "isomorphic" fluorescent nucleobases and their application for the development of discovery assays of medical and diagnostic potential. Intriguingly, tuning the photophysical characteristics of novel nucleobases to match the spectral features of other established fluorophores also facilitates the implementation of FRET-based assemblies. The chapter discusses the implementation of such tools for the development of screening assays for new antibacterial and antiviral drugs.

Complementary techniques that do not rely on tagging and labeling have evolved in recent years. The most useful ones utilize surface plasmon resonance (SPR) detection. When oligonucleotides are immobilized to thin gold surfaces creating sensor chips, ligand binding can be detected as a result of mass changes and alteration of the surface refractive index, which in turn alters the SPR angle. Chapter 4 discusses the theoretical and practical aspects of this important tool. As one chip can contain several channels, and both association and dissociation rate constants can be determined, this technique is rather effective at providing a wealth of kinetic and thermodynamic information.

Electrochemical techniques are discussed in Chapter 11, where the authors employ metal and glassy carbon electrodes to measure the diffusion properties of drugs that bind to nucleic acids. Since ligand binding to the nucleic acid reduces its mobility, the electrochemical signatures of several redox active ligands can determine their bound state from free state. Also included in this book is the use of electron paramagnetic resonance (EPR) in Chapter 10, which discusses how continuous wave (CW) EPR can be used to study RNA structural dynamics, that is, how information about motion can give insight into RNA/small molecule and RNA/protein interactions.

Approaching the problem of RNA recognition from drastically different angle, Chapter 9 describes a microarray-based method that identifies RNA motifs that bind to specific small molecules. The approach, described as two-dimensional combinatorial screening (2DCS), relies on an immobilized library of small molecules that is hybridized to an RNA library, which displays discrete secondary structural elements. By simultaneously screening the RNA and ligands spaces and statistically analyzing the results, RNA motif/ligand interactions are identified. These motifs can then be used to mine cellular RNA sequences for potential drug targets. This complements rational design approaches for RNA-friendly small molecules, which are rather challenging due to our incomplete understanding of RNA/ligand interactions. The chapter then demonstrates the utility of this approach for the development of ligands that bind tightly to RNA sequences that cause myotonic muscular dystrophy.

Finally, virtually no experimental effort is unaccompanied by a theoretical counterpart, and the two have gone hand in hand throughout scientific progress. In Chapters 13 through 15, various theoretical developments that are useful for analyzing and interpreting experimental data are discussed. Chapter 13 investigates chemical kinetics with great detail, and in particular, aspects that pertain to dye binding to nucleic acid structures. Chapter 14 gives a theoretical perspective on DNA/drug interactions that discusses various developments in the field. Finally, Chapter 15 reports on a wide array of molecular dynamics studies that investigate the interactions of various ligands with nucleic acids.

We would like to thank all the authors who have taken the time and worked hard to contribute to this book and adhere to its format. Our hope is that students, amateur scientists, and perhaps established scientists will find inspiration within the pages of this book and continue to develop new and clever ways to investigate biomolecular systems. As much as we look forward to the realization of the currently emerging tools described in this book, our hopes are that future discoveries will reshape our approaches to the study of biomolecules and provide new insights into our understanding of nucleic acid/drug interactions.

Editors

Meni Wanunu completed his PhD in 2005 at the Weizmann Institute of Science, where he specialized in supramolecular chemistry, self-assembly, and nanomaterials science. Later, he was a postdoctoral fellow at Boston University and a research associate at the University of Pennsylvania, where he developed ultrasensitive synthetic nanopores for nucleic acid analysis at the single-molecule level. Currently, he is an assistant professor at the Department of Physics and the Department of Chemistry and Chemical Biology in Northeastern University, Boston. His research interests include developing chemical approaches for investigating biomolecular structure and behavior, nucleic acid mechanics and dynamics, and probing biological processes at the single-molecule level.

Yitzhak Tor completed his PhD in 1990 at the Weizmann Institute of Science. After a postdoctoral study at the California Institute of Technology (1990–1993), he joined as a faculty at the University of Chicago. In 1994, he moved to the University of California, San Diego, where he is currently a professor of chemistry and biochemistry and the Traylor Scholar in organic chemistry. His research interests are diverse and include chemistry and biology of nucleic acids, the discovery of novel antiviral and antibacterial agents, as well as the development of cellular delivery agents and fluorescent probes. He is currently the editor-in-chief of *Perspectives in Medicinal Chemistry* (http://la-press.com/journal.php?journal_id=25) and *Organic Chemistry Insights* (http://www.la-press.com/organic-chemistry-insights-journal-j104). Apart from chemistry, his interests are predominantly in music, playing, recording, and producing his own instrumental CDs (http://www.guitormusic.net).

Contributors

Jozef Adamcik
Institut de Physique des Systèmes
  Biologiques (IPSB)
Ecole Polytechnique Fédérale
  de Lausanne (EPFL)
Lausanne, Switzerland

Janice R. Aldrich-Wright
College of Health and Science
University of Western Sydney
Penrith South, Australia

P. Bourassa
Département de Chimie-Biologie
Universite du Québec á Trois-Rivières
Trois-Rivières, Canada

Robert M. Clegg
Department of Physics and Department
  of Bioengineering
University of Illinois at
  Urbana-Champaign
Urbana, Illinois

J. Grant Collins
Strathclyde Institute of
  Pharmacy and Biomedical
  Sciences
University of Strathclyde
Glasgow, United Kingdom

Giovanni Dietler
Institut de Physique des Systèmes
  Biologiques (IPSB)
Ecole Polytechnique Fédérale
  de Lausanne (EPFL)
Lausanne, Switzerland

Matthew D. Disney
Department of Chemistry
The Scripps Research Institute
Jupiter, Florida

Dariusz Ekonomiuk
Interdisciplinary Centre for
  Mathematical and Computational
  Modelling
University of Warsaw
Warsaw, Poland

Katja Eydeler
Institute for Biochemistry and
  Molecular Biology
University of Hamburg
Hamburg, Germany

Marcia O. Fenley
Department of Physics and Institute
  of Molecular Biophysics
Florida State University
Tallahassee, Florida

Richard H. Griffey
Science Applications International
  Corporation
San Diego, California

Ulrich Hahn
Institute for Biochemistry and
  Molecular Biology
University of Hamburg
Hamburg, Germany

I. Hasni
Département de Chimie-Biologie
Universite du Québec á Trois-Rivières
Trois-Rivières, Canada

Matthew R. Hicks
Department of Chemistry
University of Warwick
Coventry, United Kingdom

B. Jayaram
Department of Chemistry

and

Supercomputing Facility for
    Bioinformatics and Computational
    Biology

and

School of Biological Sciences
Indian Institute of Technology
New Delhi, India

Volker Alexander Lenski
Institute for Biochemistry and
    Molecular Biology
University of Hamburg
Hamburg, Germany

Yang Liu
Department of Chemistry
Georgia State University
Atlanta, Georgia

Eileen Magbanua
Institute for Biochemistry and
    Molecular Biology
University of Hamburg
Hamburg, Germany

Micah J. McCauley
Department of Physics
Northeastern University
Boston, Massachusetts

Manoj Munde
Department of Chemistry
Georgia State University
Atlanta, Georgia

Rupesh Nanjunda
Department of Chemistry
Georgia State University
Atlanta, Georgia

Mary S. Noé
Department of Chemistry and
    Biochemistry
University of California San Diego
La Jolla, California

Ulai Noomnarm
Center for Biophysics and
    Computational Biology
University of Illinois at
    Urbana-Champaign
Urbana, Illinois

Nikita U. Orkey
College of Health and Science
University of Western Sydney
Penrith South, Australia

A. Ahmed Ouameur
Département de Chimie-Biologie
Universite du Québec á Trois-Rivières
Trois-Rivières, Canada

Thayaparan Paramanathan
Department of Physics
Northeastern University
Boston, Massachusetts

Michelle J. Pisani
School of Physical,
    Environmental and
    Mathematical Sciences
University of New South Wales at the
    Australian Defence Force
    Academy
Canberra, Australia

Julia Romanowska
Department of Biophysics

and

Interdisciplinary Centre for
    Mathematical and Computational
    Modelling
University of Warsaw
Warsaw, Poland

Shinobu Sato
Department of Applied Chemistry
Kyushu Institute of Technology
Kitakyushu, Japan

Snorri Th. Sigurdsson
Science Institute
University of Iceland
Reykjavik, Iceland

Tanya Singh
Department of Chemistry

and

Supercomputing Facility for
    Bioinformatics & Computational
    Biology
Indian Institute of Technology
New Delhi, India

H. A. Tajmir-Riahi
Département de Chimie-Biologie
Universite du Québec á Trois-Rivières
Trois-Rivières, Canada

Shigeori Takenaka
Department of Applied Chemistry
Kyushu Institute of Technology
Kitakyushu, Japan

T. J. Thomas
The Cancer Institute of New Jersey
University of Medicine and Dentistry
    of New Jersey
New Brunswick, New Jersey

Yitzhak Tor
Department of Chemistry and
    Biochemistry
University of California San Diego
La Jolla, California

Joanna Trylska
Interdisciplinary Centre for
    Mathematical and Computational
    Modelling
University of Warsaw
Warsaw, Poland

Sai Pradeep Velagapudi
Department of Chemistry
University at Buffalo
The State University of New York
Buffalo, New York

and

Department of Chemistry
The Scripps Research Institute
Jupiter, Florida

Meni Wanunu
Department of Physics and
    Department of Chemistry and
    Chemical Biology
Northeastern University
Boston, Massachusetts

Arne Werner
Institute for Biochemistry and
    Molecular Biology
University of Hamburg
Hamburg, Germany

Nial J. Wheate
Strathclyde Institute of Pharmacy and
    Biomedical Sciences
University of Strathclyde
Glasgow, United Kingdom

Mark C. Williams
Department of Physics

and

Center for Interdisciplinary Research
  on Complex Systems
Northeastern University
Boston, Massachusetts

W. David Wilson
Department of Chemistry
Georgia State University
Atlanta, Georgia

Yun Xie
Department of Chemistry and
  Biochemistry
University of California San Diego
La Jolla, California

Section 1

Classical Techniques

1 Using Spectroscopic Techniques to Examine Drug–DNA Interactions

Matthew R. Hicks, Nikita U. Orkey, Michelle J. Pisani, J. Grant Collins, Nial J. Wheate, and Janice R. Aldrich-Wright

CONTENTS

Since the first x-ray structures of DNA revealed that the structure was sensitive to the environment from which it was crystallized, a combination of biophysical methods has been developed to probe structural changes that occur when molecules and metal complexes interact with DNA. Metal complexes that exhibit noncoordinate interactions such as intercalation and groove binding have been used to probe the structural complexity of DNA, whereas covalent binders like cisplatin are anticancer agents. Advances in spectroscopic techniques allow DNA interactions to be examined in far greater detail, yielding important structural information on binding. With this information, innovative drugs can be designed and synthesized to incorporate more than one binding characteristic to improve their therapeutic effect. In this chapter, we report on the use of spectroscopic methods to elucidate the interactions of small molecules with DNA.

1.1   INTRODUCTION

DNA has been a fascination for many researchers since the first structure was published in 1953.[1] Understanding the cellular functions of DNA such as replication, transcription, and regulation by specific protein interactions has been the focus of intense research. Small molecules that can mimic specific interactions with DNA and that can be modulated to inhibit or even activate specific functions could be used to diagnose or treat disease. In particular, transition metal complexes are excellent molecules for biophysical studies with DNA because they offer unique structural, geometrical, and spectroscopic properties. Herein we examine the use of biophysical methods to investigate the interactions of metal complexes with DNA.

The most prevalent forms of DNA are A, B, and Z, although other conformation and structural forms have been described in the literature.[2] Differences in the conformation of A, B, and Z DNA are mainly a consequence of the different sugar puckering that sets the chirality of the helix, that is, right- or left-handed. Changes in the sugar pucker from a C3'-*endo* to C2'-*endo* alter the distance between consecutive base pairs and the degree of rotation of the helix per residue. Environmental conditions can effect changes in the sugar pucker, including salt concentration, relative humidity, base pair sequence, and the molecules that bind. Electrostatic, coordinate/covalent binding, groove binding, and intercalation are examples of drug interactions that can effect structural change.

Electrostatic binding occurs when cationic molecules are attracted to the polyanionic surface of DNA. Cations and charged metal complexes such as Na^+ and $[Co(NH_3)_6]^{3+}$, respectively, associate electrostatically with DNA by forming ionic or hydrogen bonds along the outside of the DNA double helix of the G-bases.[3,4]

The anticancer drugs cisplatin (1) and oxaliplatin (2) form coordinate bonds with DNA through intrastrand and interstrand cross-links as well as protein–DNA cross-links to produce cytotoxicity.[5–9] The complex *mer*-[Ru(terpy)Cl₃] (3) (terpy = 2,2':6',2"- terpyridine, Figure 1.1) that has activity comparable to that of *cisplatin* in L1210 murine leukemia cells also forms coordinate interstrand cross-links with DNA through two consecutive guanine residues.[10]

Groove binding, in either the major or minor grooves, involves interactions between molecules that are often round or crescent shaped (Figure 1.2). Binding affinity is

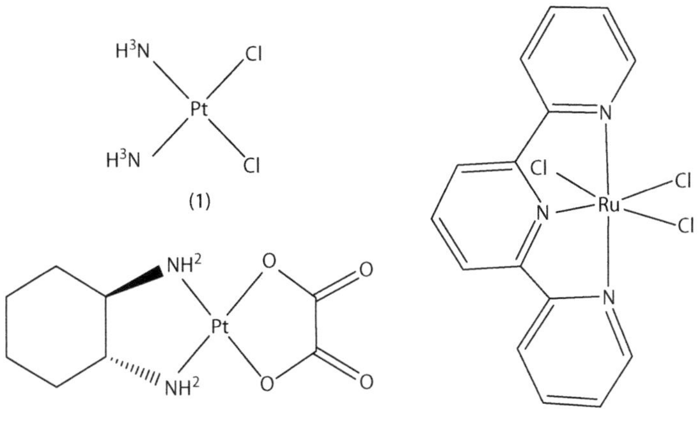

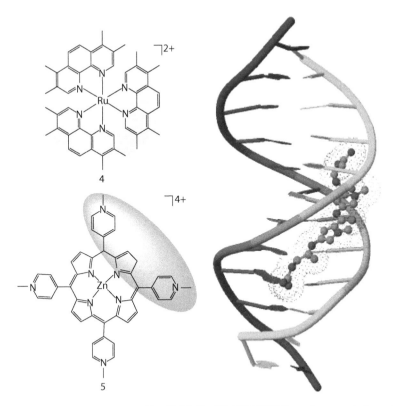

FIGURE 1.1 The coordinate DNA binding complexes: cisplatin (1), oxaliplatin (2), and *mer*-[Ru(terpy)Cl$_3$] (3).

FIGURE 1.2 Groove binders, Δ-[Ru(TMP)$_3$]$^{3+}$, [Zn(TMPYP)]$^{4+}$, and netropsin bound to B DNA. The DNA is shown with ribbons and the netropsin is colored by element and shown in the ball-and-stick representation with a dot molecular surface.[19]

dependent upon the size of the molecule, the sequences of DNA, as well as hydration and electrostatic potential of the groove. The organic antitumor drug netropsin binds within the minor groove of DNA (Figure 1.2)[11] making hydrogen bonds with adenine N-3 and thymine O-2 atoms and once bound, the groove widens slightly but not sufficiently to unwind or elongate the DNA.[11] Large octahedral complexes, such as Δ-[Ru(TMP)$_3$]$^{3+}$, which form hydrogen bonds also groove bind, favoring the major groove.[12–16] Square planar metal complexes appear to favor the minor groove, and show preference for AT sequences.[17,18]

Intercalation is described as the reversible insertion of a planar aromatic chromophore in between the base pairs of DNA. Lerman used biophysical experiments to explore intercalation,[20,21] and his initial observations have been expanded and refined over five decades.[20–23] Intercalation has been classified, such that full intercalation is the insertion of a planar aromatic chromophore between consecutive base pairs of DNA, partial intercalation is where full insertion is prevented, for example, by bulky substituents,[24,25] and quasi-intercalation is where the intercalator pushes one base pair out of the base pair stack and achieves some stacking overlap in its place with the neighboring base pairs.[26]

Intercalation is stabilized by π-electron overlap and by hydrophobic and polar interactions with polyanionic nucleic acid.[27,28] Insertion of an intercalator between adjacent base pairs produces a change in the torsion angle of the sugar–phosphate backbone. This causes a lengthening, stiffening, and unwinding of the DNA helix.[29,30] X-ray diffraction studies have demonstrated that the binding limit is one intercalator for every four nucleotides[31,32] at intervals of 10.2 Å between two intercalators.[30,33,34] This phenomena is commonly referred to as the Nearest Neighbor Exclusion Principle.[33] Figure 1.3 shows an x-ray diffraction example of the Nearest Neighbor Exclusion Principle.

1.2   SPECTROSCOPIC METHODS OF ANALYSIS

1.2.1   X-Ray Diffraction

X-ray diffraction of B-DNA, exemplified by Photo 51 taken by R. Franklin, was used to provide decisive evidence of the nature of DNA's structure.[1,35] X-ray diffraction

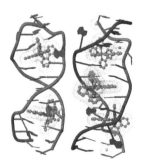

FIGURE 1.3 Examples of intercalation and Nearest Neighbor Exclusion Principle are evident in the images of the crystal structure of Δ-[Rh(bpy)$_2$(chrysi)]$^{3+}$ and d(CGGAAATTACCG)$_2$, acquired using different crystallization conditions.[36]

data remains a well-recognized method for examining DNA–drug interactions because it provides information on the DNA-binding site specificity, location, and orientation as well as any changes to the DNA helix conformation. The number of DNA x-ray crystal structures lodged with the protein data bank (>2500) is testament to the role that this technique continues to play in structure analysis. Improvements in crystallization and resolution techniques have seen ~1700 structures deposited in the past 10 years, but only a limited number of DNA/metal complex structures have been reported.[37–46]

A number of organic compounds with DNA fragments have been examined by x-ray diffraction. They include groove binders [distamycin,[47] netropsin,[19] Hoechst 33258,[46] and the polybenzamide 6-amino-1-methyl-4-(4-(4-(1-methylpyridinium-4-ylamino)phenylcarbamoyl)phenylamino) quinolinium (SN7167)[42]] and intercalators (nogalamycin[48] and proflavine[39]) (Table 1.1).

Suitable crystals of platinum complexes with DNA fragments have been produced by slow evaporation of a buffered solution,[33,49] by hanging[50] or sitting drop vapor diffusion[43,44] using a range of experimental conditions (Table 1.1).[51] Crystallization can also be promoted by the addition of cacodylic acid, glycine, polyethylene glycol (PEG), spermine, magnesium acetate, and $[Co(NH_3)_6]^{3+}$ as well as many other compounds that may be added in isolation or in concert (see Table 1.1).[37–46]

Crystal structures of octahedral metal complexes, such as $[Rh(R,R-Me_2trien)(phi)]^{3+}$, Δ-$[Rh(bpy)_2(chrysi)]^{3+}$, and $[Fe_2L_3]^{4+}$ (where R,R-Me$_2$trien = 2R,9R-2,9-diamino-4,7-diazadecane, phi = 9,10-diaminophenanthrene, bpy = 2,2′-bipyridine, chrysi = 5,6-chrysenequinone diimine, L = (NE,N′E)-4,4′-methylenebis(N-(pyridin-2-ylmethylene) aniline)) bound to DNA have also been reported.[36,38,52] $[Rh(Me_2trien)(phi)]^{3+}$ was specifically designed to intercalate, from the major groove, into the 5′-TGCA-3′ sequences of DNA. DNA-NMR and photocleavage experiments together with a crystal structure with 5′-GPU̲TGCA̲AC-3′ were used to confirm sequence-selective intercalation. This structure represented the first high-resolution view of an octahedral metal complex intercalated into the base pairs of DNA.[38] Different crystallization conditions for Δ-$[Rh(bpy)_2(chrysi)]^{3+}$ with d(CGGAAATTACCG)$_2$ resulted with two complexes intercalated in one case and three in the other (Figure 1.3 and Table 1.1).[36] A recent structure showed an iron helicate fitting perfectly into the central hydrophobic cavity of a DNA three-way junction that has not been previously reported (Figure 1.4).[37,40] This new mode of DNA recognition is without precedent but more importantly it demonstrated that a three-way DNA junction is a potential structural target for metal complexes.

1.2.2 Nuclear Magnetic Resonance Spectroscopy

Nuclear magnetic resonance (NMR) spectroscopy is arguably the most powerful and informative spectroscopic technique available for examining drug–DNA interactions, as it can provide detailed binding information under biologically relevant conditions, that is, body-like temperatures, in buffered salt solutions. These experiments can utilize single nucleotides or short, sequence-specific oligonucleotides that are of 2–12 base pairs in length. Oligonucleotides of this size are typically used because they allow good resolution of the proton resonances in ^{1}H NMR spectra of

TABLE 1.1
Drug, Action, and Mode of Binding of Some DNA-Binding Drugs

Drug	Action	Mode of Binding	PDB[a]	Conditions	Reference
Distamycin	Antitumor, antiviral	Minor groove binding	2DND	pH 6.50, vapor diffusion	47
Netropsin	Antitumor, antiviral	Minor groove binding	121D	pH 6.50, vapor diffusion	19
Hoechst 33258	Antitumor	Minor groove binding	264D	pH 7.00, vapor diffusion, sitting drop	46
SN7167	Antitumor, antiviral	Minor groove binding	328D	pH 7.00, vapor diffusion	42
Cisplatin	Anticancer antibiotic	Covalent cross-linking	3LPV	120 mM magnesium acetate, 50 mM sodium cacodylate pH 6.5, 28% w/v PEG4000, 1 mM spermine, vapor diffusion, hanging drop, at 277 K	45
Cisplatin	Anticancer antibiotic	Covalent cross-linking	1AIO	pH 6.50, vapor diffusion, sitting drop, at 277 K	44
Oxaliplatin	Anticancer antibiotic	Intrastrand cross-link	1IHH	MPD, sodium cacodylate, spermine 4HCl, $BaCl_2$, ethyl acetate, pH 7.0, vapor diffusion, hanging drop, at 277 K	50
Nogalamycin	Antitumor	Intercalation	182D	pH 6.50, vapor diffusion, sitting drop, temperature 277 K	48
Proflavine	Probe	Intercalation	3FT6	HEPES, $[Co(NH_3)_6]^{3+}$, MPD, PEG1000, pH 7.0, vapor diffusion, sitting drop, at 293 K	39
[Rh(Me$_2$trien)(phi)] and d(GPUTGCAAC)$_2$	Probe	Intercalation	454D	MPD, calcium acetate, sodium cacodylate, pH 7.0, vapor diffusion, sitting drop at 277 K	38
Δ-[Rh(bpy)$_2$(chrysi)]$^{3+}$ and d(CGGAAATTACCG)$_2$	Probe	Intercalation	3GSJ 3GSK	20 mM sodium cacodylate, 6 mM spermine, 4HCl, 40 mM KCl, 5% MPD, pH 7, vapor diffusion, at 323 K 20 mM sodium cacodylate, 6 mM spermine, 4HCl, 40 mM NaCl, 5% 2-methyl-2,4-pentanediol, pH 7, vapor diffusion, at 323 K	36
[Fe$_2$L$_3$]$^{4+}$	Metallosupramolecular helicate	Three-way junction binding	2ET0	0.08 M Mg acetate, 0.05 M TrisCl, pH 8.5, 5% PEG400, vapor diffusion, sitting drop, at 298 K	40
		Three-way junction binding	3FX8	1 μL of 10 mM cylinder, 1 μL of 3 mM DNA, 2 μL of crystallization buffer (10 mM magnesium chloride, 5% v/v isopropanol, 50 mM Tris–HCl), pH 8.5, vapor diffusion, sitting drop, at 293 K	37

a   PDB, Protein Data Bank, http://www.pdb.org/pdb/home/home.do

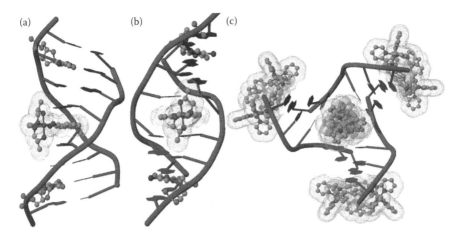

FIGURE 1.4 Images generated from x-ray data of (a) [Rh(Me$_2$trien)(phi)]$^{3+}$ intercalating into 5′-TGCA-3′ confirm sequence-selective intercalation and (b) the orthogonal view,[38] and (c) [Fe$_2$L$_3$]$^{4+}$ fitting perfectly into the central hydrophobic cavity of a DNA three way junction.[37,40]

the DNA using standard magnetic fields (300–600 MHz).[53–58] With longer oligonucle-otides, the resonances can display significant overlap making the determination of drug-binding sites, orientation, and changes to the DNA conformation difficult. The higher-field magnets (800–1000 MHz) now available, however, make experiments with large oligonucleotides more practical.

In designing drug–DNA-binding experiments, the choice of the oligonucleotide sequence can be important and the use of palindromic sequences makes the interpreta-tion of the NMR considerably easier. Palindromic oligonucleotides, for example, the Dickerson–Drew dodecamer (see Figure 1.5),[59] are symmetrical, and hence, resonances are only observed for one strand.

1.2.2.1 Assignment of the ¹H NMR Spectrum of an Oligonucleotide

Before drug–DNA binding can be studied, it is essential to assign the ¹H NMR reso-nances of the free oligonucleotide. The procedures for determining the proton chemical shift assignments and the solution conformation of oligonucleotides are well established.[53–55] Assignment of the resonances from the aromatic base protons (H8, H6, and H2) can be made from the one-dimensional spectrum. The AH2 reso-nances can be assigned from a T_1 experiment, as their T_1 time is considerably longer than that of the other aromatic resonances. Cytosine H6 peaks occur as doublets due to spin–spin coupling with the adjacent H5. Thymine H6 resonances are slightly broader than the other aromatic resonances and generally occur upfield of the other aromatic peaks, but can be unambiguously assigned in nuclear Overhauser effect

5′-GCGCAATTGCGC-3′
3′-CGCGTTAACGCG-5′

FIGURE 1.5 The Dickerson–Drew dodecamer oligonucleotide showing the palindromic nature of the sequence.

spectroscopy (NOESY) spectra (see below) due to the strong NOE to the adjacent methyl group. The specific assignment of each H6 resonance to a particular base, as well as the assignment of the adenine and guanine H8 protons (which are generally more downfield than the thymine and cytosine H6 resonances), is achieved through the analysis of NOESY spectra.

1.2.2.2 NOESY Spectra

A NOESY experiment is a two-dimensional technique that is able to determine through space interactions between protons that are within 5 Å. The intensity of the NOE is related to r^{-6}, where r is the interproton distance. Using the so-called sequential walk, it is possible to assign the resonances from the base and sugar protons from each nucleotide in a sequential manner.[53-55] For a right-handed oligonucleotide duplex, each aromatic H8 and H6 proton should exhibit an NOE to its own H1′ and H2′/H2″ sugar protons as well as to the H1′ and H2′/H2″ protons on the sugar of the nucleotide in the 5′-direction (Figure 1.6).

Furthermore, for a B-type DNA conformation, the distance between the base H8/H6 and its own H2′ is ~2 Å and ~4 Å to the H2′ proton on the flanking 5′-sugar.[55] Therefore, when the NOE cross-peak from each H8/H6 proton to its own H2′ proton is significantly larger than to the H2′ proton on the flanking 5′-sugar, it can be concluded that the oligonucleotide adopts a B-type conformation. Alternatively, for an A-type duplex, the relative distances between the H8/H6 and the H2′ protons are reversed (Figure 1.7).

1.2.2.3 COSY

A correlation spectroscopy (COSY) (double quantum filtered correlation spectroscopy (DQFCOSY), total correlation spectroscopy (TOCSY)) experiment measures through-bond ^{1}H–^{1}H coupling. The assignment of the resonances from each deoxyribose ring can be confirmed by the analysis of DQFCOSY or TOCSY spectra. Furthermore, COSY-type experiments can also be used to confirm the basic solution

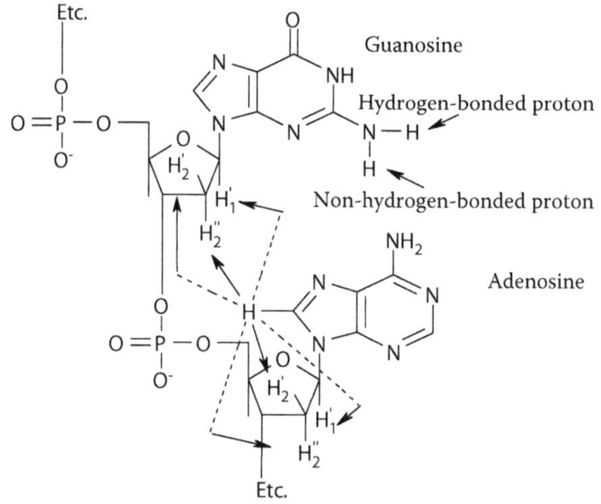

FIGURE 1.6 A schematic diagram showing the expected NOEs between the H8 of an A base to its own H1′, H2″, H2″ and to the H1′, H2′, H2″ of the base in the 5′-direction.

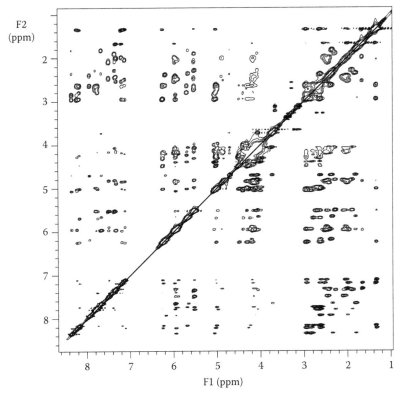

FIGURE 1.7 The full two-dimensional NOESY spectrum of the oligonucleotide d(CAATCCGGATTG)$_2$ showing the NOE cross-peaks of each resonance to itself on the diagonal, and intramolecular NOE cross-peaks between different protons on the oligonucleotide. Each side of the diagonal is a mirror image of the other side.

conformation of the oligonucleotide. For example, in normal B-type DNA, the coupling (J) and therefore the cross-peak magnitude of the H2′ resonance to the H1′ resonance is larger ($J = 10$ Hz) than for the H2″ to the H1′ ($J = 6$ Hz); however, for a Z-type DNA conformation, H2″–H1′ coupling is larger ($J = 8$ Hz) compared to the H2′–H1′ coupling ($J = 2$ Hz).

1.2.2.4   Solution Conformation

The generic solution conformation of an oligonucleotide (A-, B-, or Z-type DNA) can be established from NOESY or DQFCOSY spectra, as outlined above. A more detailed solution structure can be generated using molecular dynamics simulations that are constrained by the interproton distances obtained from NOESY spectra and torsion angles calculated from coupling constants.[60,61] Because the NOE constraints are from adjacent nucleotide residues or base pairs, the structures are less well defined than that of small proteins. The quality of the NMR-derived structure can be improved by the introduction of ^{13}C and ^{15}N labeling into the oligonucleotide or small segment of DNA. This provides additional torsion angle restraints for the molecular dynamics simulations, but again, it only aids the determination of local, rather than global, conformation.[62,63]

Owing to the development of techniques to measure and utilize residual dipolar couplings, it is now possible to obtain information on the global structure of the segment of DNA.[60,64,65] As residual dipolar couplings are a function of both distance and orientation between two nuclei, they provide additional structural data. Residual couplings average zero in isotropic solutions; however, various methods can be utilized that induce anisotropic rotational diffusion of the DNA segment that allows the measurement of the residual couplings.[64,65]

1.2.2.5 Drug–DNA Binding

The binding of a drug to DNA can be studied with the same NMR experiments used to examine free oligonucleotides. Depending upon the selectivity and kinetics of the drug binding, NMR experiments (particularly NOESY experiments) can be used to obtain a detailed high-resolution structure of the drug bound to an oligonucleotide.[66]

1.2.2.6 Covalent Drug Binding

If a drug covalently binds to an oligonucleotide at a single site, and the drug–oligonucleotide conjugate can be isolated, a full structure could be obtained from molecular dynamics simulations constrained by the proton distances obtained from NOESY spectra and torsion angles calculated from coupling constants (Figure 1.8). Alternatively, NMR spectroscopy can be used to follow the course of a reaction as a

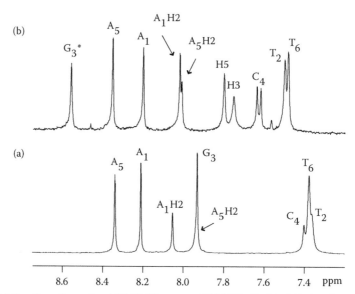

FIGURE 1.8   One-dimensional ^{1}H NMR spectra (400 MHz, D_2O) of (a) the oligonucleotide d(ATGCAT)$_2$ and (b) after reaction with one equivalent of a dinuclear platinum metal complex and HPLC purification. The spectra demonstrate the large downfield shift of the guanine H8 resonance upon binding of the platinum(II) atom at the N7 position. (Reprinted with permission from N. J. Wheate et al., *Dalton Trans.*, 2003, 3486–3492. Copyright 2003 Royal Society of Chemistry.)

function of time, or to analyze a mixture of products. For example, the covalent binding of platinum anticancer drugs to guanine bases in an oligonucleotide can be simply monitored from the chemical shift changes of the guanine H8 resonance. Direct binding of a platinum(II) atom to the guanine N7 results in the downfield shift of the adjacent H8 resonance from around 8.0 to 8.5–9 ppm depending on the solvent, temperature, and salt concentration (Figure 1.8).[67]

1.2.2.7 ^{195}Pt NMR

NMR experiments are not limited to the ^{1}H nuclei, and in instances, other nuclei can be employed. For platinum-based drugs, the formation of coordination bonds to the guanosine N7 can be directly monitored by ^{195}Pt NMR. The ^{195}Pt isotope is 33% naturally abundant, but has a receptivity compared with ^{1}H of only 3.5%. The resonances of platinum drugs can be found anywhere between 0 and −3500 ppm depending on the oxidation state of the platinum atom and the type and number of ligands attached.[68] More typically, platinum(II) drugs with oxalate, chloride, and am(m)ine ligands have resonances between −1600 and −2300 ppm. Importantly, upon binding to guanosine, these resonances shift upfield by around 100 ppm, which can be easily resolved and quantified (Figure 1.9). Platinum experiments running on medium-strength NMR magnets require between 2000 and 100,000 scans to get good signal to noise at low millimolar concentrations. If relatively high concentrations are practicable (20–30 mM), then significantly fewer scans are required (500–1000). The relatively fast-relaxing ^{195}Pt nucleus means experiments of thousands of scans can still be run in several hours. This drawback can be overcome using ^{1}H–^{195}Pt heteronuclear multiple bond correlation (HMBC) experiments where acceptable spectra can be acquired in 2 h.

1.2.2.8 HSQC Experiments

Heteronuclear single quantum coherence (HSQC) is a technique mostly applicable to platinum-based drugs and is used to examine simultaneously the hydrolysis, preassociation, and binding of a drug to DNA.[69] It is a two-dimensional experiment where one axis is ^{1}H and the other is another nucleus (typically ^{15}N). For platinum complexes that contain ^{15}N-labeled am(m)ine ligands, the HSQC cross-peak is sensitive to the ligand in the *trans*-position across from it and any hydrogen-bonding interactions in

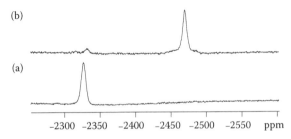

FIGURE 1.9 ^{195}Pt spectra of (a) the platinum(II) complex and (b) the approximate upfield shift of the resonance by 100 ppm upon binding to the N7 of guanosine.

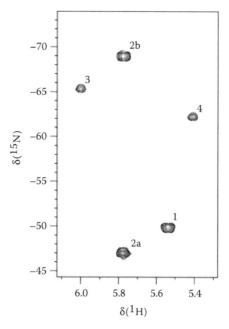

FIGURE 1.10   A two-dimensional HSQC (^{1}H–^{15}N, 90% H$_2$O/10% D$_2$O) spectrum showing simultaneous cross-peaks of a platinum drug in the (1) bis-chlorido form, (2a) amine resonance *trans* to a H$_2$O ligand, (2b) amine resonance *cis* to a H$_2$O ligand, (3) bis-aquated form, and (4) bis-hydroxy form. (Reprinted from L. Cubo et al., *Inorg. Chim. Acta*, 2009, 362, 1022–1026. Copyright 2008 with permission from Elsevier B.V.)

which the ligand is participating (Figure 1.10). Replacement of chloride or carboxylate ligands with water, hydroxide, or the N7 atom of guanosine results in significant chemical shift changes of the HSQC cross-peak.

A typical HSQC experiment will utilize a mixed solvent system of 90% H$_2$O and 10% D$_2$O so that the exchangeable protons of the am(m)ine ligands can be visualized and with minimum deuterated solvent so that a signal lock can be obtained. The HSQC technique is relatively sensitive compared to other two-dimensional experiments and can be completed with submillimolar concentrations. Measurement of the cross-peak volumes at each time point and compared to the peak volumes at zero time allows the calculation of reaction/aquation rates for all species in the experiment as they appear and disappear.

1.2.2.9   Reversible Drug Binding

NMR spectroscopy can be used to study the reversible binding (intercalation or groove binding) of a drug to an oligonucleotide. The quality of the resultant structure, however, is critically dependent upon the binding selectivity and exchange kinetics.

Where binding is reversible, the observation of resonances from the free and bound drug simultaneously (or DNA) indicates slow exchange kinetics on the NMR timescale. For groove-binding or intercalating drugs, it is rare that they bind with slow exchange kinetics (e.g., Ref. 66) and fast to intermediate kinetics is generally observed (e.g.,

Refs. 56–58). With fast or intermediate exchange kinetics, only one set of resonances is observed for the drug and oligonucleotide. The chemical shift of resonances from protons on the drug and oligonucleotide will shift as the drug is added to the oligonucleotide. The line width of the drug or DNA resonances is also indicative of binding kinetics, with increases in line width observed for intermediate exchange.

The observation of intermolecular NOEs between drug and oligonucleotide protons provides the best evidence of a drug-binding site. However, if the drug binds with fast or intermediate kinetics and at multiple-binding sites, NOEs will be detected from exchange-averaged drug resonances to a range of oligonucleotide protons. In such cases, it is not possible to obtain an accurate structure from restrained molecular dynamics simulations. Alternatively, if the drug does bind the oligonucleotide at a single site, but with selective resonances exchange broadened, it may be possible to obtain a qualitative structure. The resultant structure would be limited to some degree by imprecise distance constraints.

An indication of the drug-binding site can also be obtained from the observation of selective changes in chemical shift of resonances from the oligonucleotide as a function of added drug. For example, binding in the minor groove will generally result in selective shifts of resonances from minor groove protons (AH2, H1′, H4′, and H5′/H5″), whereas significant shifts for the base H8/H6 and sugar H2′/H2″ and H3′ resonances are indicative of major groove binding. Furthermore, if a groove-binding drug exhibits some sequence selectivity, only resonances from protons on nucleotides close to the location of binding will show significant shifts. Similarly, drugs that bind by intercalation should generally induce shifts in the proton resonances of the base pairs between which they bind.

Where the binding kinetics are fast or intermediate on the NMR timescale and where at least one drug or DNA resonance can be completely resolved, the binding constant of the drug to specific DNA sequences can be determined. At each titration point, Equation 1.1 can be used to determine the concentrations of free drug and DNA and the concentration of the drug–DNA complex,[70] where δ_{obs} is the chemical shift of the drug or DNA resonance being observed at any given titration point, δ_f is the chemical shift of the drug resonance in the absence of DNA or the chemical shift of the DNA resonance in the absence of drug, δ_b is the chemical shift of the drug resonance at the lowest observable drug: DNA ratio or the chemical shift of the DNA resonance with an excess of drug, and P_b is the population fraction of bound drug or DNA.

$$\delta_{obs} = (1 - P_b)\sigma_f + P_b\delta_b \tag{1.1}$$

The value of P_b is used to calculate the concentrations needed to solve Equation 1.2, where K_b is the binding constant, [drug] is the equilibrium concentration of free drug, [DNA] is the equilibrium concentration of free DNA, and [drug–DNA] is the equilibrium concentration of the bound drug–DNA complex.

$$K_b = \frac{[\text{drug} - \text{DNA}]}{[\text{drug}][\text{DNA}]} \tag{1.2}$$

^{1}H NMR experiments can also be used to examine the retention of Watson–Crick base pairing after drug binding. The imino protons in the Watson–Crick base pairs

G–C and A–T are exchanged with the solvent in pure D_2O and are therefore not observed in a normal ^1H NMR spectrum. If a mixed H_2O/D_2O solvent is used, however, then the imino protons can be clearly observed in the region between 10 and 14.0 ppm.[58] Disruption of the normal base pairing and/or stacking after drug binding is observed through the loss of one or more imino resonances, when compared to the unbound oligonucleotide under the same conditions.

1.2.2.10 Diffusion Experiments

Any molecule in a solution, including drugs and DNA, will travel in a random path due to Brownian motion. How fast and how far the drug or DNA moves is a function of its size, in particular its hydrodynamic radius. The self-diffusion of a drug in solution can be directly determined using NMR diffusometry using pulsed gradient spin-echo pulse sequences that are now commonly provided on modern NMR spectrometers.[71] In a typical experiment, the signal attenuation of one or more drug proton resonances is measured as a function of increasing gradient strength (usually ~16 gradient increments are employed) and the diffusion coefficient (D, $m^2 s^{-1}$) determined by plotting resonance attenuation against gradient strength.

The drug's effective hydrodynamic diameter is determined using the Stokes–Einstein equation[71]

$$D = \frac{kT}{6\pi\eta r} \tag{1.3}$$

where k is the Boltzmann's constant, T is the temperature (K), η is the viscosity of the solvent (Pa s), and r is the hydrodynamic radius (m) of the drug or DNA molecule being examined.

To calculate the binding constant of the drug, the diffusion coefficient of the drug is measured in its free form and when added to an oligonucleotide in various drug–DNA ratios.[72] A small-molecule drug in its free form will self-diffuse relatively quickly (~10^{-9} $m^2 s^{-1}$) but when bound to an oligonucleotide, it will diffuse much more slowly as it "takes on" the effective size of the oligonucleotide ($10^{-10}–10^{-12}$ $m^2 s^{-1}$). How much the drug's diffusion coefficient changes when added to the oligonucleotide will be a function of both the drug's binding constant and the drug–DNA ratio. At each drug–DNA ratio, the amount of free and bound drug can be determined from the observed diffusion coefficient

$$D_{obs} = (1 - P_b)D_f + P_b D_b \tag{1.4}$$

where D_f and D_b are the free and bound diffusion coefficients of the drug, respectively. Thus, P_b, and hence P_f, can be detected, which are then used in Equation 1.2 to determine K_b.

An NMR diffusion experiment can also be used to probe DNA conformational changes (e.g., Ref. 73). Drug binding that bends the DNA helix, for example, covalent binders and some groove binders, will result in a larger diffusion coefficient as it will have a smaller effective hydrodynamic radius. Drug binding that increases the length of the helix, for example, intercalators, will result in a smaller diffusion coefficient. While the Stokes–Einstein equation assumes that the particle being examined is spherical in shape, ellipsoids are more realistic approximations, in general, to the shapes of molecules like DNA and proteins. Modification of the equation to account for the more cylindrical/rod-like shape (i.e., a prolate ellipsoid) of DNA helices[74] allows the estimation of the oligonucleotide's length. The ratio of the friction coefficient of a geometry compared to that of the sphere of equivalent volume

$$F = \frac{f}{f_{\text{sphere}}} \tag{1.5}$$

is known as the Perrin (or shape) factor; examples are given in Table 1.2.

1.2.3 Electronic Spectroscopy

The excitation of electrons from the ground state to some excited state can give us a lot of information about molecular structure. The phenomena that are exploited include absorption, linear dichroism (LD), circular dichroism, and fluorescence. These methods give different types of information and so when they are used together they can provide a powerful toolbox with which to investigate molecular structures, including the different modes of DNA binding.

Absorbance spectroscopy is widely used for the determination of DNA purity since the absorbance ratio, A_{260}/A_{280}, should be between 1.7 and 2.0 for protein-free

TABLE 1.2
The Perrin Factor for Cylinders and Ellipsoids

Shape		Parameters	f
Sphere		$a = $ radius	$6\pi\eta a \ (= f_{\text{sphere}})$
Cylinder[a]		$d = $ diameter $l = $ length $p = l/d$ $a = l\left(\dfrac{3}{16p^2}\right)^{1/3}$	$f_{\text{sphere}} \dfrac{(2/3p^2)^{1/3}}{\ln(p) + \vartheta}$ $\vartheta = 0.312 + 0.565/p - 0.100/p^2$
Ellipsoid	Oblate	$b = $ semimajor axes $c = $ semiminor axes $p = b/c > 1$ oblate	$f_{\text{sphere}} \dfrac{\sqrt{1-p^2}}{p^{1/3} \tan^{-1}\left(\sqrt{1-p^2}/p\right)}$
	Prolate	>1 prolate $a = (b^2c)^{1/3}$ oblate $a = (bc^2)^{1/3}$ prolate	$f_{\text{sphere}} \dfrac{\sqrt{p^2-1}}{p^{1/3} \tanh^{-1}\left[\left(\sqrt{p^2-1}\right)/p\right]}$

Source: W. S. Price, In *Encyclopedia of Nuclear Magnetic Resonance*, eds. D. M. Grant and R. K. Harris, John Wiley & Sons, Chichester, 2002, pp. 364–374.

[a] This formula is only valid for p in the range of 2–20, whether the cylinders are open, closed, or capped with hemispheres.

a denotes the radius of the sphere, or for nonspherical geometries, the radius of a sphere of equivalent volume. The value of a is used to calculate the value of f_{sphere}.

DNA. However, caution should be used since DNA has a higher extinction coefficient than protein, meaning that this method is sensitive to nucleic acid contamination of protein samples but relatively insensitive to protein contamination of DNA samples.[76] This of course also means that the estimation of DNA concentration by measuring the 260 nm absorbance is also relatively unaffected by small amounts of proteins.

The characteristic UV absorbance band of DNA has a maximum around 260 nm and a minimum at 230 nm. The DNA concentration (in bases) can be calculated from the absorbance at ~260 nm. Some examples include calf thymus DNA $\varepsilon_{258} = 6600$ cm^{-1} mol^{-1} dm^3; poly[d(G–C)$_2$] $\varepsilon_{254} = 8400$ cm^{-1} mol^{-1} dm^3; poly[d(A–T)$_2$] $\varepsilon_{262} = 6600$ cm^{-1} mol^{-1} dm^3; poly[d(A)] $\varepsilon_{257} = 8600$ cm^{-1} mol^{-1} dm^3; and poly[d(T)] $\varepsilon_{264} = 8520$ cm^{-1} mol^{-1} dm^3.[77]

For metal complexes, like the ruthenium complexes^{262+}, [Ru(phen)$_3$]$^{2+}$ and [Ru(phen)$_2$(dpq)]$^{2+}$, the band systems in these spectra have been assigned. The metal-centered (MC) bands of many complexes are often dominated by more intense ligand-centered (LC) and metal-to-ligand charge transfer (MLCT) systems that are associated with the polypyridyl groups. The polypyridyl ligands, bpy, phen, and dpq, are colorless as they do not absorb in the visible region of the electromagnetic spectrum, but the ruthenium(II) polypyridyl complexes have strong and distinctive colors, as a consequence of the MLCT transitions. These MLCT transitions can be used to probe their binding interactions (Figure 1.11).

Strong absorption bands are fairly typical for metal complexes ($\varepsilon > 10,000$ M^{-1} cm^{-1}). For binding studies, longer-wavelength LC or MLCT absorption bands in the 400–500 nm (visible) region are often used. Bands of this type may be associated with

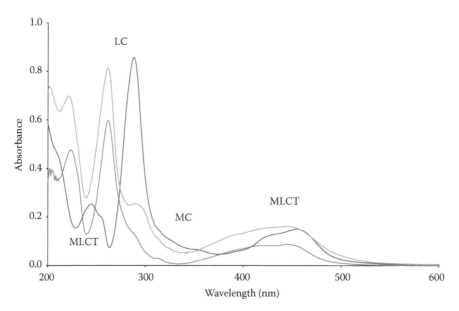

FIGURE 1.11 (**See color insert.**) UV–visible absorption spectra of [Ru(bpy)$_3$]$^{2+}$ (blue), [Ru(phen)$_3$]$^{2+}$ (red), and [Ru(phen)$_2$(dpq)]$^{2+}$ (green) in water (band assignments for [Ru(bpy)$_3$]$^{2+}$ from Campagna et al.[78]).

a cluster of excited MLCT states, which leads to temperature and solvent dependence of the fluorescence from some complexes.

Fluorescence occurs when the electrons from an excited state return to the ground state with the release of energy in the form of light. Changes in the energy of the ground state or the excited state by molecular structural changes or by environmental changes around the fluorophore can therefore change the intensity or the energy (wavelength) of the emitted light.

Different modes of DNA association can be examined through fluorescence emission spectroscopy. DNA exhibits some intrinsic fluorescence but this emission is weak, and too far down in the ultraviolet spectrum to be used. Intercalators and groove binders are often inherently fluorescent[79] and for this reason changes in the fluorescence emission spectra of these DNA binders are often used to probe their molecular interactions. For example, the DNA binding of $[Ru(phen)_2(dppz)]^{2+}$ greatly enhances the fluorescence intensities, so much so that complexes of this type are often described as a "molecular light-switch."[80] Ruthenium(II) polypyridine complexes in particular are valuable spectroscopic probes of DNA binding because of the intensity, sensitivity to chemical environment, and general lack of absorption band overlap with DNA.[10] The fluorescence emission spectrum of many ruthenium intercalators is quenched in water or buffer, but the intensity increases upon intercalation into DNA. In this case, the increase in fluorescence is a consequence of the shielding of the nitrogen atoms of dppz from aqueous environment and as such is a measure of the number of $[Ru(bpy)_2(dppz)]^{2+}$ molecules inserted into DNA.[79,81–83]

The interactions of $[Ru(phen)_2(polyamide-1)]^{2+}$, a groove binder and intercalator, can be followed using the changes in fluorescence shown in Figure 1.12. Note that the fluorescence of the Δ isomer is much higher than that of the Λ. It is apparent that there is a change in fluorescence at a ratio of around 1:1 and another change around 5 or 6:1 metal complex to DNA bases. This may correspond to intercalation at lower ratios and a change in binding mode where the polyamide binds in one of the DNA grooves by wrapping around the DNA helix. The binding constant can be determined by titration of the fluorophore into a fixed concentration of DNA or through the titration of DNA into a fixed concentration of fluorophore. Once saturation is reached, the binding constant can be determined.[84]

For nonfluorescent molecules, this can also be determined through the use of fluorescent displacement experiments. For an intercalator, a solution containing a fixed concentration of a fluorescent intercalator, such as ethidium bromide and DNA, is prepared whereupon the nonfluorescent intercalator is titrated into the solution. As the concentration of the nonfluorescent intercalator increases, it displaces the fluorescent intercalator from the DNA and the fluorescence intensity decreases until a minimum is reached.[28,85] The same strategy can be used for groove binders. Displacement experiments are to be used with some caution. The fluorescence/emission spectrum of the nonfluorescent and fluorescent intercalators together needs to be measured in order to exclude changes in the spectrum due to their association.

In summary, fluorescence emission changes in response to the environment, which can be brought about by many things, for example, changes in solvent exposure, proximity of other chemical groups on the molecule to which it binds, and the

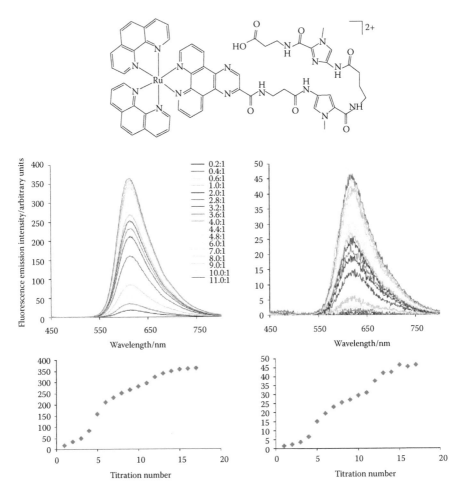

FIGURE 1.12　Fluorescence emission spectra of [Ru(phen)$_2$(polyamide-1)]$^{2+}$ showing the increase of fluorescence intensity with increasing ratio of ct-DNA concentration to intercalator. The emission spectra of Δ-[Ru(phen)$_2$(polyamide-1)]$^{2+}$ (20 × 10^{-6} M, on the left) and Λ-[Ru(phen)$_2$(polyamide-1)]$^{2+}$ (5 × 10^{+6} M, on the right) are shown as a function of successive additions of ct-DNA (300 mM). Both are in phosphate buffer (10 mM) and NaF (20 mM) with the pH adjusted to 7.0. An excitation wavelength of 430 nm was used.

conformation of the fluorophore. In addition to simple fluorescence emission experiments, it is possible to measure other properties using fluorescence techniques, for example, the fluorescence polarization anisotropy reports on the diffusion of the fluorophore. This is measured by exciting the fluorophore using polarized light and measuring the fluorescence emission in the same polarization and at right angles to this. During the fluorescence lifetime, the molecule will have changed its orientation in the solution so the polarization of the emitted light will have changed. In this way, changes in the tumbling time of the fluorophore can be measured. This is useful if the fluorophore binds to a larger molecule, thus changing its diffusion in solution.[86]

1.2.4   CIRCULAR DICHROISM

Circular dichroism spectroscopy (CD) is used to study chiral or asymmetric mole-cules. CD is simply the difference in absorbance of left- and right-handed circularly polarized light. This is usually measured by modulating the polarization of the light at around 50,000 times per second using a photoelastic modulator. A lock-in ampli-fier and signal processing can then be used to measure the difference in absorbance between the two circular polarizations of light.

CD may be used to check the chirality of a molecule or the purity of enantiomers (although to give a definitive answer the value for ε needs to be known). The CD spectra of DNA in the region that we can measure arise from the transitions in the DNA bases. However, the bases themselves are not chiral and give CD spectra because they are bound to a chiral sugar and arranged in a helical conformation. As a result of this, changes in the conformation of the DNA result in different character-istic CD spectra. For example, in Figure 1.13, we show the UV and CD spectra of calf thymus DNA and the effect of adding a molecule that intercalates between the base pairs. The intercalation changes the conformation of the DNA and the CD spectrum. In addition to the changes that occur due to the difference in the DNA conformation, there is an induced CD (ICD) signal that arises from the ligand.

CD is a sensitive probe of changes in biomacromolecular structure and may be used to probe DNA interactions with small molecules. This is also true with achiral molecules whose ICD is solely due to their interaction with DNA. Many DNA-

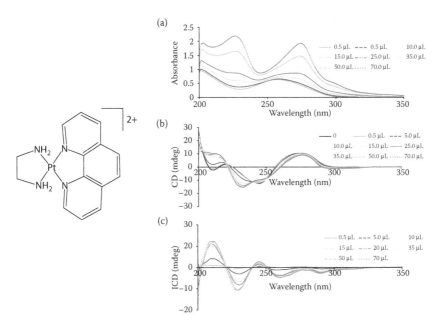

FIGURE 1.13   The absorbance (a), circular dichroism spectra (b) of ct-DNA (the solid black line indicates initial B-type DNA conformation) with increasing concentration of the interca-lator [Pt(phen)(en)]$^{2+}$, and the induced circular dichroism spectra (c) showing the net effect of intercalation on DNA.

binding molecules are achiral but upon binding to DNA they acquire an ICD that is characteristic of their interactions. The ligand also perturbs the DNA CD signal. However, it should be noted that metallocomplexes usually have transitions that occur in the same wavelength range as the DNA signals. Therefore, deductions about DNA structure based on the CD spectrum in the presence of a metal complex may not be reliable as one is viewing both DNA CD changes and metal complex CD changes overlaid.

The binding of metallointercalators to DNA usually involves a cationic metal complex binding to the anionic DNA. Before embarking on any experiment, it is important to check whether there is any precipitation, DNA condensation, or aggregation of any kind. A useful method for this is to observe the region of the absorbance spectrum at high wavelength that should have no absorbance bands. If there is a slope in this region that increases with decreasing wavelength, then this may be due to light scattering induced by particles of aggregated material. Light scattering is complex but as a qualitative diagnostic for aggregation it is sufficient to observe a change in the nonabsorbing part of the spectrum at high wavelength. Another general caveat is that we like to think of one ligand binding to DNA in one mode, such as intercalation; however, the energy difference between different DNA-binding modes may not be very large and so the possibility that multiple binding modes may be operative must be taken into account. If only one binding mode is present, then the spectrum will have isodichroic points where the spectra all cross at the same point. This is an indication that there is a two-state process occurring. This is because if there is one spectrum for state A and one for state B, then any combination of these will result in a spectrum that has the same value where the two spectra cross.

There are different units used in CD, which can be confusing. A good explanation of the units and how they are related and derived is given in Ref. 87. Many commercially available machines that measure CD report the data in millidegrees and will convert between units—some of which take into account the concentration of the sample and the path length. This allows for comparison of different concentration samples and/or different sample path lengths. It is essential that if one uses concentration in the conversion of one unit to another, then it is clearly stated what concentration is being used. It is usually the chromophore concentration but as mentioned above it is common to have both DNA and metallomolecule contributions to the spectrum so it may not be clear unless explicitly stated. We use millidegrees here.

An example of the absorbance, CD and ICD for DJ1953, a groove and covalent binder, is given in Figure 1.14. It should be noted here that we observe several isodichroic points and thus one binding mode as discussed above. This is more apparent in the ICD spectrum where the DNA spectrum has been subtracted and all of the ICD spectra cross around zero, that is, the spectral shapes are similar and are a measure of the proportion of metallomolecule that is bound to the DNA. In summary, the binding of metallomolecules to DNA can change the spectrum of the nucleic acid as well as result in CD signals from the drug. These signals may arise from the influence of the chiral environment of the DNA on the drug. For intercalators, the ICD signals that are generated by transition moments that stick out into the DNA groove tend to be positive and those that point more along the long dimension of the intercalation

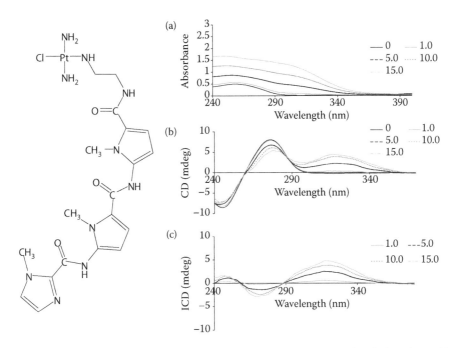

FIGURE 1.14 The absorbance (a), circular dichroism spectra (b) of ct-DNA (the solid black line indicates initial B-type DNA conformation) with increasing DJ1953 concentration, and the induced circular dichroism spectra (c) showing the net effect of groove binding on DNA.

pocket tend to be negative. However, groove binding tends to result in positive ICD signals. The reasons for this are explained in detail in Ref. 87.

1.2.5 Synchrotron Radiation Circular Dichroism

In recent years, developments in instrumentation have resulted in synchrotron radiation circular dichroism (SRCD) spectrometers suitable for examining samples under a wide range of conditions, and for the first time, spectra in aqueous solutions have been obtained to wavelengths as low as 160 nm.[84] An example of the extended wavelength range for SRCD is shown in Figure 1.15. This is due to the increased light flux produced by synchrotron radiation at lower wavelength compared to that produced by the xenon arc lamps used in benchtop spectropolarimeters. Similar advantages are gained when using synchrotron radiation for linear dichroism (SRLD) measurements. For example, see Ref. 88. The following section describes LD measurements and some of their uses in the context of DNA–drug interactions.

1.2.6 Linear Dichroism

LD measures the difference in the absorption of light linearly polarized parallel and perpendicular to an orientation axis

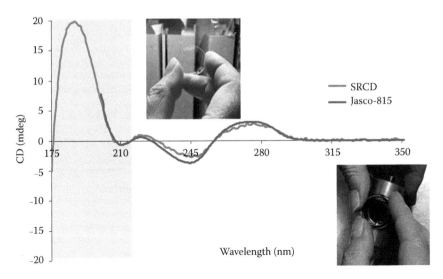

FIGURE 1.15 (**See color insert.**) The SRCD spectra of ct-DNA is shown as the solid green line (indicates B-type DNA conformation) and the CD spectra as the solid red line showing the increase in wavelength range. Insets show the cells and the cell holder required.

$$LD = A_{//} - A_{\perp} \qquad (1.7)$$

This gives us information on the orientation of absorbance transitions in the molecule. In the case of flow orientation of long molecules, the orientation axis is the long axis of the molecule. If a transition moment (the direction of net electron displacement during an electronic transition) is aligned more parallel than perpendicular to the orientation axis, a positive LD signal is observed. Conversely, if the orientation is more perpendicular than parallel to the orientation axis, a negative LD signal is observed. There are different alignment methods used, including squeezed gels, electrical, magnetic, and shear flow. For DNA alignment in solution, Couette flow is often used.

This is particularly useful for drug–DNA interactions where one is often interested in the binding mode of the drug to the DNA. Couette flow alignment is where the sample is placed in the annular gap between two concentric cylinders and one or both are rotated. This generates a shear flow and molecules that are sufficiently long and stiff align in the flow (Figure 1.16). The alignment of DNA in Couette flow is affected by many factors. As a guideline for using the microvolume Couette cells that are commercially available (Kromatek, Great Dunmow, UK), the following conditions are a good starting point: for DNA in water with salt at near physiological concentrations and the temperature around 25°C, one can obtain good LD spectra with DNA that is ~1000 base pairs or longer at concentrations of ~100 μM in base pairs or more. The sample size is around 50 μL although, with care, less can be used. Note that often the drug molecules themselves do not align in Couette flow due to their size and/or aspect ratio. This is an advantage in this case because the drug only gives an LD signal when it is aligned by binding to the DNA.

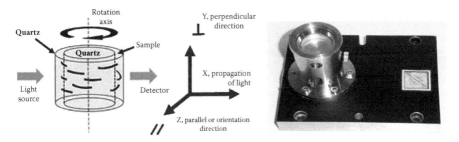

FIGURE 1.16 Schematic of a Couette flow cell showing flow orientation in a coaxial flow cell with radial incident light[96] from a laboratory fixed axis system and a standard volume Couette flow cell.

The LD signals from DNA arise mainly from the base absorbances. The bases are arranged approximately perpendicular to the helix long axis and their transition moments are in the plane of the rings of the bases. This means that the transition moments point, on average, perpendicular to the alignment direction and so give a negative LD signal (see Equation 1.7). In order to understand what the LD signals from the drug molecules mean, it is necessary to know something about the direction of the transition moments within the drug. In order to determine the direction of the transition moments in a drug molecule, one can align the molecule in a stretched film. Alignment of asymmetric molecules can be achieved using stretched films where a polymer film containing a small molecule is stretched forcing the molecule to line up with its long axis along the direction that the film is stretched.[89–95] This will be discussed in the following film LD section.

The amplitude of the LD signal depends on the angle that the transition moment makes with the alignment axis according to the following equation:

$$\text{LD} = A_{\text{iso}} \frac{3}{2} S(3\cos^2 \alpha - 1) \tag{1.8}$$

where A_{iso} is the absorbance of the nonaligned sample, S is the orientation parameter (which is 1 for a perfectly aligned transition and 0 for an unaligned one), and α is the angle that the transition moment makes with the alignment axis. This means that the LD signal gives a maximum positive signal when α is 0° (i.e., the transition polarization moment is pointing along the alignment axis) and the LD signal is maximum negative when α is 90° (i.e., the transition polarization moment is perpendicular to the alignment axis). Another consequence of this is that transitions where α is 54.7° result in zero LD signal. If one assumes that in B DNA the bases are around 86° from the helix axis, then S can be calculated from the LD signal of the bases around 260 nm.[91] Care should be taken if doing this in the presence of ligand because the base orientation may change and there may be some contribution to the LD around 260 nm if the ligand has absorbance in this region.

An example of a titration of a metallomolecule Δ-[Ru(phen)$_2$(polyamide-1)]$^{2+}$ into DNA is shown in Figure 1.17. The data shown are induced LD signal, that is, the LD signal that remains after subtraction of the signal from the DNA alone. Note that the

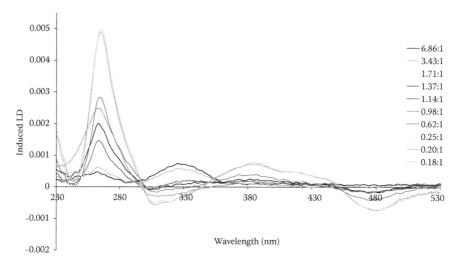

FIGURE 1.17 LD of increasing ratios of ct-DNA to Δ-[Ru(phen)$_2$(polyamide-1)]$^{2+}$. The 0.5 mm annular gap in LD cell makes a total 1 mm path length. Ratios of DNA: Δ-[Ru(phen)$_2$(polyamide-1)]$^{2+}$ were prepared from solutions of a constant ct-DNA concentration (50 µM) and aliquots of Δ-[Ru(phen)$_2$(polyamide-1)]$^{2+}$ (from a 700 µM stock solution) in phosphate buffer (10 mM) and NaF (20 mM) adjusted to pH 7.0.[52]

alignment of the DNA and therefore the contribution by the DNA to the LD signal may change at different DNA:drug ratios, so induced LD should be used only with this in mind. There are clearly two different binding modes of this molecule to DNA; at low drug loading (high DNA:drug ratio), there is a positive LD signal at 320 nm. From the film LD (see below, Figure 1.18), we know that this transition is polarized along the long axis of the drug. This means that the long axis of the drug being oriented more parallel than perpendicular to the DNA long axis. This is consistent with groove

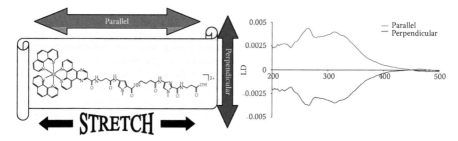

FIGURE 1.18 Molecules are aligned in a stretched film and the absorbance and LD are measured. When the film is oriented with the stretch direction in the same direction as the parallel light, the transition moments that are long-axis polarized give an LD signal that is positive. When the thin film is rotated 90°, the long-axis spectrum is equal but negative. Note that some of the minima in, for example, the parallel red spectrum may arise from short-axis polarized transitions but are not large enough to be negative on the background of large positive signals and/or a large baseline that arises from the film. The film background should be subtracted but this can be challenging and several measurements should be taken to check the reliability of the background measurements.

binding. The binding mode at high drug loading (low DNA:drug ratio) shows the opposite sign LD at 320 nm, indicating that the drug is bound in a different orientation. In addition to this, there are several other transitions that give strong LD signals, indicating that the parts of the drug that give rise to those signals have also become oriented on the DNA. The binding mode at high drug loading does not change further after a DNA:drug ratio of 0.25:1 showing that the binding sites are saturated at this ratio.

1.2.6.1   Film LD

In order to interpret the solution LD spectra, one needs to know the direction of the transition moments in the molecule. This can be measured using stretched film LD. The molecule is aligned in or on a film and the LD of the film is measured when it is placed with the stretch direction in the same direction as the parallel polarized light. The long axis of the molecule is aligned in the stretch direction so, according to Equation 1.7, long-axis polarized transitions will give a positive LD signal and short-axis polarized transitions will give a negative signal.

For a molecule to be aligned in a stretched film, it must be either an integral part of the film or associated sufficiently strongly that when the film is stretched, the molecule follows the film alignment axis. The principle of this is shown in Figure 1.18. In practice, one of two film types is appropriate for most molecules: polyethylene for nonpolar molecules (typically ligands) and polyvinyl alcohol for polar molecules (typically the metal complexes).

1.2.6.2   Alignment in Polyethylene Films

Polyethylene (PE) is well suited for orienting nonpolar molecules for spectroscopy as it has transparency in UV (above 200 nm), in the visible, and in the infrared regions. The key to success with PE film LD is the choice of PE and the degree of stretching. An example of the film LD of a ligand designed for a ruthenium metallointercalator is given in Figure 1.19. The degree of orientation in the film was high (LD^r(maximum) = 0.45) as expected for such a rod-shaped molecule.[84] The baseline is often shifted in these measurements due to the difficulty in measuring the baseline in exactly the same part of the film as the sample measurements. In this example, there is a baseline for the LD^r/10 spectrum at around 0.025 absorbance. The 230–255 nm region is essentially long-axis polarized and the 200–225 nm and 260–280 nm regions are short-axis polarized.

Polyvinylalcohol (PVA) is a good host for polar molecules; the film is transparent in the UV (above 200 nm) and visible regions of the spectrum, although it has a strong absorption over large regions of the infrared.[97] For small molecules, it has to be used in a dry form (less than a few percent water). PVA films are more difficult to prepare than PE films; however, the quality of data is often better. An example of a PVA film absorbance and LD spectrum is given in Figure 1.20.

Here we see clearly that the transitions below around 440 nm give a positive LD^r signal, indicating that they are long-axis polarized. Conversely, the higher wavelength transition around 580 nm is short-axis polarized inferred from the negative LD^r signal. Nonoverlapping transitions should give a horizontal line in the LD^r plot, which we do not see here, indicating that there are many overlapping transitions in this spectral region.

In summary, flow LD spectroscopy in combination with stretched film LD is an excellent technique for analyzing the binding of metallomolecules to DNA. It can be

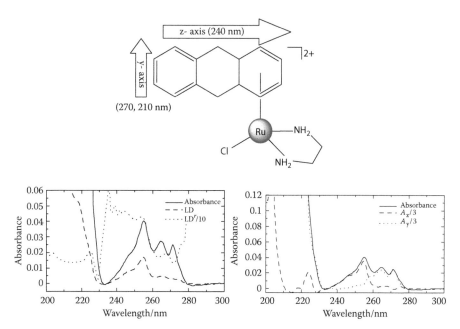

FIGURE 1.19 [(η^6-Dihydroanthracene)Ru(en)Cl]PF$_6$ indicating the energies and transition polarizations of the isolated dihydroanthracene transitions determined from the dihydroanthracene film UV–visible absorbance, LD, and LD^r spectra in a PVA-stretched film.[98]

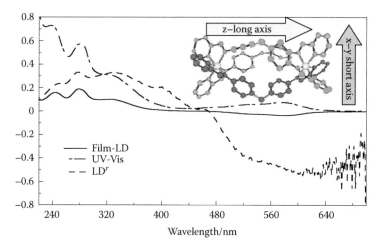

FIGURE 1.20 Absorbance, LD, and LD^r spectra for [Fe$_2$L$_3$]$^{4+}$ in PVA film prepared as described in the text.[52] Inset shows the structure of the bis iron tris-chelate helicate [Fe$_2$L$_3$]$^{4+}$ where L is (NE,N'E)-4,4'-methylenebis(N-(pyridinyl-2-methylene)aniline). (Adapted from M. J. Hannon et al., *Angew. Chem. Int. Ed.*, 2001, 40, 879–884.)

done with relatively small amounts of sample and gives information on the orientation of the molecules that can be used to elucidate the binding mode of drug molecules to DNA.

1.2.7 VISCOSITY

Viscosity is an exquisitely sensitive technique with which to measure the changes in the length of DNA and is an effective method to distinguish between the modes of DNA binding, such as coordinate/covalent binding, groove binding, and intercalation.[19,99–103] Cohen and Eisenberg[104] were the first to report the relationship between binding modes of a variety of ligands with DNA and viscosity. For molecules that form bonds with DNA, such as cisplatin, kinking/bending of the DNA helix toward either the minor or major groove (depending on the location of the ligand) produces a decrease in the DNA solution viscosity. This is generally observed as a result of reducing the end-to-end length of the DNA. Groove binders, like netropsin and Hoechst 33258, bind within the groove and do not change the end-to-end length of DNA and as a result have a very little effect on viscosity (Figure 1.21). Intercalation however results in lengthening, unwinding, and stiffening of the helix, which increases the viscosity of the DNA solution as seen for ethidium and Δ-[Ru(bpyMe$_2$)(dppz)]$^{2+}$ in Figure 1.21.[105]

For viscosity experiments, sonicated DNA (200 base pairs) is used as the DNA is almost rod-like and a change in the axial length of the helix can be easily observed. In a typical experiment, sonicated DNA, in buffer, is passed through a viscometer,

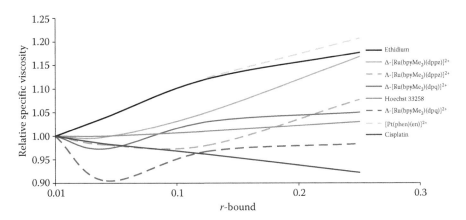

FIGURE 1.21 (**See color insert.**) A plot of *r*-bound (ratio of bound drugs per DNA nucleotide) versus $(\eta/\eta_0)^{1/3}$. An increase in viscosity is seen with intercalators like ethidium bromide, [Pt(phen)(en)]$^{2+}$,[106] a negligible effect on viscosity with the addition of the groove binder Hoechst 33258, whereas a decrease in viscosity is seen with the addition of cisplatin. Subtle difference in enantiomer binding, as the ratio increases, can also be observed. Δ-[Ru(bpyMe$_2$)$_2$dppz]$^{2+}$ increases viscosity and as such binds by intercalation; however, Λ-[Ru(bpyMe$_2$)$_2$dppz]$^{2+}$ and Δ-[Ru(bpyMe$_2$)$_2$dpq]$^{2+}$ produce a reduction in viscosity initially, due to partial intercalation, as the concentration increases, and so does the viscosity indicating that full insertion is taking place. Λ-[Ru(bpyMe$_2$)$_2$dpq]$^{2+}$ only partial intercalations even at higher concentrations.[107]

which is kept at a constant temperature by the use of a thermostatic water bath. The viscosity of the unbound DNA (η_0) is the time difference measured in seconds for the DNA in buffer (t) minus the time measured for the buffer alone (t_0):

$$\eta_0 = t - t_0 \qquad (1.9)$$

The viscosity of different ratios of metal complex and DNA solutions are prepared and measured. The viscosity (η) is the time difference measured in seconds for the DNA/metal complex sample in buffer (t) minus the time measured for the buffer alone (t_0). Viscosity is plotted against ratio bound on the x-axis (r-bound) and the relative specific viscosity on the y-axis ($\eta/\eta_0)^{1/3}$, where η is the viscosity time in seconds of the bound DNA solution (see Figure 1.21).

1.2.8 ISOTHERMAL TITRATION CALORIMETRY

Calorimetry can provide direct thermodynamic information that is essential for understanding how molecules interact with each other and aid with the design of new metal complexes to target different structural forms of DNA. Isothermal titration calorimetry (ITC) is a useful technique for studying drug–DNA interactions because it can be used to not only determine the binding affinity of the reversible association, and under physiological conditions, but also a breakdown of the thermodynamic driving forces.[108] In a typical ITC experiment, the heat absorbed or released (ΔH—the change in enthalpy) is measured as a function of added drug. From this, the binding affinity (K_a) and binding stoichiometry of the interaction can be determined. After additional manipulation of the data using the standard calorimetry, Equation 1.10 provides information on the Gibbs free energy (ΔG) and entropy (ΔS).

$$\Delta G = -RT \ln K_a = \Delta H - T\Delta S \qquad (1.10)$$

ITC has been successfully used to characterize the interaction of a huge range of biological molecules.[109] Owing to the structural changes induced in DNA by intercalative binding, it could be assumed that the thermodynamic "signature" of such binding would be significantly different compared to that where the drug binds in either the major or minor groove. Chaires was the first to propose a relationship between reversible DNA-binding modes and the observed specific thermodynamic profile.[110] In general, groove binding was found to be entropically driven, where intercalation was predominantly enthalpically driven for organic ligands (see Table 1.3).[110] Additionally, there is a favorable entropic contribution to the binding of positively charged molecules to DNA, regardless of their binding mode, as a consequence of the release of cations from the DNA structure as explained by the polyelectrolyte effect.[111,112] Studies have also been conducted to characterize other aspects of drug–DNA interactions to further supplement the binding data. As such, hydrophobic interactions between molecules have been shown to have favorable entropic processes with small enthalpic contributions to binding,[113] while interactions involving van der Waals forces tend to display large enthalpically driven processes and be entropically opposed.[114–116]

ITC has gained popularity in recent years both as a stand-alone technique to determine drug–DNA interactions and as a complement to other structural studies, in

TABLE 1.3
Thermodynamic Data of Selected Drug–DNA Interactions at 25°C

Compound	Binding Mode	ΔG (kcal mol^{-1})	ΔH (kcal mol^{-1})	$-T\Delta S$ (kcal mol^{-1})
Berenil	Groove binder	−8.0	+0.6	−8.6
DB75	Groove binder	−9.0	−2.2	−6.8
Propamidine	Groove binder	−7.0	−1.1	−5.9
Distamycin	Groove binder	−10.5	−5.8	−4.7
Netropsin	Groove binder	−8.7	−5.8	−2.9
Doxorubicin	Intercalator	−8.9	−7.4	−1.5
Propidium	Intercalator	−7.5	−6.8	−0.7
Chartreusin	Intercalator	−7.4	−7.1	+0.3
Daunorubicin	Intercalator	−7.9	−9.0	+1.1
Ethidium	Intercalator	−6.7	−9.0	+2.3

Source: Data taken from J. B. Chaires, *Arch. Biochem. Biophys.*, 2006, 453, 26–31.

particular spectrographic studies. Since each experiment directly measures the heat evolved as a result of the ligand–substrate binding, the technique is capable of measuring nanomolar-binding affinities by direct measurement and picomolar-binding affinities through competitive association experiments.[117] Unlike other techniques, it is not necessary to modify a drug (and consequently its behavior in a solution) by adding fluorescent tags or attaching it to a stationary phase for testing.[118] ITC is also particularly advantageous for samples containing colored or turbid solutions and/or precipitate suspensions.[119] Despite its many advantages, the accuracy of ITC is restricted, ironically, by its extreme sensitivity, and users must be incredibly precise with sample preparation in order to prevent large experimental errors. Regardless, ITC remains an excellent technique to obtain an entire thermodynamic profile in a single experiment. Appropriate buffers must be selected and thought given to the volatility and degradation of the buffer under the experimental conditions.[120]

1.2.9 MASS SPECTROMETRIC STUDIES

Mass spectroscopy has become more widely used in applications for large biomolecules and increasingly so since the presentation of the 2003 Nobel Prize to John Fenn and Koichi Tanaka for ion source development. The advent of soft ionization methods such as matrix-assisted laser desorption ionization (MALDI) and electrospray ionization (ESI) has revolutionized mass spectrometric analysis of biomolecules and has seen ESI-MS added to the techniques used to examine metal complex–DNA interactions. ESI-MS was used for studying oligonucleotides as early as 1988 but it was not until 1993 that the ESI-MS of duplex DNA was reported.[121,122] ESI instruments that provide "soft" ionization conditions minimize the dissociation of dsDNA but it is still necessary to carefully select the length and base sequence of the oligonucleotides to be used, and the buffer in order to ensure their stability for the process. The problem of what buffer to use has now been overcome through the use of aqueous ammonium acetate solutions, which can be adjusted to the desired pH. It is essential to choose an ammonium acetate concentration that permits the structure of dsDNA to be maintained on transfer from solution to the gas phase, without resulting in the formation of large quantities of nonspecific ammonium cation adducts. For the same reason, the concentration of other cations, such as Na^+, K^+, Mg^{2+}, and Ca^{2+}, must be minimized in the DNA sample. An ESI-MS spectrum contains several ions corresponding to multiple charge states of the same analyte; this important property enables a quick and accurate estimation of the molecular mass of large biomolecules.

Chromosomal DNA is now known to contain a variety of structures. These structures include A, B, and Z DNA; hairpins; bulge regions; and quadruplexes.[123–125] Quadruplex DNA structures, such as those in telomeres, have attracted increasing attention. Telomeres are regions at the end of chromosomes that do not code for specific proteins. In normal somatic human cells, they undergo shortening every time replication and cell division occur. Eventually, the telomeres become too short and this results in apoptosis and cell death. In contrast, the length if the telomere regions in ~85% of cancer cells are maintained by an increase in telomerase activity, rendering these cells immortal. Telomere regions and the G-quadruplex structures form has become a cellular target for anticancer drugs.[126–132]

G-quadruplex DNA is comprised of four guanine bases held together in a square planar arrangement by eight Höögsteen hydrogen bonds (G-tetrad, Figure 1.22). G-quadruplex structures include the parallel tetrameric G-quadruplex structure

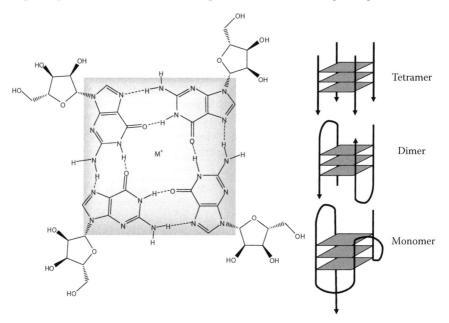

FIGURE 1.22   Structure of a G-tetrad. Schematic illustrations of different G-quadruplex structures formed from the telomere sequence. This sequence can hydrogen bond with itself (left) and fold upon itself to form quadruplex structures, each composed of three hydrogen-bonded G-quartets, a tetrameric, dimeric, and monomeric.

(tetramer, Figure 1.22), the antiparallel structure involving two DNA strands (dimer, Figure 1.22), and a single DNA form (monomer, Figure 1.22).

ESI-MS can be used to characterize quadruplex DNA and provide information about its interactions with potential drug molecules. Organic ligands, including *N,N'*-bis(2-morpholinylpropyl)-3,4,9,10-perylenetetracarboxylic acid diimide (Tel01), distamycin A, and diethylthiocarbocyanine iodide (DTC), were examined and their binding to dsDNA and the quadruplex DNA ([d(TTGGGGGT)]₄) compared.[133] Tel01 interacts with G-quadruplex DNA by stacking at the end of the G-tetrad quadruplex, whereas distamycin and DTC were shown to bind to the grooves of the quadruplex. These results corroborated the binding evidence obtained from other studies.[134,135]

1.2.9.1   Interactions between Metal Complexes and Duplex DNA

ESI-MS has been able to effectively show covalent binding of cisplatin to DNA and that sequence selectivity can be determined. This is partly because when a metal complex binds to dsDNA, it shields the nearby phosphodiester bonds from enzymatic hydrolysis. Various phosphodiesterases can be used to degrade metal/DNA complex/es into fragments that can be identified by ESI-MS.[136–138] The fragments include free nucleotides or nucleosides, depending on the enzymes used, as well as the metal complex still attached to one or two nucleotides or nucleosides, which can then be identified and the binding preference determined.[139]

Efforts to examine noncovalent-binding interactions with dsDNA using ESI-MS employed classical organic minor groove-binding agents, intercalators, and some metal complexes, including several sterically hindered metalloporphyrins and [Ru(12S4)(dppz)]²⁺ (12S4 = 1,4,7,10-tetrathiatetradecane). The metal complexes were shown to have mixed-binding modes and to form noncovalent complexes with DNA that could be characterized by ESI-MS. Experiments performed with different base sequences showed that the metalloporphyrins exhibited a preference for AT-rich DNA,[140] suggesting that metalloporphyrins function as minor groove binder whereas [Ru(12S4)(dppz)]²⁺ demonstrated a clear preference for d(GCGCGC)₂, indicating that this technique has some potential for assessing the relative DNA-binding affinities.

In experiments using a DNA 16-mer (D2, Figure 1.23) and [Ru(bpy)₂(dppz)]²⁺, the results indicate that increasing numbers of [Ru(bpy)₂(dppz)]²⁺ bind up to a maximum loading of six metal complexes per 16-mer (Figure 1.24).[141] It would seem that the Nearest Neighbor Exclusion Principle[33] can also be observed in ESI-MS experiments.

Experiments using an analogous group of ruthenium(II) complexes of the type [Ru(phen)₂(I)]²⁺ (where I = dpq, dpqC, or dppz) and three different DNA 16-mers (D1, D2, and D3, Figure 1.23) were undertaken to determine if binding affinity and preferences could be compared using ESI-MS (Figure 1.24).[142] It was expected that the degree to which they were attracted to DNA would be modulated by the size of the intercalator. The 16-mers, D1, D2, and D3 (Figure 1.23) with different central sequences were used to assess if a preference for binding mode or sequence could be determined. It was anticipated that D1 and D2 would be more attractive to intercalating metal complexes due to the greater GC content, whereas D3 would facilitate groove binding.[143–146] DNA ratios, ranging from 1:1 up to 30:1, were used.

The results clearly indicated that [Ru(phen)₂(dpqC)]²⁺ and [Ru(phen)₂(dppz)]²⁺ have significantly higher overall binding affinities toward D2 than either [Ru(phen)₂(dpq)]²⁺

D1   5'-CCT<u>CGGCCGG</u>CCGACC-3'
      3'-CCA<u>GCCGGCCG</u>GCTGG-5'

D2   5'-CCT<u>CATGGCC</u>ATGACC-3'
      3'-GGA<u>GTACCGGT</u>ACTGG-5'

D3   5'-CCT<u>CAAAATTT</u>TGACC-3'
      3'-GGA<u>GTTTTAAA</u>ACTGG-5'

FIGURE 1.23   The palindromic oligonucleotide D1, D2, and D3.

or $[Ru(phen)_3]^{2+}$. The overall affinity toward D2 decreased in the following order: $[Ru(phen)_2(dppz)]^{2+} \sim [Ru(phen)_2(dpqC)]^{2+} > [Ru(phen)_2(dpq)]^{2+} > [Ru(phen)_3]^{2+}$, which is in agreement with other spectroscopic techniques (Figure 1.25).[147,148]

ESI-MS can also provide information on the DNA sequence selectivity of a specific metal complex. This can be accomplished by plotting the relative abundances

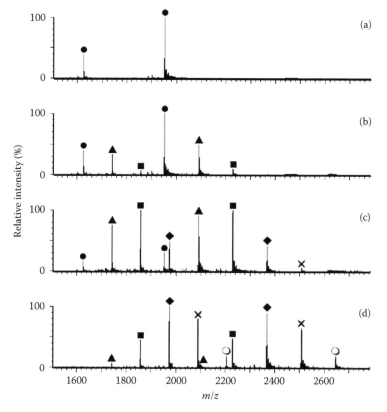

FIGURE 1.24   Negative ion ESI mass spectra of solutions containing different $[Ru(bpy)_2(dppz)]^{2+}$:D2 ratios: (a) D2; (b) Ru:D2 = 1:1; (c) Ru:D2 = 3:1; (d) Ru:D2 = 6:1. ● D2; ▲ D2 + 1$[Ru(bpy)_2(dppz)]^{2+}$; ■ D2 + 2$[Ru(bpy)_2(dppz)]^{2+}$; ◆ D2 + 3$[Ru(bpy)_2(dppz)]^{2+}$; × D2 + 4$[Ru(bpy)_2(dppz)]^{2+}$; ○ D2 + 5$[Ru(bpy)_2(dppz)]^{2+}$.

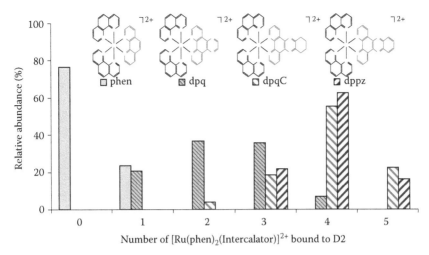

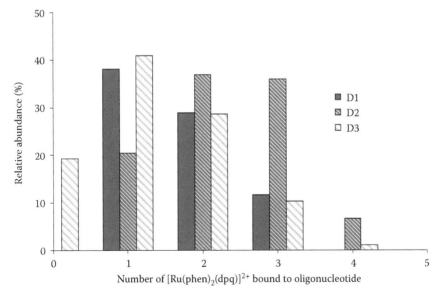

FIGURE 1.25 Relative abundances of complexes present in solutions containing a 6:1 ratio of [Ru(phen)$_2$(intercalator)]$^{2+}$ and D2.[149] The ruthenium complexes used for the binding experiments are included and the intercalator in each case is colored red.

of ions from solutions containing the same ratio of the metal complex and DNA molecules with different base sequences. Relative abundances obtained from ESI mass spectra can be used to determine the order of relative binding affinities for metal complexes. These are calculated by summing the absolute intensities of all ions assigned to a specific noncovalent complex in a given spectrum, and dividing by the combined intensity of all ions present to convert them to relative abundances. Figure 1.26 shows

FIGURE 1.26 Relative abundances of [Ru(phen)$_2$(dpq)]$^{2+}$ present in solutions in a 6:1 ratio with different oligonucleotides.[150]

that the relative abundance of $[Ru(phen)_2(dpq)]^{2+}$ with DNA was significantly greater when the DNA duplex involved was D2, than when it was either D1 or D3. In comparison with DNA-NMR experiments conducted on $[Ru(phen)_2(dpq)]^{2+}$ [150,151] and $[Ru(phenMe_2)_2(dpq)]^{2+}$ [152] and the hexanucleotide $d(GTCGAC)_2$, results suggested that intercalation was favored between the TC/GA site but this is not exclusive.[152] It is clear that both NMR and MS indicate that GC is less favored.

ESI-MS can determine the number, relative amounts, and stoichiometry of the noncovalent metal complexes interacting with DNA. Binding constants, obtained in this way, have been shown to be comparable to those obtained for the same systems by fluorescence spectroscopy[153] but the binding mode cannot be determined by ESI-MS. Other techniques such as x-ray, NMR spectroscopy, or relative viscosity measurements are required to offer information about mode for binding.[148,154]

1.2.9.2   Interactions between Metal Complexes and Quadruplex DNA

Several investigations[133,155–160] have focused on organic ligand binding to quadruplex DNA; however, recently, ESI-MS in combination with CD spectroscopy was used to examine the quadruplex $(TTGGGGGT)_4$ (Q5), and with metal complexes such as $[Ru(phen)_2(intercalator)]^{2+}$ (where intercalator = phen, dpq, dpqC, or dppz) or one of several platinum(II) intercalating complexes, including $[Pt(5,6-Me_2phen)(S,S-dach)]^{2+}$ (56MESS).[161] Both ruthenium and platinum complexes interacted with Q5 and in every case the resulting ions consisted of one or more intact molecules of the metal complex, and four ammonium ions, bound to Q5. This observation strongly suggests that the interaction does not affect the G-quadruplex structure of Q5.

These experiments showed that solutions containing the same ratio of different ruthenium complexes and Q5 gave similar spectra.[161] In this case, extending the intercalator has no effect on overall binding affinity with G-quadruplex DNA. The ruthenium complexes $[Ru(phen)_2(intercalator)]^{2+}$ (intercalator = phen, dpq, dpqC, or dppz) have a significantly higher binding affinity toward D2 than Q5. A similar conclusion was reached for the platinum(II) complexes. These differences were attributed to variations in secondary structure between duplex and quadruplex DNA molecules. These results further emphasize the role that ESI-MS can play in understanding differences between the binding interactions used by potential drugs. The unique features of ESI-MS, combined with its very high sensitivity, speed, ease of analysis, and the prospect of acquiring further information about mechanism and binding sites, would affirm that ESI-MS is one of the most useful tool available to the researcher.

1.3   CONCLUSIONS

Developing an understanding of the physical and chemical principals involved in DNA recognition has fascinated researchers for many years. Spectroscopic techniques are important methods that can be used to elucidate these principles. The use of a variety of techniques, both in isolation and in combination, has already allowed for a great deal to be learnt about how organic compounds and metal complexes interact with DNA using these spectroscopic tools.

X-ray diffraction data can provide the information that allows the structural forms of DNA to be revealed. X-ray structures of metal complexes with DNA afford unique information about these interactions; we have also seen that experimental conditions and DNA sequence can have an effect on the resulting interactions. NMR spectroscopy provides detailed binding information under biologically relevant conditions using single nucleotides or short, sequence-specific oligonucleotides in solution. NMR experiments such as ^{1}H, NOESY, COSY, RDC, HSQC, DOSY, ^{195}Pt, or diffusion can generate comprehensive structural information about metal complex–DNA interactions when used in combination.

Electronic spectroscopy, including absorption, fluorescence, CD, LD, SRCD, and SRLD, provides information about molecular structure. When techniques are used together they can offer effective methods to interrogate the different modes of DNA binding. Absorbance spectroscopy measures the change in absorption as a function of frequency or wavelength during a binding experiment, whereas fluorescence emission spectroscopy measures changes in response to the environment around a fluorophore. CD and SRCD spectroscopy are sensitive to changes in biomacromolecular structure and the interactions of metal complexes with that structure. Spectra can be used to characterize these interactions and can be used together with other spectroscopic data to understand the binding interaction more comprehensively. LD spectra inform on the orientation of a metal complex with respect to the biomacromolecule and this is used to determine the mode of binding. Viscosity measures the changes in DNA length and can be used to distinguish between DNA-binding modes.

ITC enables a comprehensive thermodynamic profile to be obtained for a binding interaction. There is a delicate balance between the favorable contributions to binding and the unfavorable that can affect an interaction. However, these contributions are not apparent from structural techniques, such as x-ray crystallography, and this demonstrates the essential role that thermodynamics plays in understanding molecular interactions.

ESI-MS offers high sensitivity, speed, and ease of analysis but does not distinguish between intercalation, electrostatic bonding, or "groove binding." However, ESI-MS can be used to probe both covalent and noncovalent systems, providing information about the covalent adducts formed in a metal complex:DNA reaction mixture or how many noncovalent metal complexes bind to DNA.

Spectroscopic techniques are some of the most versatile and powerful tools available to the bioinorganic chemist. When used in combination, they can produce a comprehensive model of binding and afford a better understanding of the physical and chemical principles involved in DNA recognition.

REFERENCES

1. J. D. Watson and F. H. C. Crick, *Nature*, 1953, 171, 737–738.
2. R. R. Sinden, *DNA Structure and Function*, Academic Press, Sydney, 1994.
3. R. V. Gessner, G. J. Quigley, A. H. J. Wang, G. A. Van der Marel, J. H. Van Boom, and A. Rich, *Biochemistry*, 1985, 24, 237–240.
4. W. J. O'Sullivan and D. D. Perrin, *Biochemistry*, 1964, 3, 18–26.
5. A. Eastman, *Chem. Biol. Interact.*, 1987, 61, 241–248.
6. A. Eastman, *Pharmacol. Ther.*, 1987, 34, 155–166.

7. P. Perego, L. Gatti, C. Caserini, R. Supino, D. Colangelo, R. Leone, S. Spinelli, N. Farrell, and F. Zunino, *J. Inorg. Biochem.*, 1999, 77, 59–64.
8. H. Rauter, R. Di Domenico, E. Menta, A. Oliva, Y. Qu, and N. Farrell, *Inorg. Chem.*, 1997, 36, 3919–3927.
9. P. K. Wu, Y. Qu, B. van Houten, and N. Farrell, *J. Inorg. Biochem.*, 1994, 54, 207–220.
10. P. M. van Vliet, S. M. S. Toekimin, J. G. Haasnoot, J. Reedijk, O. Novakova, O. Vrana, and V. Brabec, *Inorg. Chim. Acta.*, 1995, 231, 57–64.
11. M. L. Kopka, C. Yoon, D. Goodsell, P. Pjura, and R. E. Dickerson, *J. Mol. Biol.*, 1985, 183, 553–563.
12. J. K. Barton, L. A. Basile, A. Danishefsky, and A. Alexandrescu, *Proc. Natl. Acad. Sci. U.S.A.*, 1984, 81, 1961–1965.
13. J. K. Barton, A. Danishefsky, and J. M. Goldberg, *J. Am. Chem. Soc.*, 1984, 106, 2172–2176.
14. J. K. Barton and A. L. Raphael, *J. Am. Chem. Soc.*, 1984, 106, 2466–2468.
15. J. K. Barton and A. L. Raphael, *Proc. Natl. Acad. Sci. U.S.A.*, 1985, 82, 6460–6464.
16. L. A. Basile, A. L. Rapheal, and J. K. Barton, *J. Am. Chem. Soc.*, 1987, 109, 7550–7551.
17. C. A. Franklin, J. V. Fry, and J. G. Collins, *Inorg. Chem.*, 1996, 35, 7541–7545.
18. H. Xu, Y. Liang, P. Zhang, F. Du, B.-R. Zhou, J. Wu, J.-H. Liu, Z.-G. Liu, and L.-N. Ji, *J. Biol. Inorg. Chem.*, 2005, 10, 529–538.
19. L. Tabernero, N. Verdaguer, M. Coll, I. Fita, G. A. van der Marel, J. H. van Boom, A. Rich, and J. Aymami, *Biochemistry*, 1993, 32, 8403–8410.
20. L. S. Lerman, *J. Mol. Biol.*, 1961, 3, 18–30.
21. L. S. Lerman, *J. Mol. Biol.*, 1961, 3, 634–639.
22. H. M. Berman and P. R. Young, *Ann. Rev. Biophys. Bioeng.*, 1981, 10, 87–114.
23. G. Doherty and W. J. Pigram, *Crit. Rev. Biochem.*, 1982, 246, 103–132.
24. M. Cusumano, M. Letizia, and A. Giannetto, *Inorg. Chem.*, 1999, 38, 1754–1758.
25. M. Cusumano, M. L. D. Petro, A. Giannetto, F. Nicolo, and E. Rotondo, *Inorg. Chem.*, 1998, 37, 563–568.
26. P. Lincoln and B. Norden, *J. Phys. Chem. B*, 1998, 102, 9583–9594.
27. J. R. Aldrich-Wright, I. D. Greguric, C. H. Y. Lau, P. Pellegrini, and J. G. Collins, *Rec. Res. Dev. Inorg. Chem.*, 1998, 1, 13–37.
28. M. Howe-Grant, K. C. Wu, W. R. Bauer, and S. J. Lippard, *Biochemistry*, 1976, 15, 4339–4346.
29. M. Cusumano, M. L. D. Petro, A. Giannetto, and P. A. Vainiglia, *J. Inorg. Biochem.*, 2005, 99, 560–565.
30. K. W. Jennette, J. T. Gill, J. A. Sadownick, and S. J. Lippard, *J. Am. Chem. Soc.*, 1976, 98, 6159–6168.
31. V. Luzzatti, F. Masson, and L. S. Lerman, *J. Mol. Biol.*, 1961, 3, 634–639.
32. R. H. Sarma, *Nucleic Acid Geometry and Dynamics*, Pergamon Press, New York, 1980.
33. P. J. Bond, R. Langridge, K. W. Jennette, and S. J. Lippard, *Proc. Natl. Acad. Sci. U.S.A.*, 1975, 72, 4825–4829.
34. S. J. Lippard, *Acc. Chem. Res.*, 1978, 11, 211–217.
35. R. Franklin and R. G. Gosling, *Nature*, 1953, 171, 740–741.
36. B. M. Zeglis, V. R. C. Pierre, J. T. Kaiser, and J. K. Barton, *Biochemistry*, 2009, 48, 4247–4253.
37. D. Boer, J. Kerckhoffs, Y. Parajo, M. Pascu, I. Usón, P. Lincoln, M. Hannon, and M. Coll, *Angew. Chem. Int. Ed. Engl.*, 49, 2336–2339.
38. C. L. Kielkopf, K. E. Erkkila, B. P. Hudson, J. K. Barton, and D. C. Rees, *Nat. Struct. Mol. Biol.*, 2000, 7, 117–121.
39. T. Maehigashi, O. Persil, N. V. Hud, and L. D. Williams, *PDB No. 3FT6*, 2010.
40. A. Oleksi, A. G. Blanco, R. Boer, I. Usón, J. Aymamí, A. Rodger, M. J. Hannon, and M. Coll, *Angew. Chem. Int. Ed. Engl.*, 2006, 45, 1227–1231.

41. S. E. Sherman, D. Gibson, A. H.-J. Wang, and S. J. Lippard, *Science*, 1985, 230, 412–417.
42. C. J. Squire, G. R. Clark, and W. A. Denny, *Nucleic Acids Res.*, 1997, 25, 4072–4078.
43. P. M. Takahara, C. A. Frederik, and S. J. Lippard, *J. Am. Chem. Soc.*, 1996, 118, 12309.
44. P. M. Takahara, A. C. Rosenzweig, C. A. Frederick, and S. J. Lippard, *Nature*, 1995, 377, 649–652.
45. R. C. Todd and S. J. Lippard, *J. Inorg. Biochem.*, 2010, 104, 902–908.
46. M. C. Vega, I. GarcÍA SÁEz, J. AymamÍ, R. Eritja, G. A. Van Der Marel, J. H. Van Boom, A. Rich, and M. Coll, *Eur. J. Biochem.*, 1994, 222, 721–726.
47. M. Coll, C. A. Frederick, A. H. Wang, and A. Rich, *Proc. Natl. Acad. Sci. U.S.A.*, 1987, 84, 8385–8389.
48. C. K. Smith, G. J. Davies, E. J. Dodson, and M. H. Moore, *Biochemistry*, 1995, 34, 415–425.
49. Y.-S. Wong and S. J. Lippard, *J. Chem. Soc. Chem. Commun.*, 1977, 824–825.
50. B. Spingler, D. A. Whittington, and S. J. Lippard, *Inorg. Chem.*, 2001, 40, 5596–5602.
51. A. H. J. Wang, J. Nathans, G. van der Marel, J. H. van Boom, and A. Rich, *Nature*, 1978, 276, 471–474.
52. M. J. Hannon, V. Moreno, M. J. Prieto, E. Molderheim, E. Sletten, I. Meistermann, C. J. Isaac, K. J. Sanders, and A. Rodger, *Angew. Chem. Int. Ed.*, 2001, 40, 879–884.
53. R. M. Scheek, R. Boelens, N. Russo, J. H. v. Boom, and R. Kaptein, *Biochemistry*, 1984, 23, 1371–1376.
54. J. Feigon, W. Leupin, W. A. Denny, and D. R. Kearns, *Biochemistry*, 1983, 22, 5943–5951.
55. D. J. Patel, L. Shapiro, and D. Hare, *J. Biol. Chem.*, 1986, 261, 1223–1229.
56. C. M. Dupureur and J. K. Barton, *J. Am. Chem. Soc.*, 1994, 116, 10286–10287.
57. N. J. Wheate and J. G. Collins, *J. Inorg. BioChem.*, 2000, 78, 313–320.
58. N. J. Wheate, S. M. Cutts, D. R. Phillips, J. R. Aldrich-Wright, and J. G. Collins, *J. Inorg. Biochem.*, 2001, 84, 119–127.
59. H. R. Drew, R. M. Wing, T. Takano, S. Tanaka, K. Itakura, and R. E. Dickerson, *Proc. Natl. Acad. Sci. U.S.A.*, 1981, 78, 2179.
60. A. Vermeulen, H. Zhou, and A. Pardi, *J. Am. Chem. Soc.*, 2000, 122, 9638–9647.
61. H. Zhou, A. Vermeulen, F. M. Jucker, and A. Pardi, *Biopolymers*, 1999, 52, 168–180.
62. J. Van Wijk, B. D. Huckriede, J. H. Ippel, and C. Altona, *Method Enzymol.*, 1992, 211, 286–306.
63. D. P. Zimmer, J. P. Marino, and C. Griesinger, *Magn. Reson. Chem.*, 1996, 34, S177–S186.
64. N. Tjandra and A. Bax, *Science*, 1997, 278, 1111–1114.
65. N. Tjandra, S.-i. Tate, A. Ono, M. Kainosho, and A. Bax, *J. Am. Chem. Soc.*, 2000, 122, 6190–6200.
66. B. P. Hudson, C. M. Dupureur, and J. K. Barton, *J. Am. Chem. Soc.*, 1995, 9379–9930.
67. N. J. Wheate, B. J. Evison, A. J. Herlt, D. R. Phillips, and J. G. Collins, *Dalton Trans.*, 2003, 3486–3492.
68. B. M. Still, P. G. Anil Kumar, J. R. Aldrich-Wright, and W. S. Price, *Chem. Soc. Rev.*, 2007, 36, 665–686.
69. S. J. Berners-Price, L. Ronconi, and P. J. Sadler, *Prog. Nuc. Mag. Res. Spec.*, 2006, 49, 65–98.
70. M. Eriksson, M. Leijon, C. Hiort, B. Norden, and A. Graslund, *Biochemistry*, 1994, 33, 5031–5040.
71. Y. Cohen, L. Avram, and L. Frish, *Angew. Chem. Int. Ed.*, 2005, 44, 520–554.
72. W. S. Price, *Aust. J. Chem.*, 2003, 56, 855–860.
73. W. S. Price, F. Tsuchiya, C. Suzukib, and Y. Arata, *J. Biomol. NMR*, 1999, 13, 113–117.
74. A. M. Krause-Heuer, N. J. Wheate, W. S. Price, and J. Aldrich-Wright, *Chem. Commun.*, 2009, 1210.

75. W. S. Price, In *Encyclopedia of Nuclear Magnetic Resonance*, eds. D. M. Grant, and R. K. Harris, John Wiley & Sons, Chichester, 2002, pp. 364–374.
76. J. Glasel, *BioTechniques*, 1995, 18, 62–63.
77. I. Meistermann, V. Moreno, M. J. Prieto, E. Moldrheim, E. Sletten, S. Khalid, P. M. Rodger, J. C. Peberdy, C. J. Isaac, A. Rodger, and M. J. Hannon, *Proc. Natl. Acad. Sci. U.S.A.*, 2002, 99, 5069–5074.
78. S. Campagna, F. Puntoriero, F. Nastasi, G. Bergamini, and V. Balzani, In *Photochemistry and Photophysics of Coordination Compounds I*, Springer-Verlag Berlin, Berlin, Germany, 2007, pp. 117–214.
79. J. R. Lakowicz, *Principles of Fluorescence Spectroscopy*, Kluwer Academic/Plenum Publishers, New York, 1999.
80. Y. Sun, D. A. Lutterman, and C. Turro, *Inorg. Chem.*, 2008, 47, 6427–6434.
81. A. E. Friedman, J. C. Chambron, J. P. Sauvage, N. J. Turro, and J. K. Barton, *J. Am. Chem. Soc.*, 1990, 112, 4960–4962.
82. R. E. Holmlin, E. D. A. Stemp, and J. K. Barton, *Inorg. Chem.*, 1998, 37, 29–34.
83. Y. Jenkins, A. E. Friedman, N. J. Turro, and J. K. Barton, *Biochemistry*, 1992, 31, 10809–10816.
84. A. Rodger and B. Nordén, *Circular Dichroism and Linear Dichroism*, Oxford University Press, Oxford, UK, 1997.
85. T. K. Schoch, J. L. Hubbard, C. R. Zoch, G.-B. Yi, and M. Sorlie, *Inorg. Chem.*, 1996, 35, 4383–4390.
86. J. R. Lakowicz, *Principles of Fluorescence Spectroscopy*, 3rd ed., Springer, New York, 2006.
87. B. Nordén, A. Rodger, and T. R. Dafforn, *Linear Dichroism and Circular Dichroism—A Textbook on Polarized-Light Spectroscopy*, Royal Society of Chemistry, Cambridge, 2010.
88. C. Dicko, M. R. Hicks, T. R. Dafforn, F. Vollrath, A. Rodger, and S. V. Hoffmann, *Biophys. J.*, 2008, 95, 5974–5977.
89. R. Marrington, T. R. Dafforn, D. J. Halsall, M. Hicks, and A. Rodger, *Analyst*, 2005, 130, 1608–1616.
90. R. Marrington, T. R. Dafforn, D. J. Halsall, and A. Rodger, *Biophys. J.*, 2004, 87, 2002–2012.
91. Y. Matsuoka and B. Nordén, *Biopolymers*, 1982, 21, 2433–2452.
92. A. Rodger, In *Methods Enzymology*, eds. J. F. Riordan, and B. L. Vallee, Academic Press, San Diego, 1993, pp. 232–258.
93. A. Rodger, R. Marrington, M. A. Geeves, M. Hicks, L. de Alwis, D. J. Halsall, and T. R. Dafforn, *PCCP*, 2006, 8, 3161–3171.
94. A. Wada, *Biopolymers*, 1964, 2, 361–380.
95. A. Wada, *App. Spectros. Rev.*, 1972, 6, 1–30.
96. J. Aldrich-Wright, C. Brodie, E. C. Glazer, N. W. Leudtke, L. Elson-Schwab, and Y. Tor, *Chem. Commun.*, 2004, 1018–1019.
97. Y. Matsuoka and B. Nordén, *J. Phys. Chem.*, 1982, 86, 1378–1386.
98. O. Novakova, H. Chen, O. Vrana, A. Rodger, P. J. Sadler, and V. Brabec, *Biochemistry*, 2003, 42, 11544–11554.
99. H. Deng, J. Li, K. C. Zheng, Y. Yang, H. Chao, and L. N. Ji, *Inorg. Chim. Acta*, 2005, 358, 3430–3440.
100. E. C. Long and J. K. Barton, *Acc. Chem. Res.*, 1990, 23, 271–273.
101. S. Shi, J. Liu, K. C. Zheng, C. P. Tan, L. M. Chen, and L. N. Ji, *Dalton Trans.*, 2005, 2038–2046.
102. S. Tabassum and I. u. H. Baht, *Trans. Metal Chem.*, 2005, 30, 998–1007.
103. V. G. Vaidyanathan and B. U. Nair, *Dalton Trans.*, 2005, 2842–2848.
104. G. Cohen and H. Eisenberg, *Biopolymers*, 1969, 8, 45–55.
105. J. B. Chaires, N. Dattagupta, and D. M. Crothers, *Biochemistry*, 1982, 21, 3933–3940.

106. C. R. Brodie, J. G. Collins, and J. R. Aldrich-Wright, *Dalton Trans.*, 2004, 1145–1152.
107. P. P. Pellegrini and J. R. Aldrich-Wright, *Dalton Trans.*, 2003, 176–183.
108. Y. Liang, F. Du, S. Sanglier, B. R. Zhou, Y. Xia, A. Van Dorsselaer, C. Maechling, M. C. Kilhoffer, and J. Haiech, *Biol. Chem.*, 2003, 278, 30098–30105.
109. J. E. Ladbury and B. Z. Chowdhry, *Chem. Bio.*, 1996, 3, 791–801.
110. J. B. Chaires, *Arch. Biochem. Biophys.*, 2006, 453, 26–31.
111. G. S. Manning, *Q. Rev. Biophys.*, 1978, 11, 179–246.
112. M. T. Record Jr., C. F. Anderson, and T. M. Lohman, *Q. Rev. Biophys.*, 1978, 11, 103–178.
113. P. M. Wiggins, *Physica A*, 1997, 238, 113–128.
114. M. V. Rekharsky and Y. Inoue, eds., *Cyclodextrins and Their Complexes*, Wiley-VCH Verlag GmbH & Co. KGaA, Weinheim, Germany, 2006.
115. Y. Inoue, T. Hakushi, L. Y. L. Tong, B. Shen, and D. Jin, *J. Am. Chem. Soc.*, 1993, 115, 475–481.
116. M. V. Rekharsky and Y. Inoue, *Chem. Rev.*, 1998, 1875–1917.
117. E. A. Lewis and K. P. Murphy, *Methods Mol. Biol.*, 2005, 305, 1–15.
118. K. Bouchemal, *Drug Discov. Tod.*, 2008, 13, 960–972.
119. J. E. Ladbury, *Biotechniques*, 2004, 37, 885–887.
120. R. N. Goldberg, N. Kishore, and R. M. Lennen, *J. Phys. Chem. Ref. Data*, 2002, 31, 231–370.
121. B. Ganem, Y. T. Li, and J. D. Henion, *Tetrahedron Lett.*, 1993, 34, 1445–1448.
122. K. J. Light-Wahl, D. L. Springer, B. E. Winger, C. G. Edmonds, D. G. Camp, II, B. D. Thrall, and R. D. Smith, *J. Am. Chem. Soc.*, 1993, 115, 803–804.
123. D. E. Gilbert and J. Feigon, *Curr. Opin. Struct. Biol.*, 1999, 9, 305–314.
124. S. M. Nelson, L. R. Ferguson, and W. A. Denny, *Cell Chromosome*, 2004, 3, 1–26.
125. P. C. Champ, S. Maurice, J. M. Vargason, T. Camp, and P. S. Ho, *Nucleic Acids Res.*, 2004, 32, 6501–6510.
126. N. V. Anantha, M. Azam, and R. D. Sheardy, *Biochemistry*, 1998, 37, 2709–2714.
127. H. Han, C. L. Cliff, and L. H. Hurley, *Biochemistry*, 1999, 38, 6981–6986.
128. H. Han, D. R. Langley, A. Rangan, and L. H. Hurley, *J. Am. Chem. Soc.*, 2001, 123, 8902–8913.
129. R. J. Harrison, J. Cuesta, G. Chessari, M. A. Read, S. K. Basra, A. P. Reszka, J. Morrell, S. M. Gowan, C. M. Incles, F. A. Tanious, W. D. Wilson, L. R. Kelland, and S. Neidle, *J. Med. Chem.*, 2003, 46, 4463–4476.
130. G. S. Minhas, D. S. Pilch, J. E. Kerrigan, E. J. LaVoie, and J. E. Rice, *Bioorg. Med. Chem. Lett.*, 2006, 16, 3891–3895.
131. K. Shin-ya, K. Wierzba, K. Matsuo, T. Ohtani, Y. Yamada, K. Furihata, Y. Hayakawa, and H. Seto, *J. Am. Chem. Soc.*, 2001, 123, 1262–1263.
132. D. Sun, B. Thompson, B. E. Cathers, M. Salazar, S. M. Kerwin, J. O. Trent, T. C. Jenkins, S. Neidle, and L. H. Hurley, *J. Med. Chem.*, 1997, 40, 2113–2116.
133. W. M. David, J. Brodbelt, S. M. Kerwin, and P. W. Thomas, *Anal. Chem.*, 2002, 74, 2029–2033.
134. T-M. Ou, Y-J. Lu, J-H. Tan, Z. s. Huang, K. Y. Wong, and L-Q. Gu, *ChemMedChem*, 2008, 3, 690–713.
135. E. W. White, F. Tanious, M. A. Ismail, A. P. Reszka, S. Neidle, D. W. Boykin, and W. D. Wilson, *Biophys. Chem.*, 2007, 126, 140–153.
136. F. Gonnet, F. Kocher, J. C. Blais, G. Bolbach, J. C. Tabet, and J. C. Chottard, *J. Mass Spectrom.*, 1996, 31, 802–809.
137. R. Gupta, J. L. Beck, M. M. Sheil, and S. F. Ralph, *J. Inorg. Biochem.*, 2005, 99, 552–559.
138. H. Troujman and J.-C. Chottard, *Anal. Biochem.*, 1997, 252, 177–185.
139. J. L. Beck, M. L. Colgrave, S. F. Ralph, and M. M. Sheil, *Mass Spectrom. Rev.*, 2001, 20, 61–87.

140. K. X. Wan, T. Shibue, and M. L. Gross, *J. Am. Chem. Soc.*, 2000, 122, 300–307.
141. J. Talib, C. Green, K. J. Davis, T. Urathamakul, J. L. Beck, J. R. Aldrich-Wright, and S. F. Ralph, *Dalton Trans.*, 2008, 1018–1026.
142. R. Gupta, J. L. Beck, S. F. Ralph, M. M. Sheil, and J. R. Aldrich-Wright, *J. Am. Soc. Mass Spectrom.*, 2004, 15, 1382–1391.
143. L. M. Wilhelmsson, F. Westerlund, P. Lincoln, B. Norden, *J. Am. Chem. Soc.*, 2002, 124, 12092–12093.
144. C. Rajput, R. Rutkaite, L. Swanson, I. Haq, and J. A. Thomas, *Chem. Eur. J.*, 2006, 12, 4611–4619.
145. J. A. Smith, J. G. Collins, B. T. Patterson, and F. R. Keene, *Dalton Trans.*, 2004, 1277–1283.
146. J. A. Smith, J. L. Morgan, A. G. Turley, J. G. Collins, and F. R. Keene, *Dalton Trans.*, 2006, 3179–3187.
147. I. Haq, P. Lincoln, D. Suh, B. Norden, B. Z. Chowdhry, and J. B. Chaires, *J. Am. Chem. Soc.*, 1995, 117, 4788–4796.
148. S. Satyanarayana, J. C. Dabrowiak, and J. B. Chaires, *Biochemistry*, 1992, 31, 9319–9324.
149. J. L. Beck, R. Gupta, T. Urathamakul, N. L. Williamson, M. M. Sheil, J. R. Aldrich-Wright, and S. F. Ralph, *Chem. Commun.*, 2003, 626–627.
150. I. Greguric, J. R. Aldrich-Wright, and J. G. Collins, *J. Am. Chem. Soc.*, 1997, 119, 3621–3622.
151. J. G. Collins, A. D. Sleeman, J. R. Aldrich-Wright, I. Greguric, and T. W. Hambley, *Inorg. Chem.*, 1998, 37, 3133–3141.
152. J. G. Collins, J. R. Aldrich-Wright, I. D. Greguric, and P. A. Pellegrini, *Inorg. Chem.*, 1999, 38, 5502–5509.
153. F. Rosu, V. Gabelica, C. Houssier, and E. De Pauw, *Nucleic Acids Res.*, 2002, 30, e82/81–e82/89.
154. S. Satyanarayana, J. C. Dabrowiak, and J. B. Chaires, *Biochemistry*, 1993, 32, 2573–2584.
155. C. Carrasco, F. Rosu, V. Gabelica, C. Houssier, E. De Pauw, C. Garbay-Jaureguiberry, B. Roques, W. D. Wilson, J. B. Chaires, M. J. Waring, and C. Bailly, *ChemBioChem*, 2002, 3, 1235–1241.
156. W.-H. Chen, Y. Qin, Z. Cai, C.-L. Chan, G.-A. Luo, and Z.-H. Jiang, *Bioorg. Med. Chem. Lett.*, 2005, 13, 1859–1866.
157. K. C. Gornall, S. Samosorn, J. Talib, J. B. Bremner, and J. L. Beck, *Rapid Commun. Mass Spectrom.*, 2007, 21, 1759–1766.
158. F. Rosu, E. De Pauw, L. Guittat, P. Alberti, L. Lacroix, P. Mailliet, J.-F. Riou, and J.-L. Mergny, *Biochemistry*, 2003, 42, 10361–10371.
159. F. Rosu, V. Gabelica, C. Houssier, P. Colson, and E. De Pauw, *Rapid Commun. Mass Spectrom.*, 2002, 16, 1729–1736.
160. F. Rosu, V. Gabelica, K. Shin-ya, and E. De Pauw, *Chem. Commun.*, 2003, 2702–2703.
161. L. Cubo, D. S. Thomas, J. Zhang, A. G. Quiroga, C. Navarro-Ranninger, and S. J. Berners-Price, *Inorg. Chim. Acta*, 2009, 362, 1022–1026.

2 Probing DNA and RNA Interactions with Biogenic and Synthetic Polyamines
Models and Biological Implications

H. A. Tajmir-Riahi, A. Ahmed Ouameur, I. Hasni, P. Bourassa, and T. J. Thomas

CONTENTS

In this chapter, we report the applications of several analytical methods and the molecular modeling to determine the binding sites of biogenic and synthetic polyamines on DNA and tRNA. The analytical methods evaluated here are capillary electrophoresis, Fourier-transform infrared (FTIR), UV–Visible, circular dichroism (CD) spectroscopic techniques, as well as molecular modeling.

2.1 INTRODUCTION

The biogenic polyamines putrescine, spermidine, and spermine (Scheme 2.1) are small aliphatic polycationic compounds present in almost all living organisms. They are essential for normal cell growth and their concentrations increase with cell proliferation [1–5]. In mammalian cells, they are found in millimolar concentrations [1–3]. Of the many hypotheses advanced to explain their biological effects, the most important one concerns their interaction with nucleic acids [1,2,4,6,7]. However, it seems that a large fraction of polyamines exist in polyamine–RNA adducts, and the major part of their cellular function may be explained through structural changes of RNA by polyamines [2,3,8]. On the other hand, the potential effectiveness of polyamine analogs (Scheme 2.1) as antiproliferative agents against many tumor cell lines and infectious diseases provides evidence for nucleic acid interaction with biogenic polyamines [9–15]. Polyamine analogs bind oligonucleotides 15mer GC and 15mer AT via major and minor grooves, inducing B to Z DNA conformational changes for G–C base pairs and causing DNA aggregation [16,17]. Analogs also induce

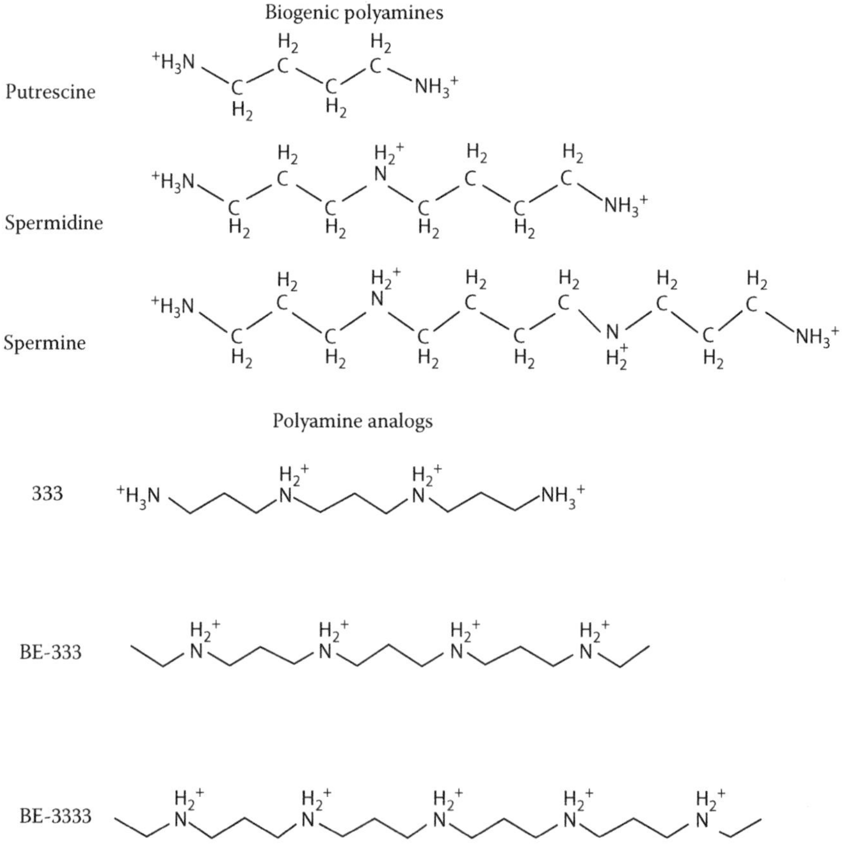

SCHEME 2.1

aggregation and precipitation of highly polymerized DNA [18]. The effects of polyamine analogs on the aggregation, precipitation, and conformations of single-, double-, and triple-stranded DNA have been reported [19,20]. DNA condensation by polyamine analogs is known and the effects of analogs on the stability of duplex structures, such as RNA–DNA, DNA–RNA, RNA–RNA, and DNA–DNA, have been investigated [21,22].

Interactions of polyamines with tRNA are less understood than with DNA, and thus more specific theoretical and experimental studies of polyamine–tRNA binding are required. The interactions of both spermine and spermidine with tRNA have been studied using solution ^{1}H-NMR [23,24], ^{13}C-NMR [25], and ^{15}N-NMR [26,27]. The authors observed that in all polyamines, internal $-NH_2^+-$ groups bind to tRNA more strongly than terminal $-NH_3^+$ groups. They suggest that factors other than the electrostatic interaction are responsible for this phenomenon since the primary amines ($-NH_3^+$) possess a higher density of positive charge than the secondary ones ($-NH_2^+-$). The ^{15}N-NMR experiments suggest that the specificity by which the nitrogen atoms in polyamines bind to tRNA is a consequence of the various hydrogen bonds that can be established between both tRNA and polyamines [26,27]. Interestingly, Bibillo et al. [28] found that biogenic polyamines are required for the nonenzymatic cleavage of tRNA and oligoribonucleotides at UA (uracil and adenine bases) and CA (cytosine and adenine bases) phosphodiester bonds. This suggests that at least two linked protonated amino groups are necessary to mediate hydrolysis. They hypothesized that one ammonium group binds to the nucleotide while the second one participates in the activation of the labile phosphodiester bond [28]. However, in spite of these extensive studies, the binding sites of biogenic polyamines on tRNA are still not clearly established.

A spermine–tRNA^{Phe} complex has been crystallized and its structure resolved at 2.5 Å resolution [29,30]. The polyamine binds at two major sites, the first in the major groove at the end of the anticodon stem and the second near the variable loop around phosphate 10 at a turn in the RNA backbone. However, ^{1}H-NMR studies found that spermine and related polyamines bind to additional sites of tRNA, in particular at the junction of the TψC and D loops, which were not seen in the crystal structure [23,24]. Binding of biogenic polyamines to tRNA has been shown to be relevant for optimal translational accuracy and efficiency [3]. They are found to modulate protein synthesis [31,32] and influence the binding of deacylated and acylated tRNA to the ribosomes [33,34]. These effects may be explained by the ability of polyamines to bind and influence the secondary structure of tRNA, mRNA, and rRNA [3,31–34]. Recent reports showed results of infrared, CD, UV–Visible spectroscopy, and capillary electrophoresis as well as molecular modeling on the interaction of DNA and tRNA with biogenic and synthetic polyamines [35–37].

In this chapter, we examine different analytical methods used for the structural characterization of DNA and RNA complexes with biogenic polyamine, spermine, spermidine, putrescine, and polyamine analogs, 1,11-diamino-4,8-diazaundecane (333), 3,7,11,15-tetrazaheptadecane.4HCl (BE-333), and 3,7,11,15,19-pentazahenicos ane.5HCl (BE-3333) (Scheme 2.1), and the advantages and disadvantages of each method are discussed here.

2.2 ANALYTICAL METHODS USED FOR STRUCTURAL ANALYSIS OF POLYAMINE–NUCLEIC ACID COMPLEXES

2.2.1 FOURIER-TRANSFORM INFRARED SPECTROSCOPY

Infrared spectra were recorded on an FTIR spectrometer (Impact 420 model), equipped with DTGS (deuterated triglycine sulfate) detector and KBr beam splitter, using AgBr windows. Spectra were collected after 2 h incubation of polyamine with the polynucleotide solution and measured. Interferograms were accumulated over the spectral range 4000–600 cm^{-1} with a nominal resolution of 2 cm^{-1} and a minimum of 100 scans. The difference spectra [(polynucleotide solution + polyamine)–(polynucleotide solution)] were obtained, using a sharp band at 968 (DNA) or 966 cm^{-1} (RNA) as internal reference. This band, which is due to sugar C–C stretching vibrations, exhibits no spectral changes (shifting or intensity variation) upon polyamine–polynucleotide complexation, and canceled out upon spectral subtraction [35,38].

2.2.2 CIRCULAR DICHROISM SPECTROSCOPY

The CD spectra of DNA and RNA and their polyamine adducts were recorded at pH 7.3 with a Jasco J-720 spectropolarimeter. For measurements in the Far-UV region (200–320 nm), a quartz cell with a path length of 0.01 cm was used. Six scans were accumulated at a scan speed of 50 nm/min, with data being collected at every nm from 200 to 320 nm. Sample temperature was maintained at 25°C using a Neslab RTE-111 circulating water bath connected to the water-jacketed quartz cuvette. Spectra were corrected for buffer signal and conversion to the Mol CD ($\Delta\varepsilon$) was performed with the Jasco Standard Analysis software [36].

2.2.3 ABSORPTION SPECTROSCOPY

The absorption spectra were recorded on a Perkin Elmer Lambda 40 spectrophotometer with a slit of 2 nm and scan speed of 240 nm/min. Quartz cuvettes of 1 cm were used. The absorbance assessments were performed at pH 7.3 by keeping the concentration of DNA or RNA constant (125 μM), while varying the concentration of polyamine (5–100 μM). The binding constants of polyamine–polynucleotide complexes were calculated as reported [39,40].

It is assumed that the interaction between the ligand L and the substrate S is 1:1; for this reason, a single complex SL (1:1) is formed. It was also assumed that the sites (and all the binding sites) are independent and finally the Beer's law is followed by all species. A wavelength is selected at which the molar absorptivities ε_S (molar absorptivity of the substrate) and S_{11} (molar absorptivity of the complex) are different. Then, at total concentration S_t of the substrate, in the absence of ligand and the light path length $b = 1$ cm, the solution absorbance is

$$A_o = \varepsilon_S b S_t \tag{2.1}$$

In the presence of ligand at total concentration L_t, the absorbance of a solution containing the same total substrate concentration is

$$A_L = \varepsilon_S b[S] + \varepsilon_L b[L] + \varepsilon_{11} b[SL] \tag{2.2}$$

where [S] is the concentration of the uncomplexed substrate, [L] the concentration of the uncomplexed ligand, and [SL] the concentration of the complex, which, combined with the mass balance on S and L, gives

$$A_L = \varepsilon_S b S_t + \varepsilon_L b L_t + \Delta\varepsilon_{11} b[SL] \tag{2.3}$$

where $\Delta\varepsilon_{11} = \varepsilon_{11} - \varepsilon_S - \varepsilon_L$ (ε_L is the molar absorptivity of the ligand). By measuring the solution absorbance against a reference containing ligand at the same total concentration L_t, the measured absorbance becomes

$$A = \varepsilon_S b S_t + \Delta\varepsilon_{11} b[SL] \tag{2.4}$$

Combining Equation 2.4 with the stability constant definition $K_{11} = [SL]/[S][L]$ gives

$$\Delta A = K_{11}\Delta\varepsilon_{11} b[S][L] \tag{2.5}$$

where $\Delta A = A - A_0$. From the mass balance expression $S_t = [S] + [SL]$, we get $[S] = S_t/(1 + K_{11}[L])$, which is Equation 2.5, giving Equation 2.6 at the relationship between the observed absorbance change per centimeter and the system variables and parameters.

$$\frac{\Delta A}{b} = \frac{S_t K_{11}\Delta\varepsilon_{11}[L]}{1 + K_{11}[L]} \tag{2.6}$$

Equation 2.6 is the binding isotherm, which shows the hyperbolic dependence on free ligand concentration.

The double reciprocal form of plotting the rectangular hyperbola $(1/y) = (f.d) \cdot (1/x) + (e/d)$ is based on the linearization of Equation 2.6 according to the following equation:

$$\frac{b}{\Delta A} = \frac{1}{S_t K_{11}\Delta\varepsilon_{11}[L]} + \frac{1}{S_t\Delta\varepsilon_{11}} \tag{2.7}$$

Thus, the double reciprocal plot of $1/\Delta A$ versus $1/[L]$ is linear and the binding constant can be estimated from the following equation:

$$K_{11} = \frac{intercept}{slope} \tag{2.8}$$

2.2.4 Capillary Electrophoresis

Affinity capillary electrophoresis (ACE) was used to detect a shift in mobility when polyamine binds to nucleic acids. The binding constants for the polyamine–polynucleotide complexes can be determined by Scatchard analysis, using mobility shift of DNA or RNA complexes [41]. The extent of saturation (R_f) of the polynucleotides was determined from the change of the migration time of polynucleotide in the presence of various concentrations of polyamine by the following equation:

$$R_f = \frac{\Delta m}{\Delta m_s} \tag{2.9}$$

where Δm is the difference in migration time between pure DNA or RNA and its complexes at different polyamine concentrations and Δm_s corresponds to the difference in migration time between pure polynucleotide and polynucleotide-saturated complex.

The binding constant K_b given by

$$K_b = \frac{[\text{polyamine} - \text{tRNA adduct}]}{[\text{tRNA}][\text{polyamine}]} \tag{2.10}$$

was determined by fitting the experimental values of R_f and polyamine concentration to the equation:

$$R_f = \frac{K_b[\text{polyamine}]}{1 + K_b[\text{polyamine}]} \tag{2.11}$$

Rearrangement of this gives a convenient form for Scatchard analysis:

$$\frac{R_f}{[\text{polyamine}]} = K_b - K_b R_f \tag{2.12}$$

Capillary electrophoresis has become a useful technique for measuring binding constants [42–45]. The cooperativity of the binding can be analyzed using a Hill plot. Assuming one binding site for polyamine (Equation 2.5), the equation below (Hill equation) can be established:

$$\log \frac{R_f}{1 - R_f} = n \log[\text{polyamine}] + \log K_d \tag{2.13}$$

where n (Hill coefficient) measures the degree of cooperativity and K_d is the dissociation constant. The linear plot of $\log\{R_f/(R_f - 1)\}$ versus $\log[\text{polyamine}]$ has a slope of n and an intercept on the $\log[\text{polyamine}]$ axis of $\log K_d/n$. The quantity n increases with the degree of cooperativity of a reaction and thereby provides a convenient and simplistic characterization of a ligand-binding reaction [46].

2.2.5 MOLECULAR MODELING

The docking studies were performed with ArgusLab 4.0.1 software (Mark A. Thompson, Planaria Software LLC, Seattle, WA, http://www.arguslab.com). tRNA and DNA structures were obtained from the Protein Data Bank (PDB) (ID: 6TNA) [28,47,48] and the polyamine three-dimensional structures were generated from PM3 semiempirical calculations using Chem3D Ultra 6.0. According to x-ray [29] and ¹H-NMR [22] data, three possible binding sites (major groove, variable loop, and TψC loop) were searched. The docking runs were performed on the ArgusDock docking engine using high precision with a maximum of 150 candidate poses. The conformations were ranked using the Ascore scoring function, which estimates the free binding energy. Upon docking of polyamines to tRNA, the current configurations were optimized using a steepest decent algorithm until convergence, using 40 iterations and nucleobase residues within a distance of 3.5 Å relative to polyamines involved in complexation [35].

2.3   CASE STUDIES: RESULTS AND DISCUSSION

2.3.1   FTIR SPECTRA

FTIR spectroscopy is widely used to characterize the nature of biogenic and synthetic polyamine bindings with DNA and RNA [1–4]. Figures 2.1 and 2.2 present the infrared

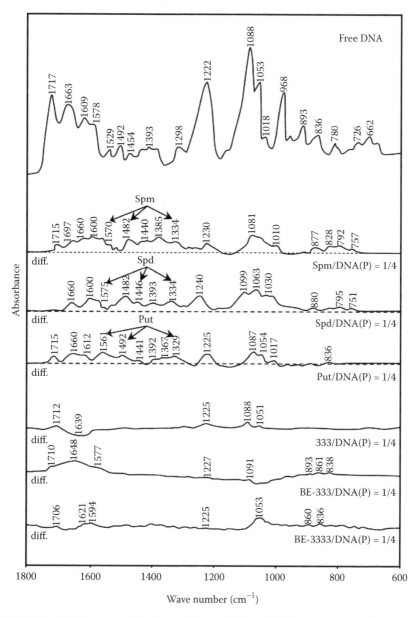

FIGURE 2.1   FTIR spectra of the free calf-thymus DNA and difference spectra of polyamine–DNA complexes in the region of 1800–600 cm⁻¹ with polyamine/DNA molar ratio 1/4.

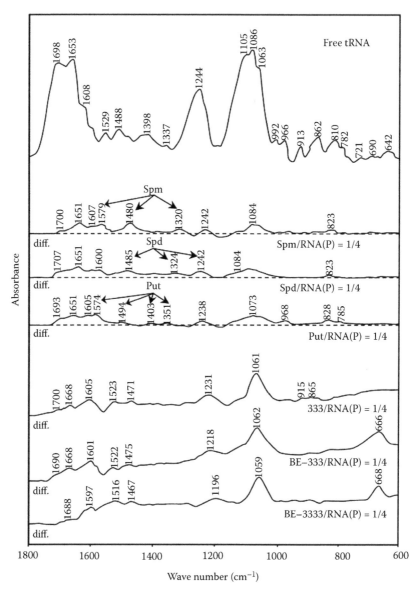

FIGURE 2.2 FTIR spectra of the free yeast tRNA and difference spectra of polyamine–tRNA complexes in the region of 1800–600 cm⁻¹ with polyamine/tRNA molar ratio 1/4.

spectra and difference spectra of DNA and RNA complexes with spermine, spermidine, putrescine (biogenic), 333, BE-333, and BE-3333 (synthetic) polyamines. The assignments of typical IR bands related to DNA and tRNA are given in Tables 2.1 and 2.2. Marker infrared bands were selected at 1717 (guanine), 1663 (thymine), 1609 (adenine), 1492 (cytosine), 1222 (PO₂), and 1088 cm⁻¹ (PO₂) (Table 2.1) to detect polyamine–base and polyamine–phosphate bindings for both biogenic and synthetic polyamine complexes with DNA (Figure 2.1). The difference spectra showed spectral changes (shifting

TABLE 2.1

Principal Infrared Absorption Bands, Relative Intensities, and Assignments for Calf-Thymus DNA

Wave Number (cm^{-1})	Intensity[a]	Assignment
1717	vs	Guanine (C=O stretching)
1663	vs	Thymine (C2=O stretching)
1609	s	Adenine (C7=N stretching)
1578	sh	Purine stretching (N7)
1529	w	In-plane vibration of cytosine and guanine
1492	m	In-plane vibration of cytosine
1222	vs	Asymmetric PO_2^- stretch
1088	vs	Symmetric PO_2^- stretch
1053	s	C–O deoxyribose stretch
968	s	C–C deoxyribose stretch
893	m	Deoxyribose, B-marker
836	m	Deoxyribose, B-marker

[a] Relative intensities: s = strong, sh = shoulder, vs = very strong, m = medium, w = weak.

and intensity changes) for the marked bands related to DNA bases and the backbone phosphate group, indicating polyamine interaction with both DNA bases and the PO_2 group (Figure 2.1). However, the spectral changes were more pronounced for biogenic polyamines spermine, spermidine, and putrescine than those of polyamine analogs 333, BE-333, and BE-3333 (Figure 2.1). Based on the spectral changes observed, the major polyamine-binding sites were guanine and adenine N7 sites as well as the backbone PO_2 group (Figure 2.1). Similarly, RNA marker bands at 1698 (guanine), 1653 (uracil), 1608 (adenine), 1488 (cytosine), 1244 (PO_2), and 1086 cm^{-1} (PO_2) (Table 2.2) were examined to detect the binding of polyamines to RNA bases and the backbone phosphate group (Figure 2.2). The difference spectra showed major spectral changes (shifting and intensity increase) for RNA marker bands related to RNA bases and the backbone phosphate group (Figure 2.2). Based on the spectral changes, polyamine interaction was mainly with guanine (N7 site), uracil (O_2), and adenine (N7 site) as well as the backbone PO_2 group (Figure 2.2). The intensity increases of the RNA marker bands were similar for both biogenic and synthetic polyamine–RNA complexes (Figure 2.2).

2.3.2   CD Spectra

CD spectroscopy is a powerful tool used to analyze the conformational aspects of DNA, RNA, and proteins. The CD spectrum of the free DNA is composed of four major peaks at 210 (negative), 221 (positive), 245 (negative), and 280 nm (positive) (Figure 2.3). This is consistent with the CD spectrum of double-helical DNA in B conformation [49,50]. Upon addition of biogenic and synthetic polyamine, no major shifting of CD bands was observed (Figure 2.3). The band at 280 nm showed no shifting upon biogenic polyamine (spermine) complexation, while it appeared at

TABLE 2.2

Measured Wave Numbers, Relative Intensities, and Assignments for the Main Infrared Bands of Baker's Yeast tRNA

Wave Number (cm^{-1})	Intensity[a]	Assignment[b]
1698	vs	Guanine (C=O, C=N stretching)
1653	vs	Uracil (C=O stretching)
1608	s	Adenine (C=N stretching)
1529	w	In-plane ring vibration of cytosine and guanine
1488	m	In-plane ring vibration of cytosine
1398	s	In-plane ring vibration of guanine in *anti* conformation
1244	vs	Asymmetric PO_2^- stretch
1086	vs	Symmetric PO_2^- stretch
1063	s	C–O ribose stretch
966	m	C–C ribose stretch
913	m	C–C ribose stretch
862	m	Ribose phosphodiester, A-marker
810	m	Ribose phosphodiester, A-marker

[a] Relative intensities: s = strong, sh = shoulder, vs = very strong, m = medium, w = weak.

[b] Assignments have been taken from the literature and relevant references are given in the results section.

273–269 nm in the spectra of polyamine analog–DNA complexes (Figure 2.3). This was due to no DNA conformational changes upon biogenic polyamine interaction, while a partial B to A transition occurred for the synthetic polyamine–DNA adducts [36]. The alterations of the intensity of the CD band at 210, 221, 245, and 280 nm were due to DNA aggregation and condensation in the presence of both biogenic and synthetic polyamines (Figure 2.3). This is consistent with the aggregation and supramolecular formation of DNA by biogenic polyamines [51].

The CD spectrum of free tRNA is composed of four major peaks at 210 (negative), 222, 237 (negative), and 269 nm (positive) (Figure 2.4). This is consistent with the CD spectrum of double-helical RNA in A conformation [49,50]. Upon addition of polyamines, no major shifting of the bands was observed, while an increase of the molar ellipticity of the band at 210 nm was observed [37]. However, as the polyamine concentration increased, a major increase in molar ellipticity of the band at 210 nm was observed (Figure 2.4). Since there was no major shifting of the bands at 210 and 269 nm, tRNA remains in A conformation. However, the major alterations of the intensity of the band at 210 nm can be attributed to the reduction of the base-stacking interaction and tRNA aggregation upon protein complexation [37].

2.3.3 UV–VISIBLE SPECTRA

UV–Visible spectroscopic methods used to monitor the structural changes occur for DNA and RNA upon synthetic polyamine complexation [36,37]. The binding constants of polyamine complexes with DNA and RNA were determined as described in the methods of analysis (UV–Visible spectroscopy). An increasing polyamine

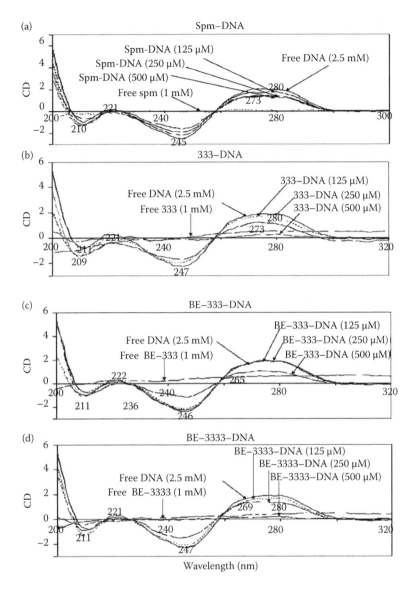

FIGURE 2.3 CD spectra of highly polymerized calf-thymus DNA (pH ~ 7.2) at 25°C (2.5 mM) and spermine (a), 333 (b), BE-333 (c), and BE-3333 (d) with 125, 250, and 500 μM polyamine concentrations.

concentration resulted in an increase in UV light absorption, as can be observed in Figures 2.5 and 2.6. This is consistent with a reduction of base-stacking interaction due to polyamine–DNA complexation (Figure 2.5A and B). The double reciprocal plot of $1/(A–A_0)$ versus $1/$(polyamine concentration) is linear and the binding constant (K) can be estimated from the ratio of the intercept to the slope. A_0 is the initial absorbance of the free DNA at 260 nm and A is the recorded absorbance of complexes at different

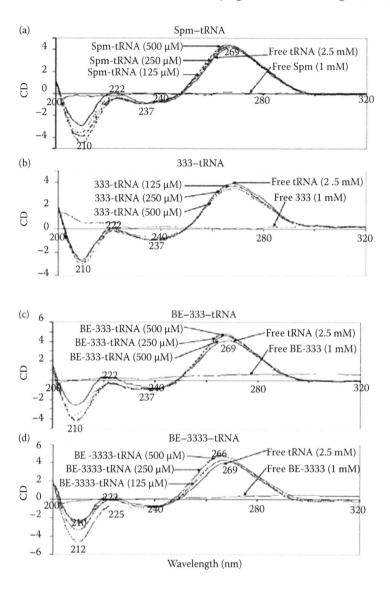

FIGURE 2.4   CD spectra of yeast tRNA (pH ~ 7.2) at 25°C (2.5 mM) and spermine (a), 333 (b), BE-333 (c), and BE-3333 (d) with 125, 250, and 500 μM polyamine concentrations.

polyamine concentrations. The overall binding constants for polyamine analog–DNA complexes are reported to be $K_{333} = 1.9 \times 10^4$ M^{-1}, $K_{BE-333} = 6.4 \times 10^4$ M^{-1}, and $K_{BE-3333} = 4.7 \times 10^4$ M^{-1} (Table 2.3). The binding constant for polyamine analog–tRNA is calculated to be $K_{333} = 2.8 \times 10^4$ M^{-1}, $K_{BE-333} = 3.7 \times 10^4$ M^{-1}, and $K_{BE-3333} = 4.0 \times 10^4$ M^{-1} (Table 2.4). The order of stability of synthetic polyamine–tRNA was BE-3333 > BE-333 > 333. However, the stability of the biogenic polyamine complexes with DNA and RNA is much higher than those of the synthetic polyamine complexes (Tables 2.3 and 2.4). The binding constants estimated are mainly due to

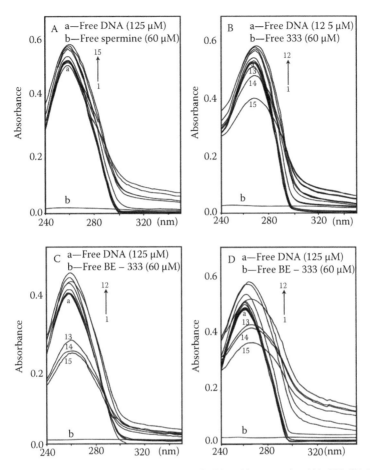

FIGURE 2.5   UV–Visible results of calf-thymus DNA and its spermine (A), 333 (B), BE-333 (C), and BE-3333 (D) complexes. Spectra of (a) free DNA (125 μM); (b) free polyamine (60 μM); (1–15) polyamine–DNA complexes: 1 (2), 2 (4), 3 (6), 4 (8), 5 (10), 6 (12), 7 (15), 8 (20), 9 (25), 10 (30), 11 (35), 12 (40), 13 (45), 14 (50), and 15 (60 μM).

the polyamine-base binding and not related to the polyamine–PO_2 interaction, which is largely ionic and can be dissociated easily in aqueous solution.

2.3.4 CAPILLARY ELECTROPHORESIS

Capillary electrophoresis is widely used for the separation and complex formation of DNA and RNA with polyamines [34,35]. Figure 2.5 shows the saturation curves for both DNA and RNA complexes with biogenic polyamines. The saturation curve (Figure 2.7a) shows a change in the migration time of DNA after addition of different polyamine concentration. As the polycation concentration reached a high level, no appreciable changes were measured for the polycation–DNA complexes. Based on these results, binding constants of the polycation–DNA complexes were determined by Scatchard analysis. Scatchard analysis for the biogenic polyamine–

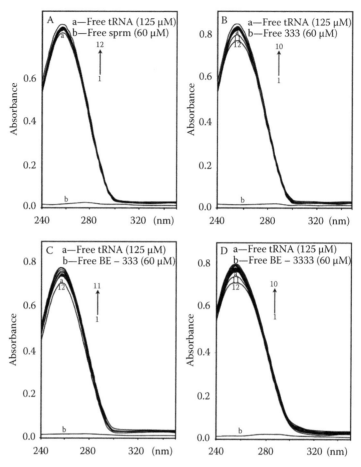

FIGURE 2.6 UV–Visible results of yeast tRNA and its spermine (A), 333 (B), BE-333 (C), and BE-3333 (D) complexes. Spectra of (a) free DNA (125 µM); (b) free polyamine (60 µM); (1–12) polyamine–DNA complexes: 1 (2), 2 (4), 3 (6), 4 (8), 5 (10), 6 (12), 7 (15), 8 (20), 9 (25), 10 (30), 11 (35), 12 (40 µM).

DNA adducts showed one overall binding constant for each polyamine, namely $K_{Spm} = 2.3 \times 10^5$ M^{-1}, $K_{Spd} = 1.4 \times 10^5$ M^{-1}, and $K_{Put} = 1.0 \times 10^5$ M^{-1} (Figure 2.7b and Table 2.3). The stability was in the following order: spermine > spermidine > putrescine. This finding is in agreement with the NMR studies [18–22] that found the secondary amino groups ($-NH_2^+-$) of polyamine bind more strongly than the primary amino groups ($-NH_3^+$). In fact, spermine contains two $-NH_2^+-$ groups, spermidine one group, while putrescine does not contain secondary amino group. Similar behaviors were observed for polyamine–tRNA complexes (Figure 2.7c). Scatchard analysis for the biogenic polyamine–DNA adducts showed one overall binding constant for each polyamine, namely $K_{Spm} = 8.7 \times 10^5$ M^{-1}, $K_{Spd} = 6.1 \times 10^5$ M^{-1}, and $K_{Put} = 1.0 \times 10^5$ M^{-1} (Figure 2.7d and Table 2.3). The stability was in the following order: spermine > spermi-

TABLE 2.3
Binding Constants (K) for Polyamine Analog–DNA
Adducts Compared with Biogenic Polyamines

Biogenic Polyamines	Binding Constant (M^{-1})
Spermine	2.3×10^5
Spermidine	1.4×10^5
Putrescine	1.0×10^5
Polyamine analogs	
333	1.9×10^4
BE-333	6.4×10^4
BE-3333	4.7×10^4

dine > putrescine. Stronger polyamine–tRNA complexes than polyamine–DNA adducts were formed and the reason is attributed to the fact that most of the poly-amines are RNA bound *in vivo*. However, no attempt has been made to calculate the binding constant of polyamine analogs with DNA and RNA using capillary electro-phoresis. The binding of synthetic polyamines with DNA and RNA was estimated by UV spectroscopy as discussed earlier in Section 2.3.3.

2.3.5 DOCKING STUDIES

Molecular models are reported based on the spectroscopic data for polyamine–DNA and polyamine–tRNA complexes [34,35]. The dockings results presented in Figure 2.8 and Table 2.5 show that putrescine is surrounded by T10, C11, C12, G14, A15, G16, and A17 with a binding energy of −4.0 kcal/mol (Figure 2.8a and Table 2.5). Spermidine is located in the vicinity of C9, T10, C11, C12, G14, A15, G16, A17, and C18 with a binding energy of −3.80 kcal/mol (Figure 2.8b and Table 2.5). However, spermine is positioned near C4, T5, T6, A20, G21, A22, and G23 with a binding energy of −3.52 kcal/mol (Figure 2.8c and Table 2.5). The binding energy (ΔG) shows the

TABLE 2.4
Binding Constants (K) for tRNA–Polyamine Analogs
Compared with Biogenic Polyamines

Biogenic Polyamines	Binding Constant (M^{-1})
Spermine	8.7×10^5
Spermidine	6.1×10^5
Putrescine	1.0×10^5
Polyamine analogs	
333	2.8×10^4
BE-333	3.7×10^4
BE-3333	4.0×10^4

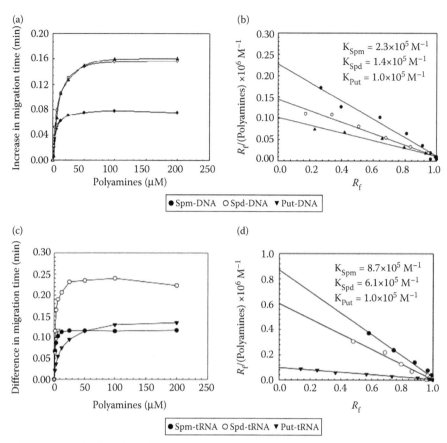

FIGURE 2.7　A plot of the difference in migration time (in minutes) of polyamine–DNA and polyamine–tRNA complexes from capillary electrophoresis following incubation of a constant concentration of DNA or tRNA (1.25 mM) with various concentrations of polyamines. The difference in migration time of the polyamine–DNA or polyamine–tRNA complexes was determined by subtracting the migration time of pure DNA or tRNA from that of each polyamine complexes (a and c). Scatchard plots and binding constants (K values) for biogenic polyamine–DNA and polyamine–tRNA complexes (b and d).

stability of the complexes formed: putrescine > spermidine > spermine in the biogenic polyamine–DNA complexes (Table 2.5).

Similarly, a docking study was carried out for putrescine, spermidine, and spermine cations docked to tRNA [35]. The models showed that putrescine is surrounded by A9, A23, G24, A25, G26, G43, A44, and G45 with a binding energy of −4.04 kcal/mol (Figure 2.9A and Table 2.6). Spermidine is located in the vicinity of A23, G24, A25, G26, C27, G43, A44, and G45 with a binding energy of −3.81 kcal/mol (Figure 2.9B and Table 2.6). However, spermine is positioned near A23, G24, A25, G26, C27, G42, G43, A44, and G45 with a binding energy of −3.57 kcal/mol (Figure 2.9C and Table 2.6). The binding energy (ΔG) shows the stability of the complexes formed: putrescine > spermidine > spermine in the polyamine–tRNA adducts (Table 2.6).

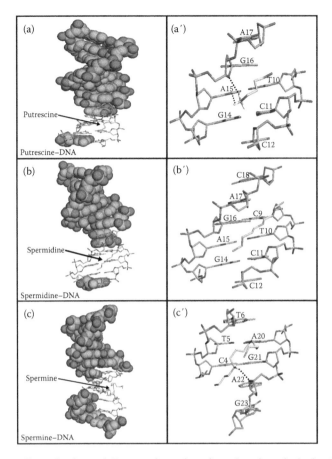

FIGURE 2.8 **(See color insert.)** Best conformations for polyamines docked to DNA (PDB entry 6TNA). The polyamines are shown in green color. (a) shows DNA in sphere-filling model with the putrescine-binding site in sticks and (a′) shows putrescine-binding sites represented in sticks with the corresponding base residues. (b) shows DNA in sphere-filling model with the spermidine-binding site in sticks and (b′) shows spermidine-binding sites represented in sticks. (c) shows DNA in sphere-filling model with the spermine-binding sites represented in sticks and (c′) shows the binding sites represented in sticks.

2.4 CONCLUDING REMARKS

We compared the applications of FTIR, CD, UV–Visible spectroscopic methods, and capillary electrophoresis as well as molecular modeling in the structural characterization of biogenic and synthetic polyamine complexes with DNA and tRNA. FTIR spectroscopy is a powerful tool to detect the binding sites of polyamine on DNA and RNA, while CD spectroscopy is used for conformational analysis of polyamine–polynucleotide complexes. However, UV–Visible spectroscopy and capillary electrophoresis were applied to determine the stability of the polyamine complexes with nucleic acids. Molecular modeling located the positions of each

TABLE 2.5

Nucleobases in the Vicinity of Putrescine, Spermidine, and Spermine with DNA (PDB 2K0V) and the Free Binding Energies of the Docked Complexes

Complex	Nucleobases Involved in Interactions	$\Delta G_{binding}$ (kcal/mol)
Putrescine–DNA	T10, C11, C12, G14, A15, G16, A17	−4.00
Spermidine–DNA	C9, T10, C11, C12, G14, A15, G16, A17, C18	−3.80
Spermine–DNA	C4, T5, T6, A20, G21, A22, G23	−3.52

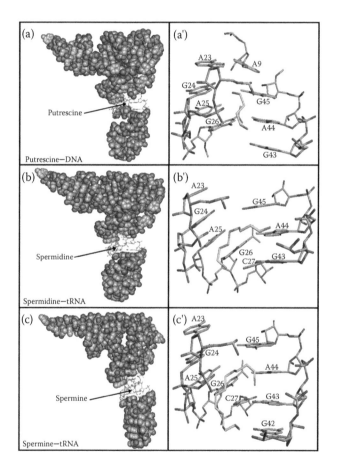

FIGURE 2.9 (See color insert.) Best conformations for polyamines docked to tRNA (PDB entry 6TNA). The polyamines are shown in green color. (a) shows tRNA in sphere-filling model with the putrescine-binding site in sticks and (a′) shows putrescine-binding sites represented in sticks with the corresponding base residues. (b) shows tRNA in sphere-filling model with the spermidine-binding site in sticks and (b′) shows spermidine-binding sites represented in sticks. (c) shows tRNA in sphere filling model with the spermine-binding sites represented in sticks and (c′) shows the binding sites represented in sticks.

TABLE 2.6

Ribonucleotides in the Vicinity of Putrescine, Spermidine, and Spermine with tRNA (PDB 6TNA) and the Free Binding Energies of the Docked Complexes

Complex	Ribonucleotides Involved in Interactions	$\Delta G_{binding}$ (kcal/mol)
Putrescine–tRNA	A9, A23, G24, A25, G26, G43, A44, G45	− 4.04
Spermidine–tRNA	A23, G24, A25, G26, C27, G43, A44, G45	− 3.81
Spermine–tRNA	A23, G24, A25, G26, C27, G42, G43, A44, G45	− 3.57

polyamine in the vicinity of DNA and RNA bases. These analytical methods were essential in elucidating the nature of polyamine–DNA and polyamine–RNA bindings and to clarify the major differences between biogenic and synthetic polyamine interactions with nucleic acids. The structural information generated can be correlated to the biological implications of biogenic polyamines in protecting and preventing the DNA damage, while polyamine analogs alter DNA structure and are used as anticancer drugs.

REFERENCES

1. Tabor, H. and Tabor, C. W. Polyamines. *Annu. Rev. Biochem.* 53, 749–790, 1984.
2. Cohen, S. S. *A Guide to the Polyamines.* Oxford University Press, New York, 1998.
3. Igarashi, K. and Kashiwagi, K. Polyamines: Mysterious modulators of cellular functions. *Biochem. Biophys. Res. Commun.* 271, 559–564, 2000.
4. Thomas, T. and Thomas, T. J. Polyamines in cell growth and cell death: Molecular mechanisms and therapeutic applications. *Cell. Mol. Life Sci.* 58, 244–258, 2001.
5. Childs, A. C., Mehta, D. J. and Gerner, E.W. Polyamine-dependent gene expression. *Cell. Mol. Life Sci.* 60, 1394–1406, 2003.
6. Feuerstein, B. G., Williams, L. D., Basu, H. S. and Marton, L. J. Implications and concepts of polyamine-nucleic acid interactions. *J. Cell. Biochem.* 46, 37–47, 1991.
7. Marton, L. J. and Pegg, A. E. Polyamines as targets for therapeutic intervention. *Ann. Rev. Pharm. Toxic.* 35, 55–91, 1995.
8. Watanabe, S., Kusama-Eguchi, K., Kobayashi, H. and Igarashi, K. Estimation of polyamine binding to macromolecules and ATP in bovine lymphocytes and rat liver. *J. Biol. Chem.* 266, 20803–20809, 1991.
9. Li, Y., Eiseman, J. L., Sentz, D. L., Rogers, F. A., Pan, S. S., Hu, L.-T., Egorin, M. J. and Callery, P. S. Synthesis and antitumor evaluation of a highly potent cytotoxic DNA cross-linking polyamine analogue, 1,12-diaziridinyl-4,9-diazadodecane. *J. Med. Chem.* 39, 339–341, 1996.
10. Frydman, B. and Valasinas, A. Polyamine-based chemotherapy of cancer. *Exp. Opin. Ther. Patents.* 9, 1055–1068, 1999.
11. Bacchi, C. J., Weiss, L. M., Lane, S., Frydman, B., Valasinas, A., Reddy, V., Sun, J. S., Marton, L. J., Khan, I. A., Moretto, M., Yarlett, N. and Wittner, M. Novel synthetic polyamines are effective in the treatment of experimental microsporidiosis, an opportunistic AIDS-associated infection. *Antimicro. Agents Chemother.* 46, 55–61, 2002.

12. Frydman, B., Porter, C. W., Maxuitenko, Y., Sarkar, A., Bhattacharya, S., Valasinas, A., Reddy, V. K., Kisiel, N., Marton, L. J. and Basu, H. S. A novel polyamine analog (SL-11093) inhibits growth of human prostate tumor xenografts in nude mice. *Cancer Chemother. Pharmacol.* 51, 488–492, 2003.

13. Fernández, C. O., Buldain, G. and Samejima, K. Probing the interaction between N^1,N^4-dibenzylputrescine and tRNA through ^{15}N NMR: Biological implications. *Biochim. Biophys. Acta.* 1476, 324–330, 2000.

14. Thomas, T., Balabhadrapathruni, S., Gallo, M. A. and Thomas, T. J. Development of polyamine analogs as cancer therapeutic agents. *Oncol. Res.* 13, 123–135, 2002.

15. Thomas, T. and Thomas, T. J. Polyamine metabolism and cancer. *J. Cell Mol. Med.* 7, 113–126, 2003.

16. Remirez, F. J., Thomas, T. J., Ruiz-Chica, A. J. and Thomas, T. Effects of aminooxy analogues of biogenic polyamines on the aggregation and stability of calf-thymus DNA. *Biopolymers (Biospectroscopy).* 65, 148–157, 2002.

17. Ruiz-Chica, A. J., Medina, M. A., Snachez-Jimenz, F. and Ramirez, F. J. On the interpretation of Raman spectra of 1-aminooxy-spermine/DNA complexes. *Nucleic Acids Res.* 32, 579–589, 2004.

18. Saminathan, M., Antony, T., Shirahata, A., Sigal, L. H., Thomas, T. and Thomas, T. J. Ionic and structural specificity effects of natural and synthetic polyamines on the aggregation and resolubilization of single-, double-, and triple-stranded DNA. *Biochemistry.* 38, 3821–3830, 1999.

19. Antony, T., Thomas, T., Shirahata, A., and Thomas, J. J. Selectivity of polyamines on stability of RNA-DNA hybrids containing phosphodiester and phosphorothioate oligodeoxyribonucleotides. *Biochemistry.* 38, 10775–10784, 1999.

20. Venkiteswaran, S., Vijayanathan, V., Shirahata, A., Thomas, T. and Thomas, T. J. Antisense recognition of the HER-2 mRNA: Effects of phosphorothioate substitution on polyamines on DNA-RNA, RNA-RNA and DNA-DNA stability. *Biochemistry.* 44, 303–312, 2005.

21. Vijayanathan, V., Thomas, T., Shirahata, A. and Thomas, T. T. DNA condensation by polyamines. A laser light scattering study of structural effects. *Biochemistry.* 40, 13644–13651, 2001.

22. Frydman, B., Westler, W. M. and Samejima, K. Spermine binds in bolution to the TψC loop of tRNA^{Phe}: Evidence from a 750 MHz ^{1}H-NMR analysis. *J. Org. Chem.* 61, 2588–2589, 1996.

23. Frydman, B., Westler, W. M., Valasinas, A., Kramer, D. L. and Porter, C. W. Regioselective binding of spermine, N^1,N^{12}-bismethylspermine, and N^1,N^{12}-bisethylspermine to tRNA^{Phe} as revealed by 750 MHz ^{1}H-NMR and its possible correlation with cell cycling and cytotoxicity. *J. Braz. Chem. Soc.* 10, 334–340, 1999.

24. Frydman, B., de los Santos, C. and Frydman, R. B. A ^{13}C NMR study of [5,8-^{13}C$_2$]spermidine binding to tRNA and to *Escherichia coli* macromolecules. *J. Biol. Chem.* 265, 20874–20878, 1990.

25. Fernandez, C. O., Frydman, B. and Samejima, K. Interactions between polyamine analogs with antiproliferative effects and tRNA: A ^{15}N NMR analysis. *Cell. Mol. Biol.* 40, 933–944, 1994.

26. Frydman, L., Rossomando, P. C., Frydman, V., Fernandez, C. O., Frydman, B. and Samejima, K. Interactions between natural polyamines and tRNA: An ^{15}N-NMR analysis. *Proc. Natl. Acad. Sci. USA.* 89, 9186–9190, 1992.

27. Bibillo, A., Figlerowicz, M. and Kierzek, R. The non-enzymatic hydrolysis of oligoribonucleotides VI. The role of biogenic polyamines. *Nucleic Acids Res.* 27, 3931–3937, 1999.

28. Quigley, G. J., Teeter, M. M. and Rich, A. Structural analysis of spermine and magnesium ion binding to yeast phenylalanine transfer RNA. *Proc. Nat. Acad. Sci. USA.* 75, 64–68, 1978.

29. Shi, H. and Moore, P. B. The crystal structure of yeast phenylalanine tRNA at 1.93 Å resolution. A classic structure revisited. *RNA.* 6, 1091–1105, 2000.

30. Shimogori, T., Kashiwagi, K. and Igarashi, K. Spermidine regulation of protein synthesis at the level of initiation complex formation of Met-tRNA, mRNA and ribosomes. *Biochem. Biophys. Res. Commun.* 223, 544–548, 1996.

31. Igarashi, K. and Kashiwagi, K. Polyamine modulon in *Escherichia coli*: Genes involved in the stimulation of cell growth by polyamines. *J. Biochem.* 139, 11–16, 2006.

32. Agrawal, R. K., Penczelk, P., Grassucci, R. A., Burkhardt, N., Nierhaus, K. H. and Frank, J. Effect of buffer conditions on the position of tRNA on the 70 S ribosome as visualized by cryoelectron microscopy. *J. Biol. Chem.* 274, 8723–8729, 1999.

33. Yoshida, M., Meksuriyen, D., Kashiwagi, K., Kawai, G. and Igarashi, K. Polyamine stimulation of the synthesis of oligopeptide-binding protein (OppA). Involvement of a structural change of the Shine-Dalgarno sequence and the initiation codon AUG in oppa mRNA. *J. Biol. Chem.* 274, 22723–22728, 1999.

34. Ahmed Ouameur, A. and Tajmir-Riahi, H. A. Structural analysis of DNA interactions with biogenic polyamines and cobalt(III)hexamine studied by Fourier transform infrared and capillary electrophoresis. *J. Biol. Chem.* 279, 42041–42054, 2004.

35. Ahmed Ouameur, A., Bourassa, P. and Tajmir-Riahi, H. A. Probing tRNA interaction with biogenic polyamines. *RNA.* 16, 1968–1976, 2010.

36. N'soukpoé-Kossi, C. N., Ahmed Ouameur, A., Thomas, T., Shirahata, A., Thomas, T. J. and Tajmir-Riahi, H. A. DNA interaction with antitumor polyamine analogues: A comparison with biogenic polyamines. *Biomacromolecules.* 9, 2712–2718, 2008.

37. N'soukpoé-Kossi, C. N., Ahmed Ouameur, A., Thomas, T., Thomas, T. J. and Tajmir-Riahi, H. A. Interaction tRNA with antitumor polyamine analogues. *Biochem. Cell Biol.* 87, 621–630, 2009.

38. Alex, S. and Dupuis, P. FTIR and Raman investigation of cadmium binding by DNA. *Inorg. Chim. Acta.* 157, 271–281, 1989.

39. Connors K. 1987. *Binding Constants: The Measurement of Molecular Complex Stability*, John Wiley & Sons, New York.

40. Stephanos, J. J. Drug-protein interactions: Two-site binding of heterocyclic ligands to a monomeric hemoglobin. *J. Inorg. Biochem.* 62, 155–169, 1996.

41. Klotz, M. I. Numbers of receptor sites from Scatchard graphs: Facts and fantasies. *Science.* 217, 1247–1249, 1982.

42. Foulds, G. J. and Etzkorn, F. A. A capillary electrophoresis mobility shift assay for protein-DNA binding affinities free in solution. *Nucleic Acids Res.* 26, 4304–4305, 1998.

43. Li, G. and Martin, M. L. A robust method for determining DNA binding constants using capillary zone electrophoresis. *Anal. Biochem.* 263, 72–78, 1998.

44. Guszcynski, T. and Copeland, T. D. A binding shift assay for the zinc-bound and zinc-free HIV-1 nucleocapsid protein by capillary electrophoresis. *Anal. Biochem.* 260, 212–217, 1998.

45. Arakawa, H., Neault, J.-F. and Tajmir-Riahi, H. A. Silver (I) complexes with DNA and RNA studied by Fourier transform infrared spectroscopy and capillary electrophoresis. *Biophys. J.* 81, 1580–1587, 2001.

46. Guijt-van Duijn, R. M., Frank, J., van Dedem, G. W. K. and Baltussen, E. Recent advances in affinity capillary electrophoresis. *Electrophoresis.* 21, 3905–3918, 2000.

47. Privé, G. G., Yanagi, K. and Dickerson, R. E. Structure of the B-DNA decamer C-C-A-A-C-G-T-T-G-G and comparison with isomorphous decamers C-C-A-A-A-G-A-T-T-G-G and C-C-A-G-G-C-C-T-G-G. *J. Mol. Biol.* 217, 177–199, 1991.

48. Sussman, J. L., Holbrook, S. R., Warrant, R. W., Church, G. M. and Kim, S-H. Crystal structure of yeast phenylalanine transfer RNA: I. Crystallographic refinement. *J. Mol. Biol.* 123, 607–630, 1978.

49. Vorlickova, M. Conformational transitions of alternating purine-pyrimidine DNAs in the perchlorate ethanol solutions. *Biophys. J.* 69, 2033–2043, 1995.
50. Kypr, J. and Vorlickova, M. Circular dichroism spectroscopy reveals invariant conformation of guanine runs in DNA. *Biopolymers.* 67, 275–277, 2002.
51. D'Agostino, L., Di pietro, M. and Di Luccia, A. Nuclear aggregates of polyamines are supramolecular structures that play a crucial role in genomic DNA protection and conformation. *FEBS J.* 272, 3777–3787, 2005.

Section 2

Emerging Techniques

3 Mass Spectrometry–Based Techniques for Studying Nucleic Acid/Small Molecule Interactions

Richard H. Griffey

CONTENTS

This chapter describes the utility of electrospray ionization mass spectrometry for studying complexes between small molecules and nucleic acids. This method is unique for the investigation of potential compounds targeting RNA and DNA, as it enables the determination of the identity, stoichiometry, and binding position along the nucleic acid molecule.

3.1 INTRODUCTION

As an analytical tool, mass spectrometry offers many advantages for the study of noncovalent complexes between nucleic acids and small molecules. Electrospray ionization (ESI) provides a rapid method to acquire a "snapshot" of the various species present in a solution sample, as the nm-sized droplets are instantly freeze-dried in the high-vacuum region of the MS instrument. Of interest are observations that the solution structures of large nucleic acids and proteins are maintained during the desolvation process. MS techniques allow the study of mixtures without the need for additional deconvolution as the intrinsic mass of a small molecule (derived from the elemental composition) acts as the "tag" for unambiguous identification. Titration experiments can be performed to determine the solution dissociation constants (K_D), as the MS detection process preserves binding stoichiometries for cationic, anionic, and neutral small molecules. The MS-derived results correlate with structures determined from conventional structural methods such as NMR. In addition, binding of multiple small molecules to the same nucleic acid can be observed, facilitating medicinal chemistry approaches to create higher-affinity compounds. MS fragmentation can be induced using infrared laser-induced heating or collisional activation of the complex. Both RNA and DNA have well-defined fragmentation patterns in the gas phase whose abundances are governed by sugar pucker and base conformation. This low-resolution structural information can be used to compare ligand-binding modes without the need to determine high-resolution structures using hard-to-grow crystals or isotopically enriched RNAs for NMR. ESI-MS methods are compatible with high-throughput screening methods and automated data analysis. This facilitates screening against libraries of small molecules and HPLC-purified fractions of bacterial broths.

The major sample preparation issue for ESI-MS studies is the need to exchange the solution nonvolatile cations such as sodium and magnesium for volatile cations (ammonium, imidazolium, trimethylammonium) that are lost from the nucleic acid or protein during desolvation [1]. As a result, volatile buffers such as ammonium acetate or ammonium formate are preferred for RNA/DNA sample preparation prior to all ESI-MS analyses. Low levels (1–200 μM) of $Mg^{2\pm}$ (or other metals) can be added back in studies of ion-binding stoichiometry and location. ESI-MS has good sensitivity compared to NMR or fluorescence-based approaches, and 50 nM to 5 μM concentrations of nucleic acid typically are adequate for screening. The capital cost and site requirements for the ESI-MS instrumentation are less than (ion trap, quadrupole) or comparable to NMR (FT-ICR).

3.2 BASICS OF ESI-MS SCREENING AND DETECTION

The utility of ESI-MS for parallel screening of small molecules against nucleic acid targets is based on the ability to transfer low-affinity noncovalent complexes from solution into the gas phase, where the masses of noncovalent complexes can be measured accurately within 1–10 ppm. No isotopic, radioactive, or fluorescent tag is required to detect the formation of a complex, as the signals from the "free" nucleic acid and the complex are readily resolved by the measured difference in mass-to-charge (m/z) ratio. As shown in Figure 3.1, the mass of the bound ligand can be measured with great precision, allowing identification of the ligand based on the exact mass determined from the elemental composition. The measured mass of the ligand can be determined to 0.1 Da with ion trap and quadrupole instruments, and to 0.0006 Da with an FT-ICR instrument. A typical collection of 5–20 small molecules will have masses differing by 1–15 Da, which produces a difference of 0.2–3.0 m/z for the [M−5H]$^{5-}$ charge state of a complex with an RNA or DNA target. The ESI-MS technique is sensitive, and consumes only pmole quantities of nucleic acid per assay [2]. The hit rates from ESI-MS screening are 10–100 times higher than conventional HTS functional assays, *because the ESI-MS measures noncovalent-binding interactions directly.* This type of a binding assay discriminates strongly against false-positives compared to functional assays where the cumulative activity from several compounds can generate a false-positive result. The ESI-MS-binding affinities of individual compounds have been shown in multiple instances to correlate with activity in functional assays [3].

A robotics-based high-throughput ESI-MS screening protocol can be fast, requiring <40 s to obtain spectra following an ~7.5 µL injection from a plate well. The

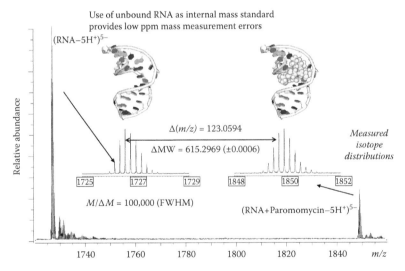

FIGURE 3.1   **(See color insert.)** ESI-MS spectrum for a mixture of a 27mer RNA analog of the 16S rRNA A site with paromomycin. The signal from the [M−5H]$^{5-}$ion of the free RNA appears at m/z 1746, while the signal from the [M−5H]$^{5-}$ion of RNA bound to paromomycin appears at m/z 1849. The accurate measurement of the difference in mass using an FT-ICR mass spectrometer allows the identity of the ligand to be determined from the limited set of elemental compositions consistent with the measured mass. For paromomycin, the measured mass of 615.2969 + 0.0006 Da is only consistent with one chemically plausible elemental composition.

ESI-MS-binding assay has robust dynamic range, and can detect complexes with dissociation constants (K_D) ranging from <10 nM to ~1 mM. The scientist should adjust the concentrations of target nucleic acid and small molecule in the sample to allow signals from both free nucleic acid target and the noncovalent complex(es) to be detected. This facilitates using the signal from the free nucleic acid as the internal mass standard and for proper estimation of solution dissociation constants [4].

A range of ESI-MS equipment can be used to examine different properties of noncovalent complexes. Trades for the choice of MS instrument type include desired mass accuracy, m/z resolving power, vacuum system type, electrospray source type and geometry, use of intermediate ion collection methods such as hexapole traps to improve sensitivity, and different methods for gas-phase dissociation and interrogation of complexes (collisional dissociation, laser-induced dissociation, black-body heating). It is beyond the scope of this chapter to examine all potential instrument configurations. However, noncovalent complexes of nucleic acids have been detected using quadrupole ion trap instruments, quadrupole instruments, Fourier transform ion cyclotron resonance instruments, and TOF instruments. As noted, this chapter focuses on electrospray ionization processes, as matrix-assisted laser desorption methods dissociate weak noncovalent complexes with small molecules through the heating required to desorb ions from the matrix surface. Studies of noncovalent complexes with proteins and peptides are also not covered in this chapter.

As noted previously, the charge of the ligand does not impact the ability of ESI-MS to detect the complex or measure affinity for the target. While positively charged ligands tend to have higher affinity for RNA and DNA, the ESI-MS technique is not biased against negatively charged ligands binding to the target. We have detected complexes between 20 μM adenosine monophosphate (AMP) and a 40mer duplex RNA aptamer (0.5 μM) designed to bind ATP. As shown in Figure 3.2, signal from the free aptamer duplex and adducted Mg ions (500 μM added Mg) is detected between m/z 2040 and 2045. Signal from the complex with AMP is detected at m/z 2098 to 2105. The ESI-MS-measured affinity (~10 μM) and stoichiometry (1:1) match observations from previous solution studies. Binding of Mg ions was not required to detect the complex, but it stabilized the duplex and increased the fraction of the duplex relative to single-stranded RNA detected using ESI-MS. Further gas-phase dissociation studies showed that the predominant-binding site for the Mg ion is at a single phosphate residue (Figure 3.2).

3.2.1 RNA Target Selection

The RNA structure is critical to function as these structures are recognition elements for proteins and nucleic acids. RNAs are comprised of independently folded helical regions, and these folded domains can typically be studied as unique entities even when removed from larger RNAs [5]. RNA domains contain only four nucleotides, but they can be arranged in a variety of ways that generate bulges, nonstandard base pairs, and internal mismatch pairs. Good (potentially druggable) RNA targets include regions with base pair mismatches and bulges, which have local deformations in helical geometry that provide exposed surfaces for stacking, H-bonding, and charge–charge interactions with small molecules. RNA loops have proven more difficult to target, as

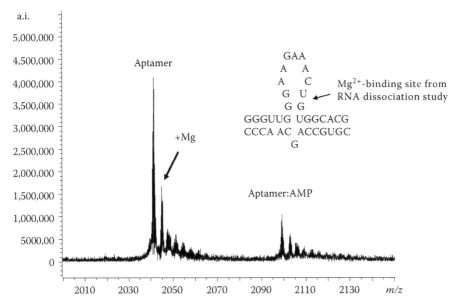

FIGURE 3.2 ESI-MS detection of noncovalent complexes between RNA and ligands is independent of ligand charge and tolerant of nonvolatile divalent cations. A 0.5 μM solution of a 40mer RNA aptamer duplex (sequence shown in inset) previously shown to bind ATP was prepared in 50 mM ammonium acetate buffer with 0.5 mM Mg ions. The signals from the "free" RNA, the Mg ion adduct, and a series of Mg and ammonium adducts appear between m/z 2040 and m/z 2060. Signals from the AMP adduct with the aptamer duplex and associated Mg and ammonium ions appear between m/z 2098 and m/z 2115. Gas-phase selection and dissociation of the aptamer–Mg ion complex showed the Mg ion preferentially bound at a single phosphate highlighted in the sequence diagram.

flexibility in loops provide induced fit in protein–RNA interactions but are too dynamic to produce stable interactions with small ligands.

The upper RNA size limit for ESI-MS screening is ~100 nucleotides, with smaller RNAs of 25–35 nucleotides being preferred. The larger RNAs are harder to desalt and the shorter sequences produced by synthesis failures are difficult to remove, and they add additional low m/z peaks to the spectrum. Complexes between these sequences and ligands produce complex spectra that can be challenging to deconvolute. Counterion condensation reduces the number of charged phosphates present in solution, causing the mass-to-charge ratio for large RNAs (98 nt) to reach m/z 3200–3800, outside the optimal detection range of most mass spectrometers. Smaller regions of RNA targets can be studied, with an assumption that the structure of the RNA subdomain mimics the structure present in the full-length RNA. This has proven to be true for many RNAs but cannot be assumed. The structure of subdomains can be stabilized through the addition of folded RNA tetraloops (or triloops on DNA) at one end to prevent breathing and helical dissociation. RNA sequences may also be circularly permuted to create a stable fold [6]. The structure of the subdomain can be confirmed using conventional methods (NMR, CD, nuclease digestion, etc.) and confirmed in the ESI-MS with collisional activation (see Section 2.6). In many cases, the structures of subdomains containing large internal bulges or loops are conformationally dynamic in the absence of stabilizing metal ions or ligands. Care should be taken to avoid switching AU and GC pairs that may alter base-stacking interactions or angles along the phosphate backbone. Magnesium ions can be added up to ~100 µM to stabilize folding of larger RNAs and duplexes (see Figure 3.2).

3.2.2 SAMPLE PREPARATION

Since ESI-MS can detect impurities below 0.2% levels, very pure RNA or DNA samples are required for screening. Typically, nucleic acids are HPLC-purified following synthesis, and then ethanol-precipitated from 1 M ammonium acetate solutions. Multiple ammonium acetate precipitations may be required to remove traces of nonvolatile metal ions (Na, K, Fe, etc.). Addition of low levels of RNAse inhibitors such as ammonium azide does not impact ESI-MS performance. However, nonvolatile phosphate and TRIS buffers should be avoided. Ligands can be added from diluted aqueous solutions or from stock solutions of dimethylsulfoxide (DMSO). We have observed that final DMSO concentrations of <2.5% are well tolerated. Care should be taken to insure that the wells in a screening plate are completely mixed, as not all robotic pipetting devices mix small-volume samples equivalently. In high-throughput screening studies, one or more wells are set up with a "standard" ligand of known affinity such that the formation of noncovalent complexes can be checked automatically by software to identify an instrument failure and stop a long run without losing samples [7]. Custom computer software may be written to measure the relative MS signal strength along every row in a 96- or 384-well plate. This effort is always important to prevent loss of valuable compounds, RNA, and data due to problems with instrument power, robotic equipment, electrospray sources, or the mass spectrometer.

3.2.3 ESI-MS OPERATING PARAMETERS

Initial experimental setup to detect noncovalent complexes with an ESI-MS instrument can be challenging. Hence, it is important to use a "known" noncovalent interaction to tune instrumental operating parameters [2]. Among the most critical operating parameters are the electrospray ionization potential, vacuum pressure in the interface region, hexapole/quadrupole storage voltages and vacuum pressures, and ion beam skimmer potentials. In general, "harsh" sources such as orthogonal electrospray units with high (µL/min) flow rates should be avoided. We have had the most success with low-flow (nL/min) ESI sources operating at high intermediate vacuum pressures. Under these conditions, the desolvation process is very mild, and collisional cooling keeps even low-affinity noncovalent complexes intact. We have observed "weak" complexes with dissociation constants of ~200 to >1000 µM. Tuning of the vacuum pressure and capillary skimmer potentials, countercurrent gas flow rate, and electrospray ionization potential are all required to adequately desolvate noncovalent complexes. Additional care must be taken to adjust the buffer concentration to reduce the formation of nonspecific electrostatic interactions between nucleic acids and positively charged small molecules. Low (10–50 mM) concentrations of ammonium acetate buffer at pH 6–8 generate stable electrospray ionization conditions, while higher concentrations of buffer (>50 mM) can be more challenging to tune and optimize and can reduce sensitivity.

Short DNA or RNA duplexes are good quality control systems for instrument setup and optimization. Duplexes of 10–12 base pairs with melting temperatures near 25°C can be studied to optimize buffer and instrument parameters. The collisional energy required to dissociate a duplex or to fragment a nucleic acid strand is well above that required to dissociate bound small molecules. Similarly, the desolvation energy can be adjusted so low (via vacuum pressure and skimmer applied potential) that multiple ammonium ion adducts ($M + 17$) on the phosphate backbone are detected. The objective of the tuning process is to find an operating point where the majority of the ammonium adducts have been dissociated (freeze-dried) down to protons, but the noncovalent complexes between DNA and RNA strands and small molecules are still detected.

For our Bruker FT-ICR instrument, example conditions and results are presented in Figure 3.3. The operating temperature for the heated capillary is set at 115°C, well

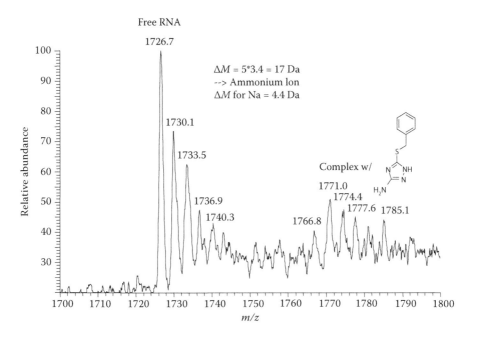

FIGURE 3.3   ESI-MS detection of very weak complexes from a sample of 5 µM of a 27mer RNA with 100 µM of a ligand in 50 mM ammonium acetate buffer. The temperature of the desolvation capillary has been reduced to 115°C and the voltage potential at capillary exit reduced to 50 V. The signal from the [M−5H]$^{5-}$ charge state of the "free" RNA with protons on 22 of the 27 phosphate groups appears at m/z 1726.7. Under these desolvation conditions, the ammonium ions adducted to phosphates are not fully dissociated in the gas phase into ammonia (lost to the vacuum system) and bound protons. The ammonium ion adducts appear at m/z +3.4 intervals from the free RNA. The signals from the complex and associated ammonium adducts appear at m/z 1771.0 to 1777.6, with signals from additional unidentified ligands present at lower abundances.

below the normal operating point of ~170°C. In addition, the capillary exit potential has been reduced to 50 V. Under these conditions, the signal for the $(M-5H)^{5-}$ charge state of a 27mer RNA hairpin is detected at m/z 1726.7, with a series of ammonium ion adducts detected at intervals of +3.4 m/z (ammonium ions adding a net mass of $M + 17$ Da through binding to phosphate groups). In contrast, adducted sodium ions would be observed at intervals of +4.4 m/z. In the presence of 100 μM of an amino-triazole ligand, a series of ions are detected from m/z 1771.0 to 1785.1, corresponding to the noncovalent complex with the ligand and combinations of adducted ammonium ions.

3.2.4 DETERMINATION OF DISSOCIATION CONSTANTS

The measurement of solution dissociation constants using ESI-MS is facilitated by the ability to directly detect the relative concentrations of the bound and unbound nucleic acids [4,8]. The RNA or DNA target is held at a concentration below or near the anticipated K_D, and multiple experiments can be performed by varying the concentration of the ligand over 1–3 orders of magnitude. The relative intensities of the MS signals from the free and bound target (and higher-order complexes) are integrated and fit mathematically to estimate the K_D values [8]. Given that the range of ligand affinities can vary from nM to mM, the concentrations of the RNA and ligands should be adjusted based on initial binding studies. In a typical experiment, the concentration of the RNA is held ~10-fold below the anticipated K_D, and a series of measurements are performed with increasing ligand concentrations. The amount of "free" RNA is determined from the relative abundance of the signal from unbound RNA, and the concentration of the complex is estimated from the relative abundance of the RNA–ligand complex. Abundances of ammonium adducts may be included in these determinations. The absolute concentration of the ligand is unknown, but can be estimated based on the initial solution concentration. This data analysis format is followed as the concentration of the ligand is increased. A K_D value can be fit mathematically using the appropriate equations. An advantage of ESI-MS is that the dissociation constants for several ligands can be determined in parallel, or the affinity of a single ligand for different RNAs can be determined from a series of experiments [9]. Rapid estimates of the RNA–ligand K_D can be determined at a single concentration of RNA and ligand in high-throughput screening.

3.2.5 HIGH-THROUGHPUT SCREENING

With modern robotics for automated sample plate preparation, it is possible to screen 2000–10,000 compounds per week against multiple nucleic acid targets using ESI-MS [10]. The ESI-MS instrument is rugged for continuous operation, and typically requires weekly service to clean the electrospray ionization source. Ancillary robotics are required for adjusting RNA and buffer concentrations from stock solutions, addition of compounds to wells, plate mixing and sealing, and sample introduction into the ESI-MS source. Small volumes (2–10 μL) are introduced via syringe drive, followed by 2–4 wash steps to rinse the compound and RNA from the fluidics system. Stacking of plates is preferred to allow completion of studies in a single run, as the ESI-MS process is most stable during continuous operation. ESI-MS data may be acquired and summed for 5–30 s per well. Associated informatics programs can be written and executed on the raw ESI-MS data to provide peak identification, mass calibration based on the free RNA, determination of mass for bound ligands, estimation of relative intensities, and calculation of relative or absolute K_D values.

Compound collections may be screened at 5 to >50 per well, as a function of the desired concentration and complexity of the set. Collections may be created from individual compounds via robotic mixing, from combinatorial syntheses, from HPLC fractions of broths derived from specific microbial strains, or from extracts of natural products. The ability to measure exact mass for the RNA adducts to <0.1 Da allows direct deconvolution of mixtures and identification of the active ligand. Additional MS fragmentation studies may be performed on natural products for the determination of their chemical structure.

3.2.6 LIGAND-BINDING SITE DETERMINATIONS

Three approaches have been demonstrated to identify sites where small molecules bind to RNA targets. The first approach exploits the ability to cleave the RNA backbone in the gas phase through mild collisional activation. The second approach incorporates modified nucleotides into the RNA, and measures changes in affinity for small molecules. The third approach uses solution enzymes to cleave the backbone with MS readout and quantitation of the results. The ability to activate the gas-phase RNA ions for fragmentation is a valuable method for determining local changes in RNA structure. The gas-phase fragmentation of RNA occurs most readily at sites where the sugar pucker has been shifted from a strong C3′-endo conformation to a C3′-exo conformation where the 2′-OH has an axial, in-line orientation to the 3′-phosphate group. This conformation is readily cleaved in the gas phase to generate a 2′,3′-cyclic phosphate and a free 5′-OH with the production of two fragment ions. The masses of the ions produced by collisional activation or IRMPD heating can be measured and the location of the cleavage ascertained from the abundances and mass/charge ratios of the product ions [11,12].

Given that many of the interesting RNA targets contain internal loops or bulges with altered sugar conformations, the ESI-MS screening methods can be used to examine conformational protection created by ligand binding, in addition to induced changes in RNA conformation and sugar pucker. RNA strand fragmentation is favored from C3′-exo sugar puckers. It should be noted that RNA is more resistant to fragmentation at uridine residues, possibly due to charge localization on the base rather than the phosphate backbone. ESI-MS fragmentation studies are rapid and provide insight into local nucleotide conformation. Fragmentation is favored at bulged or mismatch sites compared to base-paired regions, and ligand-induced changes in fragmentation can be used to locate binding sites with high confidence [13].

3.2.7 GAS-PHASE DISSOCIATION OF COMPLEXES

Griffey et al. took advantage of the bulged purine residues in an RNA analog of the 16S rRNA A site to map binding contacts for neamine-derived polysaccharides [9,14]. A solution containing a stoichiometric amount of the saccharide was mixed with the RNA, and collision-induced dissociation of the complex was produced by increasing the capillary exit voltage prior to MS detection. The fragmentation pattern and abundances of individual RNA fragments for the complex could be compared directly to the RNA in the absence of ligand. In complexes with aminoglycosides, the energy required to fragment the RNA at various locations is less than the energy required to dissociate the ligand. The bulged adenosine residues in the A site are labile to gas-phase cleavage compared to the base-paired nucleotides. Binding of the aminoglycoside sugars at the A site reduces the abundance of fragment ions produced by cleavage at G1405, U1495, and A1492 compared to the free RNA by 20–60%, with increasing protection from cleavage directly correlated with increasing numbers of sugar rings in the compound (paromomycin > ribostamycin > tobramycin). Increased protection of U1406 and U1495 is also observed with increasing numbers of sugars. This result suggests that the binding site is quite large, and the multiring compounds may provide a scaffold that binds the RNA and prevents the conformational changes required to effect cleavage of the phosphate backbone.

3.2.8 INCORPORATION OF MODIFIED NUCLEOTIDES

We have also examined the utility of incorporating modified nucleotides into the RNA to characterize small-molecule-binding sites. Incorporation of 5-iodouridine into the 27mer RNA model of the 16S A site has been used to map contacts between uridine residues and the aminoglycosides gentamicin and ribostamycin. The additional mass of the iodine relative to a proton (+125.9 Da) provides a convenient mass tag that allows the wild-type RNA and the iodinated RNA to be screened simultaneously. As shown in Figure 3.4a, a model of the structure of gentamicin with the 27mer A site RNA analog suggests that incorporation of a 5-iodouridine at U1406 should disrupt the binding of the aminoglycoside. A 62% loss in abundance has been observed for the gentamicin-5-I-U1406 complex compared to the abundance of the complex for the wild-type 27mer RNA (5.0 µM total RNA, 2.5 µM gentamicin, 50 mM ammonium acetate buffer). In comparison, incorporation of 5-iodouridine in the UUCG tetraloop does not affect binding of gentamicin relative to the wild-type RNA. Incorporation of 5-iodoU1495 reduces binding of ribostamycin by 56% (Figure 3.4b). These studies suggest that the substitution of ring C off hydroxyls 4 (gentamicin) and 5 (ribostamycin) of the 2-deoxystreptamine ring leads to contacts with different uridine residues at the U1406–U1495 mismatch base pair in the binding pocket. We have also examined the incorporation of phosphorothioate residues at specific sites. The magnitude of the steric effect is reduced as the sulfur cannot be readily incorporated stereospecifically. However, the incorporation of a phosphorothioate at the 3′-phosphate of A1492 leads to no change in the abundance of the complex with gentamicin, while the relative intensity of the ribostamycin complex is reduced by 20%. These results are consistent with a different orientation of the amines in the 2-DOS ring around A1492 in the two complexes.

3.2.9 ENZYMATIC DIGESTION STUDIES WITH MS DETECTION

Conventional solution nuclease digestion and protection assays can be performed in a rapid, high-throughput format using ESI-MS for the readout. The degradation

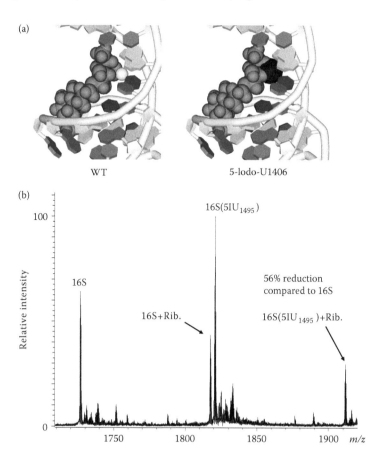

FIGURE 3.4   (a) Model of the structure of gentamicin bound to the 27mer RNA analog of the bacterial rRNA A site (left) and a 27mer RNA containing 5-iodouridine at position 1406. The proton and iodine are presented as CPK atoms in white (left) and black (right), respectively. Introduction of iodine creates a significant steric interaction with the bound aminoglycoside. (b) ESI-MS spectrum from a mixture of the 27mer RNA and iodinated analog mixed together with ribostamycin. Introduction of the iodinated U1495 residue reduces ribostamycin binding by 56% compared to the parent.

three RNAs from regions of the HCV IRES Domain II helix has been studied using ESI-MS as a function of added Ibis 528363, a small molecule shown to bind to the lower region of Domain IIA. As shown in Figure 3.5, RNAse A cleaves the 3′-phosphate at single-stranded C/N or U/N sites, while RNAse T1 cleaves the 3′-phosphate at single-stranded G/N sites. Studies of small-molecule binding to the 29mer RNA incorporating the Domain IIA internal loop suggested that the molecule contacted or significantly reduced the availability of the U, C, and G residues for cleavage. All of the protected residues were on the same face of the bent RNA helix (from comparison with structure determined using NMR) [15]. The protection by the ligand affects fewer residues in the larger RNAs, consistent with a more

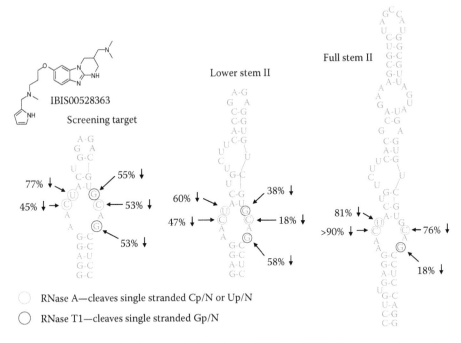

FIGURE 3.5   Analysis of enzymatic digestions of RNA and RNA–ligand complexes using ESI-MS. The screening targets are different subdomains from the HCV IRES Stem II. The RNAs were digested with RNase A and RNase T1 in the presence and absence of Ibis-00528363, a ligand known to bind in the lower internal bulge region of the RNA. The digestion pattern for the RNA used in the MS assay to identify the hit and develop the compound is presented on the left. Ligand binding protects the pyrimidines and guanines in the bulge from cleavage by both RNase A and RNase T1 by 45–77%. The bulge in a larger RNA surrogate (center) is also protected, but the pattern of protection is altered, with less protection on the 3′ side. The digestion pattern for the full Stem II is presented on the right, showing continued strong protection of the 5′-pyrimidines, and strong protection of the 3′-cytidine with loss in protection of one guanosine residue. Subsequent NMR analysis showed this cytidine to undergo a major conformational change upon ligand binding.

stable structure or a more defined conformational change induced by ligand binding.

3.3   CASE STUDIES

A series of case studies are presented that highlight the power of ESI-MS screening methods for the determination of ligand-binding properties. The most studied system is the ribosomal A site, where a 27 nt RNA has demonstrated surprising conformational similarity with the ribosomal A site in bacteria and eukaryotes. Studies of aminoglycosides and small-molecule binding to the RNA are presented. This is followed by studies of subdomains from the internal ribosomal entry site of hepatitis C virus, where a large-scale screening effort was conducted that converted an ESI-MS hit into a promising class of lead compounds that inhibit viral replication. Finally, studies of ligand identification and optimization targeted against a very large RNA target derived from the L11-binding site in the 23S rRNA are presented. These studies highlight the benefits of ESI-MS methods for identification and optimization of hits, leads, and for understanding the binding characteristics of known therapeutics.

3.3.1   16S AND 18S RIBOSOMAL A SITE

The sequence of the bacterial ribosomal RNA A site is highly conserved, and this location is the characterized-binding site for the aminoglycoside class of antibiotics. The A site binds the mRNA during translation and is proximate to the location where tRNAs bind the message. The crystal structure of the 16S complex has been solved, and NMR structures determined for subdomains of the A site RNA are similar to those observed in the intact ribosome. As a result, this has been an attractive (although largely unproductive) target for antimicrobial drug discovery efforts using a variety of screening approaches. It also has served to validate new methods where binding of the aminoglycosides can be detected with good sensitivity and specificity.

3.3.2   AMINOGLYCOSIDES

Binding of cognate tRNAs to bacterial and eukaryotic mRNAs occurs near the A site in the small (16S and 18S) ribosomal subunit (Figure 3.6a). This site has been demonstrated to be a site of action for aminoglycoside antibiotics. Griffey et al. measured the binding affinities and characterized the RNA contacts for six aminoglycosides against eight RNA subdomain variants of the 16S and 18S ribosomal A site [9]. The 16S and 18S RNA targets were screened simultaneously by introducing a short PEG linker on the 5′-terminus of the 18S analog, as shown in Figure 3.6b. This neutral mass tag shifted the signal from the 18S analog out beyond the masses for the noncovalent complexes between the 16S RNA and the aminoglycosides, allowing each complex to be monitored in parallel. Significant differences in binding affinities and specificities were observed between the paromomycin-type and the apramycin-type aminoglycosides, with paromomycin-type compounds binding preferentially to the 16S RNA surrogate and apramycin binding with higher affinity to the 18S RNA surrogate. Single-base substitutions at positions 1408 and 1409 produced changes in the conformation of the RNA and binding characteristics that could reduce or enhance the affinity for the target. Of interest were changes at position 1409, a C−G base pair in prokaryotes and a C−A mismatch pair in eukaryotes. Conversion of C1409 to A1409 enhanced the affinities of tobramycin and bekanamycin, while dramatically reducing the affinities of ribostamycin and paromomycin. Introducing an AU base pair in place of the CG base pair also reduced the affinities of paromomycin and lividomycin without altering the affinities of tobramycin and apramycin. Converting the 16S A1408−A1493 mismatch pair into the G1408−A1493 mismatch pair while retaining the rest of the 16S sequence reduced the affinities of paromomycin and lividomycin but improved the affinity of apramycin. These results showed that the specificity of aminoglycoside binding was governed by multiple RNA contacts, which independently contributed to the observed affinities and specificities.

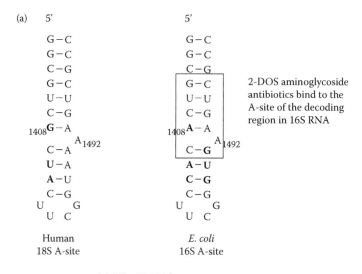

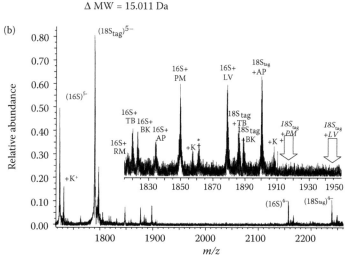

FIGURE 3.6 (a) Sequences for RNA analogs of the 18S and 16S rRNA A sites. (b) ESI-MS spectrum from a mixture containing 18S-tag and 16S rRNA A site analogs with a mixture of six aminoglycosides. The inset spectrum highlights the lack of binding observed between the 18S target, paromomycin, and lividomycin.

3.3.3 SYNTHETIC AMINOGLYCOSIDE DERIVATIVES

Hanessian and coworkers have examined a series of synthetic aminoglycoside derivates for RNA-binding affinity, functional inhibition of transcription/translation, and antimicrobial activity [16,17]. The tetracyclic aminoglycoside paromomycin was systematically altered through ring removal, hydroxyl functionalization, and amine alkylation. A series of >50 compounds were studied. The MS assay was used to measure the affinities for the consensus bacterial rRNA A site target. It was observed that affinity for the RNA target was not correlated with inhibition of protein synthesis or antimicrobial activity. This is not surprising since the compounds can bind with modest (μM) affinity in modes that are not appropriate to inhibit protein synthesis at the full ribosomal A site. However, the compounds with antimicrobial activity all had good affinity for the RNA target, and the RNA-binding assay was a better predictor of antimicrobial activity than the *in vitro* transcription/translation assay. Crystallography studies showed the disubstituted paromomycin analogs bound at the A site and at an interface site as expected for the mRNA codon and the tRNA anticodon [18].

3.3.4 SMALL MOLECULES

We have characterized the binding of the ribosomal A site target with low-affinity small-molecule ligands through direct binding and competition studies [19,20]. A common limitation of ESI-MS screening is the nonspecific binding of ligands at sites other than the desired active site. This is a particular problem for the A site RNA target, whose large-binding site can fit multiple ligands into and around the pocket. Two approaches have been evaluated to determine the binding sites. Competitive-binding studies can be performed using the ESI-MS methods where low-affinity hits are displaced by higher-affinity molecules. This provides confirmation that ligands bind in the same site. Another approach, using modified RNA nucleotides that have steric groups attached off the nucleoside, sugar, or phosphate, has been described in Section 2.6.2. For the A site RNA target, a series of low-affinity ligands were discovered through high-throughput screening. These ligands were classified as specific or nonspecific based on the ability of 2-deoxystreptamine or glucosamine to displace them from the RNA or to be displaced from the RNA [19]. As shown in Figure 3.7, Ibis-326732 is a substituted aminobenzimidazole ligand that binds the A site RNA target with ~70 μM affinity. When the A site target is bound to glucosamine, addition of Ibis-326732 displaces glucosamine from the RNA. This demonstrates that they bind at (or adjacent) the same site on the RNA. Further elaboration of Ibis-326732 provided no improvement in binding affinity or the ability to displace paromomycin or ribostamycin from the A site target. The inability to take hits that bind the A site and turn them into superior leads may reflect the influence of RNA conformational dynamics and the special properties of the aminoglycosides. The A site is physically large, and the aminoglycosides are large, preorganized, and relatively rigid. Their binding may induce a conformational change in the RNA that cannot be created with smaller, more flexible ligands that have a low probability of adopting a single low-energy structure that can induce (or prevent) a conformational change in the A site RNA [21].

3.3.5 HEPATITIS C INTERNAL RIBOSOMAL ENTRY SITE DOMAIN IIA

Translation of the hepatitis C virus (HCV) RNA occurs through an internal ribosomal entry site (IRES) located at the 5′-terminus. This highly structured region of the RNA has four helical domains and a pseudoknot that bind to eIF2, the 40S ribosome, and the 60S subunit. Alignment and analysis of HCV mutants support a strongly conserved structure for Stem II and the Domain IIA, which has a 19 base pair helix with two internal loops. A 29mer RNA surrogate for the lower internal

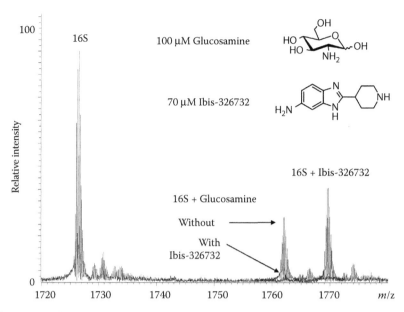

FIGURE 3.7 Competitive displacement of glucosamine from the 16S rRNA model RNA by Ibis-326732. Addition of 100 μM glucosamine produces a single noncovalent complex (m/z 1762; dotted lines). Addition of 70 μM Ibis-326732 to this solution reduces the intensity of the glucosamine adduct to near zero.

loop and surrounding helical domains of Stem II has been prepared and used for ESI-MS screening [22,23].

A total of 270,901 compounds from commercial sources have been screened against a 29mer surrogate of HCV Domain IIA. The estimated dissociation constants are presented in Table 3.1. Only 31 compounds (0.01%) had starting K_D values <50 μM, and the majority of these compounds (intercalating dyes, highly positively charged molecules) were not suitable for further elaboration. The screening results were linked to chemical structures and the chemical subdomains present in the molecules. Structurally similar compounds were clustered together for SAR analysis. Of particular interest for hit to lead optimization was a series of *N*-substituted

---

TABLE 3.1

Dissociation Constants for 270,901 Compounds with HCV Domain IIA RNA Surrogate Determined Using ESI-MS

Estimated K_D (μM)	Number of Compounds	Percent of Compounds
500–1000	10,667	3.94
250–500	4776	1.76
100–250	1153	0.43
50–100	137	0.05
<50	31	0.01

benzimidazoles that had threefold selectivity for the HCV domain over a control RNA and estimated affinities of 100–800 µM (Figure 3.8).

Initial SAR on the parent benzimidazole demonstrated that modifications at positions C6 and N1 were tolerated, but substitution at C5 and N2 were not tolerated. The C6 SAR studies provided the key breakthrough, as incorporation of a methoxy group improved affinity 2.5-fold while simultaneously increasing target selectivity threefold. Further elaboration of the C6 oxygen yielded a dimethypropoxy-substituted compound with a 10 µM K_D and 15-fold selectivity for the Domain IIA target. Variation in the length of the C6 side chain demonstrated that a 3 carbon linker provided the best selectivity with similar binding affinity to 2 and 4 carbon chains. Systematic removal of the C6, N1, and N2 moieties each resulted in >10-fold loss in affinity for the target RNA, showing that all three groups are required for target binding [22].

Substitution of small heterocycles off the C6 amine improves the binding affinity for the target and increases the lipophilicity of the compound, with a furan providing a twofold improvement in affinity. We reasoned that preorganizing the position of the positive charge would be critical for improving affinity and specificity, and examined a series of cyclic and chiral C6 side chains. None provided the desired effect.

A second breakthrough in the SAR occurred when the N1 side chain is constrained to N2. This provides a threefold improvement in affinity (3.5 µM) and demonstrates that N2 substitution could be tolerated when the amine proton is located on the same face as the lone pair of the benzimidazole nitrogen. Furthermore, in the constrained compound, attachment of alkylpyridyl or alkylpyrrole side chains off the C6 amine provides compounds with high affinity (<500 nM) and >50-fold selectivity for the Domain IIA target.

The effect of constraining the C6 propoxyl side chain to the benzimidazole ring has been examined. While constraining the C6 chain through a furan or pyran ring produces a 6–11-fold improvement in affinity, the doubly constrained molecule only has modestly better affinity. This result is surprising, and may reflect the differences in affinity for the diastereomers present in the mixture. The three types of constrained molecules have been screened in an HCV replicon assay (Figure 3.9). The inhibitory activity in the replicon assay (3.9, 1.5, and 5.4 µM) matched the ESI-MS K_D (0.5, 1.1, and 0.6 µM), and none of the three demonstrated toxicity at 100 µM concentrations

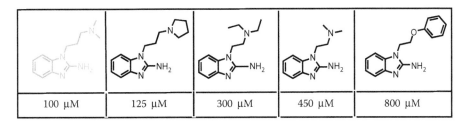

| 100 µM | 125 µM | 300 µM | 450 µM | 800 µM |

FIGURE 3.8 ESI-MS screening results for a series of substituted benzimidazoles against an HCV RNA Stem II domain. The presence of a positive charge on the side chain enhanced affinity, while the aromatic group suggested that the binding site had sufficient space for further elaboration. K_D values were estimated from screening at two ligand concentrations.

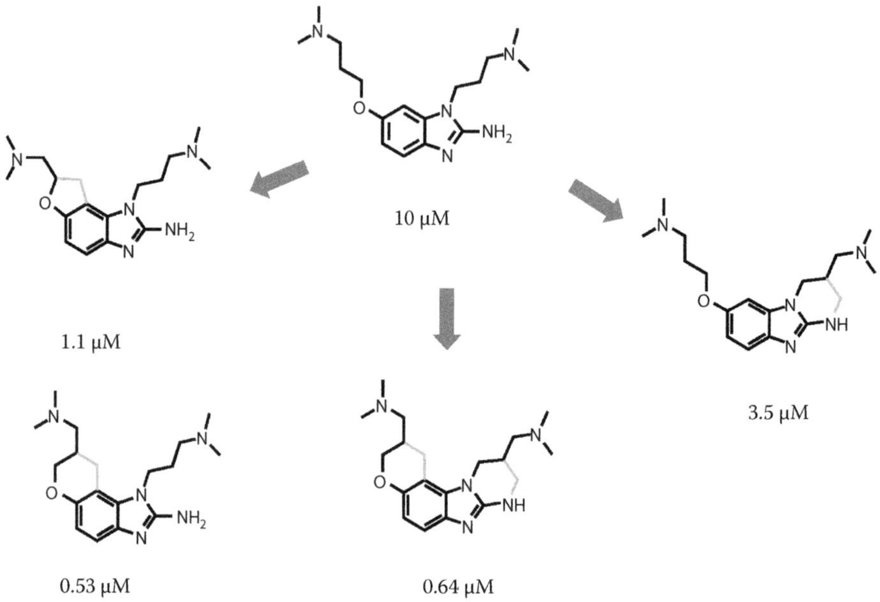

FIGURE 3.9   Structures of three classes of constrained benzimidazoles elaborated from an ESI-MS hit with their respective inhibitory activities in an HCV replication assay. The activity of the parent compound is improved by three- to eight-fold through constraint of either cationic side chain. However, the doubly constrained analog showed minimally increased activity relative to compounds with constrained side chains on the aromatic side.

in Huh-7 cells after 96 h. These results match a more detailed study of 10 compounds where the rank order of affinity for Domain IIA correlates with the replicon IC_{50}.

As described in Section 2.6.3, the binding location for Isis-528363 has been mapped near C8 and U9 in Domain IIA and the full Stem II using RNAse A and RNAse T1 digestions. Paulsen et al. have solved the structure for Isis-11 bound to the Domain IIA RNA using NMR and molecular dynamics refinement with explicit solvent [15,23]. Structures have been solved for both enantiomers of the compound, and energetic analysis suggests that both enantiomers bind with similar affinity and mode to the RNA. The compound induces a major conformational change in the RNA, displacing two conserved adenosines from the helical stack. These changes remove the 90° bend observed at the internal loop in the free RNA and may prevent appropriate contact between the HCV RNA and the 80S ribosome. The bound and free RNA structures have similar free energies, indicating that the inhibitor may selectively bind to a rare conformation of the RNA, as opposed to a mechanism where inhibitor binding induces a conformation change [15].

The structure of the bound inhibitor provides insight into design principals for RNA-targeted therapeutics. An electrostatic surface map of the RNA–Isis-11 complex showed that the two dimethylamino groups interact with local regions of partial negative charge within the RNA interior rather than making direct contacts with phosphate groups. This is counterintuitive, and explains why molecules in the SAR series with aliphatic and aromatic cationic groups have higher affinity than primary amines.

These charged groups can participate in cation–pi interactions with RNA bases and localized regions of negative charge produced by the folded RNA backbone. The benzimidazole-binding pocket of Domain IIA may not look like a traditional hydrophobic location in a protein, but is replete with hydrophobic surfaces and polar functional groups that could be recruited to enhance the inhibitor affinity and specificity [23].

3.3.6 L11-23S rRNA Target

The L11 ribosomal protein binds a conserved domain of the 23S rRNA in bacteria. A 78 nt RNA subdomain undergoes a conformational change coupled to GTP-dependent tRNA translocation. This subdomain is the known-binding site for thiostrepton-class antibiotics. Thiostrepton may work by locking the conformation of the RNA and/or ribosomal protein in one conformation. The conformation change requires binding of Mg^{2+} ions. Substitution of adenosine at position 1061 stabilizes the conformation with thiostrepton [24].

We have studied the binding of thiostrepton and small-molecule libraries to the wild-type (U1061) and A1061 mutant 58mer model RNA duplex in the presence of Mg^{2+} ions (Figure 3.10) [25]. As shown in Figure 3.11, ESI-MS of 5 µM thiostrepton, 1 µM RNA, and 100 µM Mg ions demonstrates that thiostrepton binds weakly to the wild-type RNA, but binds with high affinity to the stabilized A1061 RNA duplex. The A1061 RNA duplex requires only 100 µM Mg ions to fold and bind solution thiostrepton as monitored using ESI-MS (Figure 3.12), compared to the U1061 RNA that was only 90% bound at 500 µM Mg. Incorporation of a 2′-O-methyl

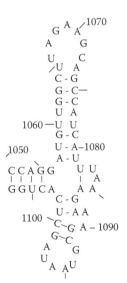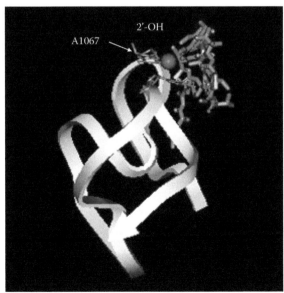

FIGURE 3.10 (See color insert.) RNA sequence of the 58mer RNA analog for the L11-binding site in the 23S rRNA and a model structure with bound thiostrepton. The large thiostrepton molecule makes multiple surface contacts with the RNA.

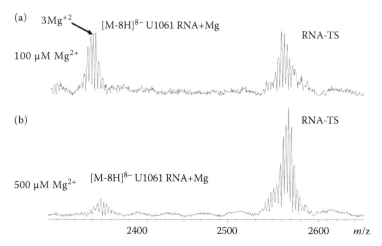

FIGURE 3.11 ESI-MS spectra from the U1061 variant of a 58mer RNA analog with 5 μM added thiostrepton (TS). (a) The [M−8H]⁸⁻ charge state in the presence of 100 μM added Mg and (b) in the presence of 500 μM added Mg ions. The fraction of thiostrepton bound to the RNA increases twofold at the higher Mg concentration.

at A1067 dramatically reduces thiostrepton binding in the MS assay, consistent with solution observations.

Two small-molecule motifs have been identified that bind the 58mer RNA duplex. Motif A contains a thiophene or a furan linked to a cationic group with ESI-MS affinities of 2–6 mM. Motif B has a quinoxalinedione with a pendant amide func-

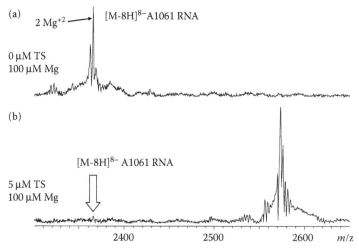

FIGURE 3.12 ESI-MS spectra from the A1061 variant of a 58mer RNA analog with and without 5 μM added thiostrepton (TS). (a) The [M−8H]⁸⁻ charge state in the absence of TS and the presence of 100 μM added Mg. The stabilized RNA binds only two Mg ions and (b) in the presence of TS with 100 μM added Mg ions. The thiostrepton K_D for the stabilized RNA is well below 1 μM even at the lower Mg ion concentration.

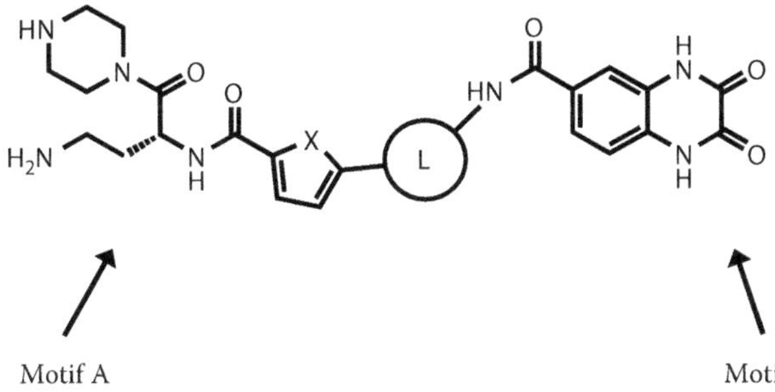

FIGURE 3.13 Chemical structures of Motif A and Motif B linked to generate a compound with improved antimicrobial activity. The motifs were identified from ESI-MS screening and fused using a linker identified through competitive binding assays with substituted derivatives.

tional group. Motif A could be functionalized off the furan ring, with methyl and phenyl groups not impacting binding affinity. The amino acid spacer between the furan Motif A and the piperidine chain has been optimized, with a short cationic tail demonstrating the highest-binding affinity relative to phenylalanine, histidine, serine, or lysine (D or L). The optimized Motif A bound with a ~360 µM affinity to the RNA target. The optimized Motif B bound with a ~700 µM affinity, and tolerated substitutions of aromatic groups from the carbonyl without a significant loss in affinity. An extensive range of linking strategies between the motifs was analyzed, including rigid groups (phenyl), flexible groups (alkyl and alkene spacers), and amino acid spacers. ESI-MS screening shows that Motif A with a phenyl group and Motif B with a methyl group could bind the RNA target simultaneously, while a longer linker off the Motif B amide nitrogen blocked simultaneous binding. Fusing the two motifs through a metasubstituted phenyl spacer produced a 20-fold improvement in binding affinity (23 µM), and the new compound inhibited bacterial transcription/translation at 10 µM. In addition, the compound inhibited bacterial growth at ~50 µM (Figure 3.13). Further elaboration of this lead and optimization in antimicrobial assays produced a compound with a bicyclic aromatic ring that inhibits bacterial translation at <10 µM and has a 3–6 µM MIC against *Escherichia coli* and *Staphylococcus aureus* [25].

3.4 CONCLUSIONS

High-throughput ESI-MS screening of RNA targets for low-affinity ligands can provide hits suitable for elaboration into lead compounds from large compound collections. The speed and information content of the assay are superior to any other screening format, as noncovalently complexed compounds are directly identified using their exact molecular mass as the label. The gas-phase dissociation of complexes can be used to map contacts with the RNA. Relative binding affinities can be determined from direct competition assays, and distances between concurrently binding ligands can be mapped through SAR studies of substituted analogs. ESI-MS screening has been used to identify molecules that bind viral and bacterial RNA targets with high and low affinities. Elaboration of these hits generates leads that inhibit viral replication and bacterial translation. There is a strong correlation between measured affinities and biological activity *in vitro* and in cells highlighting the utility of ESI-MS for drug discovery against a range of RNA targets.

3.5   PROSPECTS AND OUTLOOK

RNA targets continue to be discovered and implicated in human disease. Screening and small-molecule drug discovery against RNA targets present formidable challenges, as the hit rates are low compared to protein targets, few classes of prototypical lead molecules are known, and the "rules" for the generation of contacts with the RNA are largely unexplored. Recent efforts have developed leads from 2-deoxystreptamine or aminoglycoside cores. Disney et al. have discovered aminoglycoside-linked molecules that bind the myotonic muscular dystrophy RNA repeat regions with low nM affinities [26]. This work has yielded molecules with selectivity for U–U internal loops versus CU–UC internal loops. The lack of specificity for aminoglycoside analogs can be a toxicology drawback or a therapeutic benefit, as evidenced by work with paromomycin analogs. Kondo et al. have shown from crystallography studies that a disubstituted paromomycin derivative binds the ribosomal A site at two locations in two different conformations [18]. One molecule binds at the A site, while a second molecule binds at the crystal-packing interface where mRNA–tRNA complexes are formed.

Future drug development efforts will require identification of new motifs and mechanistic paradigms for small molecules that bind and inhibit the function of an RNA target. Examples include modeling and identification of negatively charged surface "pockets" not directly linked to phosphate groups, motifs for improved stacking with purine and pyrimidine bases, and core ligand structures that bind and limit the flexibility of RNA domains. SAR elaboration of hits for RNA targets is difficult, as the small molecules often start with low affinities and specificities that limit the utility of structural techniques such as crystallography and NMR. Hence, ESI-MS techniques will always be the most efficient methods to screen large number of compounds against a target and control to identify classes of compounds with structure and activity that warrant further medicinal chemistry and development. It is anticipated that therapeutically important RNA targets will continue to emerge, and that MS-based screening methods will be used to rapidly identify hits suitable for further elaboration via medicinal chemistry SAR and more refined structural analysis [7].

ACKNOWLEDGMENTS

I wish to acknowledge my collaborators and coauthors whose efforts made this work possible. I also acknowledge funding from NIST and DARPA that supported these research efforts.

REFERENCES

1. Greig, M.J., and Griffey, R.H. Utility of organic bases for improved electrospray mass spectrometry of oligonucleotides. *Rapid Commun. Mass Spec.*, **9**, 97–102, 1995.
2. Sannes-Lowery, K.A., Drader, J.J., Griffey, R.H., et al. Fourier transform ion cyclotron resonance mass spectrometry as a high-throughput affinity screen to identify RNA-binding ligands. *Trends Anal. Chem.*, **19**, 281–291, 2000.

3. Hofstadler, S.A. and Griffey, R.H. Analysis of non-covalent complexes of DNA and RNA by mass spectrometry. *Chem. Rev.*, **101**, 377–390, 2001.

4. Greig, M.J., Gaus, H., Cummins, L.L., et al. Measurements of macromolecular binding using electrospray mass spectrometry. Determination of dissociation constants for oligonucleotide:serum albumin complexes. *J. Am. Chem. Soc.*, **117**, 10765–10766, 1995.

5. Holbrook, S.R. Structural principals from large RNAs. *Ann. Rev. Biophys.*, **37**, 445–464, 2008.

6. Pan, T. and Uhlenbeck, O.C. Circularly permuted DNA, RNA and proteins—A review. *Gene*, **125**, 111–114, 1993.

7. Hofstadler, S.A. and Griffey, R.H. Mass spectrometry as a drug discovery platform against RNA targets. *Curr. Opin. Drug Disc. Devel.*, **3**, 423–431, 2000.

8. Sannes-Lowery, K.A., Griffey, R.H. and Hofstadler, S.A. Measuring dissociation constants of RNA and aminoglycoside antibiotics by electrospray ionization mass spectrometry. *Anal. Biochem.*, **280**, 264–271, 2000.

9. Griffey, R.H., Sannes-Lowery, K., Hofstadler, S.A., et al. A mass spectrometry-based parallel high-throughput screen for identification of RNA-binding compounds. *Proc. Natl. Acad. Sci. USA*, **96**, 10129–10133, 1999.

10. Hofstadler, S.A., Sannes-Lowery, K.A., Crooke, S.T. et al. Multiplexed screening of neutral mass tagged RNA targets against ligand libraries with electrospray ionization FTICR MS: A paradigm for high-throughput affinity screening. *Anal. Chem.*, **71**, 3436–3440, 1999.

11. McLuckey, S.A. and Habibi-Goudarzi, S. Gas-phase cleavage of oligonucleotides anions. *J. Am. Chem. Soc.*, **115**, 12085–12095, 1993.

12. Hofstadler, S.A., Sannes-Lowery, K.A. and Griffey R.H. Infrared multiphoton dissociation in an external ion reservoir. *Anal. Chem.*, **71**, 2067–2076, 1999.

13. Griffey, R.H. and Greig, M.J. Detection of base pair mismatches in duplex DNA and RNA oligonucleotides using electrospray mass spectrometry. *Ultrasensitive Biomolecular Devices SPIE*, **2985**, 82–86, 1997.

14. Griffey, R.H., Greig, M.J., An, H., et al. Targeted site-specific gas-phase cleavage of oligoribonucleotides. Application in mass spectrometry-based identification of ligand binding sites. *J. Am. Chem. Soc.*, **121**, 474–475, 1999.

15. Paulsen, R.B., Seth, P.P., Swayze, E.E., et al. Inhibitor induced structural changes in the HCV IRES Domain IIa RNA. *Proc. Natl. Acad. Sci. USA*, **107**, 7263–7268, 2010.

16. Francois, B., Szychowski, J., Adhikari, S.S., et al. Antibacterial aminoglycosides with a new mode of binding to the ribosomal RNA decoding site. *Angew. Chem.*, **43**, 6735–6738, 2004.

17. Hanessian, S., Adhikari, S., Szychowski, J., et al. Probing the ribosomal RNA A-site with functionally diverse analogs of paromomycin- synthesis of ring I mimetics. *Tetrahedron*, **65**, 827–846, 2007.

18. Kondo, J., Pachamuthu, K., Francois, B., et al. Crystal structure of the bacterial ribosomal decoding site complexed with a synthetic doubly functionalized paromomycin derivative: A new specific binding mode to an a-minor motif enhances antibacterial activity. *ChemMedChem*, **2**, 1631–1638, 2007.

19. Griffey R.H., Sannes-Lowery, K.A., Drader, J.J., et al. Characterization of low affinity complexes between RNA and small molecules using electrospray ionization mass spectrometry. *J. Am. Chem. Soc.*, **122**, 9933–9938, 2000.

20. He, Y., Yang, J., Wu, B., et al. Synthesis and evaluation of novel bacterial rRNA-binding benzimidazoles by mass spectrometry. *Bioorg. Med. Chem. Lett.*, **14**, 695–699, 2004.

21. Ding, Y., Hofstadler, S.A., Risen, L., et al. Design and synthesis of aminoglycoside-related heterocycle-substituted aminoglycoside mimetics based on a mass spectrometry RNA-binding assay. *Angew. Chem. Int. Ed.*, **42**, 3409–3412, 2003.

22. Seth, P., Miyaji, A., Jefferson, E.A., et al. SAR by MS: Discovery of a new class of RNA-binding small molecules for the hepatitis C virus internal ribosome entry site IIA subdomain. *J. Med. Chem.*, **48**, 7099–7102, 2005.
23. Davis, D.R. and Seth P.P. Therapeutic targeting of the HCV internal ribosomal entry site RNA. *Antivir. Chem. Chemother.*, **21**, 117–128, 2011.
24. Blyn, L.B., Risen, L.M., Griffey, R.H., et al. The RNA-binding domain of ribosomal protein L11 recognizes a rRNA tertiary structure stabilized by both thiostrepton and magnesium ion. *Nucleic Acid Res.*, **28**, 1778–1784, 2000.
25. Jefferson, E.A., Seth, P.P., Robinson, D.E., et al. Optimizing the antibacterial activity of a lead structure discovered by SAR by MS technology. *Bioorg. Med. Chem. Lett.*, **14**, 5257–5261, 2004.
26. Lee, J.L., Childs-Disney, A., Pushechnikov, JM., et al. Controlling the specificity of modularly assembled small molecules for RNA via ligand module spacing: Targeting the RNAs that cause myotonic muscular dystrophy. *J. Am. Chem. Soc.*, **131**, 17464–17472, 2009.

4 Real-Time Monitoring of Nucleic Acid Interactions with Biosensor-Surface Plasmon Resonance

Rupesh Nanjunda, Manoj Munde, Yang Liu, and W. David Wilson

CONTENTS

Specific biosensor surfaces with immobilized nucleic acids can be created in a hydrophilic matrix on a thin layer of gold for surface plasmon resonance detection that responds to mass changes. Nucleic acids linked to the hydrophilic layer respond in real time to binding events for kinetic and equilibrium analysis.

4.1   INTRODUCTION

Biological systems require thousands of overlapping biomolecular interactions to maintain their basic metabolism as well as to carry out any specialized cellular functions. The interactions cover the spectrum from macromolecules, including protein–protein and protein–DNA to small molecule–macromolecule interactions, and include complex molecular systems such as ribosomes and membranes. Important applications in medicinal chemistry include the design of therapeutic motifs that selectively target and interact with specific cells and/or macromolecules to exert a clinically useful biological effect. Such motifs vary from macromolecules such as antibodies to small synthetic drugs.

To understand and characterize this multitude of intrinsic biochemical and designed interactions, there is a major research effort to develop robust methods to monitor biomolecular interactions with small amounts of material while providing the maximum amount of information. For a reaction such as a small molecule, S, binding to a nucleic acid, NA, to give a complex C:

$$S + NA \xrightleftharpoons[k_d]{k_a} C \tag{4.1}$$

the basic desired information about the reaction is the equilibrium constant, K, the association, k_a, and dissociation, k_d, kinetic rate constants. This can obviously be extended to reactions with stoichiometry greater than one and such systems are presented in Section 4.2.10. In the more complex multisite case, both the stoichiometry of the interaction and any cooperativity in binding among sites must be characterized to obtain a full understanding of the complex formed.

The biosensor-surface plasmon resonance (SPR) method has emerged over the past 20 years as a powerful approach to determine all of the characteristic features of a molecular interaction in real time with very little material and without requirement of an external probe to monitor the interaction [1–8]. In this method, one component of an interaction, such as NA in Equation 4.1, must be immobilized on a sensor chip to create the biomolecular interaction surface. The other component(s) of the interaction is then injected over the flow cell surface in a solution at the desired pH and ionic strength. As will be described in detail below, the amount of complex formed is monitored in real time by SPR changes. As shown in Figure 4.1, injection of the compound can be replaced by buffer injection at a designated time and the dissociation of the complex can also be monitored by SPR. Since current research instruments have multiple channels, several nucleic acids can be immobilized on the same flow system (sensor chip) and questions about interaction specificity can be addressed. In this chapter, each of the steps in the biosensor-SPR method will be described and this will be followed with some example applications of the method to nucleic acid interactions.

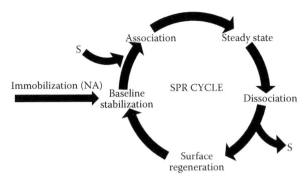

FIGURE 4.1   Outline of a single concentration injection and the steps in a surface plasmon resonance (SPR) cycle following the immobilization step.

SPR detection relies on a resonance interaction between a light beam and a thin metal film. The light beam strikes the metal, generally gold in biosensor-solution studies, at an angle such that total reflection of the light occurs. At a specific angle and light wavelength that depends on the materials used and solution conditions, resonance occurs and light is absorbed in the reflected beam at that angle, the SPR angle (Figure 4.2). The gold surface can be functionalized for linkage of a target macromolecule to provide an immobilized biosensor detection surface. Changes in the refractive index at the gold surface affect the SPR angle. As a complex is formed,

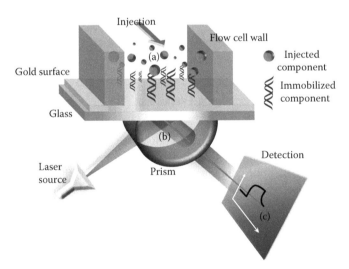

FIGURE 4.2   (See color insert.) Schematic representation of biomolecular interactions observed in a flow cell and SPR angle change with some of the critical components labeled. The flow of a compound (a) over the immobilized nucleic acid (blue) on the biosensor surface will result in a change in the refractive index with a change in the light beam–gold metal interaction and a change in the SPR angle (b) that can be monitored in real time by an array detector (c). The sensorgram overlaid on the array detector is simply to illustrate how the measured SPR angle change is converted into a time-dependent output signal.

the characteristics at the gold surface change, including the refractive index, and this can be quantitatively detected by changes in the SPR angle. The unbound concentration of the immobilized component decreases with binding and the concentration of complex increases by an equivalent amount. This leads to a change in SPR angle that is directly proportional to the concentration of the complex and allows determination of the reaction equilibrium constant, as will be described in this chapter.

4.2   BASIC COMPONENTS AND STEPS IN A BIOSENSOR-SPR EXPERIMENT

4.2.1   Materials

Biacore biosensor-SPR instruments were the first commercial instruments [4] and they currently account for the vast majority of publications in this field [9]; therefore, this chapter will lean toward those instruments, but the presentation will be general enough to apply to any of the high-sensitivity research instruments currently available. The high-sensitivity instruments of interest are able to accurately monitor small molecule–macromolecule, as well as, macromolecule–macromolecule interactions. The Biacore T100/200 research instruments are especially useful for experiments with small molecules binding to a macromolecule.

As is true in any method to monitor biomolecular interactions, it is essential to start with well-characterized, active, and pure reagents in their biologically relevant conformations [10]. It cannot be overemphasized that the data quality from the biosensor will directly depend on the purity as well as the activity of the reagents used. There are well-established procedures for immobilizing macromolecules, such as DNA, RNA, and protein, and they are generally immobilized on the sensor surface when studying macromolecule complex formation with a set of small molecules [3,5].

4.2.2   Overview of a Biosensor Experiment

The biosensor-SPR system to monitor molecular interactions consists of the following three modules: (i) an optical detection module that can detect changes in an SPR signal that are caused by binding; (ii) a replaceable, multichannel sensor chip for immobilizing one of the reaction components; and (iii) an internal flow module for liquid handling of the injected buffer and binding reagents that flow over the sensor chip in a precise manner [2,4,7]. An overview of the steps in a single concentration injection of the nonimmobilized compound in an SPR experiment to monitor a biomolecular interaction is shown in Figure 4.1 and the components of a biosensor-SPR system are illustrated in Figure 4.2.

The first step in the experimental process is immobilization of one component of a reaction to create the biospecific surface. In general, sensor chips have multiple flow cells, four in the Biacore research level instruments, which are functionalized to allow reagent immobilization (Section 4.2.3). The chip and flow cells consist of a glass plate base with a thin layer of gold on one side for the detection of the SPR signal change on complex formation (Section 4.2.4) [5]. The gold can react with a variety of thiol derivatives to give an immobilized hydrophilic matrix, such as carboxymethyl dextran, which can be activated for linkage of a target macromolecule that provides an immobilized biosensor surface [1–8]. Other surface matrices are also available [11]. Reagent immobilization raises another very critical point; however, any group that is used in immobilization cannot block or distort the macromolecular binding/active site in any significant way that perturbs the interaction. In other words, the interaction on the sensor surface must mirror the interaction in free solution as closely as possible. A method that has been very useful for nucleic acid immobilization involves streptavidin–biotin capture [3,5]. Biotin is easily linked to synthetic oligonucleotides at either the 5′ or 3′ and such synthetic strands are available from many suppliers. A biotin-linked nucleic acid can be immobilized on a streptavidin surface and because the linked biotin is at the terminus of the nucleic acid chain, it rarely perturbs the interaction to be monitored. Other groups, such as $-NH_2$, can be linked to the 5′ or 3′ terminus to allow covalent attachment of a nucleic acid [11].

As shown in Figure 4.1, immobilization is followed by buffer flow and baseline stabilization so that a constant signal with baseline drifts <2 response units (RU) is obtained as a function of time. Next, the remaining component(s) of the reaction is injected and the association to form a complex on the sensor surface is followed in real time until a steady state is reached, if the association occurs rapidly enough (Figure 4.3; Section 4.2.5). This is followed by reinjection of buffer to remove the bound small molecules from the complex and can be monitored by an SPR signal decrease. Finally, if the complex does not completely dissociate in a reasonable time period of a few minutes, a regeneration solution can be used to clean the final complex from the surface (Section 4.2.6). The process is restarted by reinjection of buffer to again yield a stable baseline, and another concentration of the nonimmobilized reaction component can be injected. The cycle in Figure 4.1 is repeated for as many injected concentration points as needed to completely define the binding curve and for each concentration a sensorgram, as shown in Figure 4.3, is obtained.

4.2.3 IMMOBILIZATION: WHICH COMPONENT AND HOW MUCH

Immobilization of one of the reaction components in Equation 4.1, creation of the selective-biosensing surface, is the single most important step in a biosensor-SPR experiment. If this step is not done properly, then no other part of the experiment can provide correct results in the experiment. As noted above, the key point in immobilization is not to perturb the binding/active site of the immobilized molecule. In this case, the binding on the sensor surface should match that in solution within an acceptable error limit. Fortunately, the usual conformation of nucleic acids is robust and relatively easily maintained or generated on the surface. Nucleic acid conformations also generally provide 5′ and/or 3′ terminal groups for immobilization to maintain the active conformation without blocking the binding site. With DNA double-helical duplexes or folded-back hairpin duplexes, both 5′ and 3′ groups are readily available. The same is true of multistrand conformations such as quadruplexes, which are becoming increasingly important in cellular chemistry and chemical biology. Folded RNAs such as tRNA, riboswitches, tetraloops, and control units such as HIV Rev and Tat RNAs also have terminal groups available that can be used to immobilize the RNA.

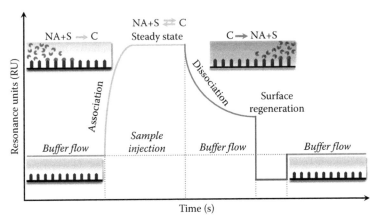

FIGURE 4.3 (**See color insert.**) Diagram of the steps and observed signal for buffer and compound injections over a sensor chip surface with immobilized DNA. These are typical responses observed in each cycle of a biomolecular interaction experiment. The immobilized NA is actually in a 3D hydrophilic matrix. In the sensorgram, the SPR angle change from Figure 4.2 has been converted to resonance units (RU) where a 1000 RU response is equivalent to binding of 1 ng/mm² for most biomolecules. In the 3D matrix of most sensor chips, this would be equivalent to approximately 6 mg/mL amount of bound material.

In order to carry out the immobilization reaction, some groups on the sensor surface that can be used to link biological receptors to create the sensor surface are needed. The goal of the surface chemistry is to provide an immobilization matrix for connection in surroundings that allow the reaction to take place as closely as possible to the solution reaction conditions. The best surfaces provide some type of hydrophilic three-dimensional matrix with chemical groups that can be used for immobilization. The primary surface in Biacore technology is composed of dextran polymers with carboxymethyl groups for immobilization chemistry [11]. The dextran is essentially a linear polymer of glucose units that is well hydrated and about 100 nm in length on average for the standard Biacore CM-5 sensor chip. The length of the dextran and number of carboxyl groups is varied on different sensor chips. This 3D surface provides space for a much greater level of immobilization, approximately 10 times more than the flat gold surface. The matrix with its hydrated dextran and solvent components provides an aqueous-like environment. This matrix has proved successful in reproducing quantitative results that are very close to those in solution for a wide variety of reactions. Since the SPR detection method effectively senses the environment to over a 100 nm from the surface, the increased immobilization on the 3D surface provides a dramatic increase in sensitivity to the reaction [11]. Sensor chips provided by other companies for their instruments, such as the six-channel Proteon instrument by BioRad, use other polymers that also perform well as aqueous reaction matrices.

The carboxyl groups in the sensor matrix are negatively charged at pH 7 and this can lead to problems that must be considered. Nucleic acids are negatively charged and are repelled by the surface. This can lead to problems with low immobilization levels, but can be relatively easily minimized by working at as low a $-COO^-$ density as possible, by lowering the pH during immobilization to partially protonate the –COO⁻ groups, and/or by increasing the salt concentration to screen the charges. Many compounds that bind to nucleic acids are positively charged and this can lead to the reverse case of nonspecific binding to the negatively charged surface. Again, this is best dealt with by using a low –COO⁻ density, and by doing experiments at 0.1 M salt concentration or higher.

The next key point in immobilization is to link one of the reaction components to the surface in such a way that it is not removed during the experimental time period. With the carboxymethyl dextran surface, this can be done through various carboxyl-linking chemistries [11]. The most common linking reaction is amide formation. The carboxyl group is activated through standard chemistry and reacts with a free amine to give a stable amide linkage [11]. An amine on the surface of a protein is useful in this respect but as already mentioned, the linkage should not change the conformation or block the binding site of the protein. Nucleic acids do not have reactive amine groups that can be used directly in this reaction, but such amines can be incorporated in oligonucleotide synthetic schemes at the 5′ or 3′ terminus and such derivatives are available commercially.

A more common way to link nucleic acids is through noncovalent capture. Streptavidin or one of its modified analogs can be linked to the surface through amide formation. Streptavidin-coated chips (Biacore SA chip) that are ready to be immobilized with biotin-labeled DNA are also commercially available. Biotin can be incorporated at the 3′ or 5′ ends of a nucleic acid and captured on the surface through the very strong and stable biotin–streptavidin complex. This immobilization method provides rapid kinetics and high-affinity binding of the nucleic acid to the surface. The concentration of DNA and the flow rate are very important during immobilization. A 100 μL solution at a low concentration (around 50–100 nM) of a 5′-biotin nucleic acid in an appropriate buffer, such as HBS-EP buffer (pH 7.4, 0.01 M 4-(2-hydroxyethyl)-1-piperazineethanesulfonic acid (HEPES), 0.15 M NaCl, 3 mM ethylenediaminetetraacetic acid (EDTA), and 0.005% surfactant P20), is loaded in the injection syringe. Typically, only about 5–10 μL of the oligonucleotide solution is consumed during actual immobilization [3]. It may be necessary to increase the concentration when using larger nucleic acids. A flow rate of 1–2 μL/min is generally a good practice to use during DNA immobilization to help control the immobilization rate and thus the level. A concentration or flow rate that is too high will make control over the amount of nucleic acid immobilized very difficult.

Generally, an immobilization amount of 300–450 RU (Figure 4.3) of a hairpin nucleic acid of 20–40 total bases in length is immobilized for running standard experiments. For cases where very accurate kinetic constants are needed, the immobilized RU should be decreased as much as possible to minimize the mass transfer effects (Section 4.2.8). Newer instruments, such as the Biacore T100 and T200, are able to provide excellent S/N with less immobilized nucleic acid. The details of the immobilization procedure are available in numerous references (e.g., [3,5]). Specific immobilized antibodies that form very strong complexes can also be used to capture nucleic acids, proteins, and other molecules to create a biospecific surface. It is important to reemphasize that the immobilization should be very stable through the binding experiment and preferably the biosensor surface should be stable enough so that it can be regenerated/cleaned after each experiment and reused a number of times (Section 4.2.6). In summary, it is relatively straightforward to immobilize various nucleic acid structures in their active conformation on sensor chip flow cells to create a biospecific reaction surface of high sensitivity that mirrors reaction results from solution.

4.2.4 DETECTION: SPR

While an entire chapter could easily be devoted to SPR signal detection, only the basic information necessary for the detection of a biomolecular interaction will be described here. The detection is based on an interaction at a gold layer of approximately 50 nm thickness on a glass surface of a sensor chip (Figure 4.2). Monochromatic light of a specific wavelength strikes the side of the gold film that is away from the functionalized surface such that the light is totally reflected [1–8]. At a certain angle and wavelength, a resonance condition is created between the reflected incident light and the electrons in the gold layer such that light at that angle creates a surface plasmon wave in the gold. At the specific incident angle where the light–metal coupling occurs, the SPR angle, the light intensity is reduced and this angle can be monitored accurately by an array detector (Figure 4.2).

The key to the detection of complex formation is that the local characteristics of the solution and matrix, specifically the refractive index, at the gold surface affect the SPR angle. The complex formed after reactant injection thus leads to a change in the SPR angle that can be quantitatively determined [1–8]. By subtracting the signal from a blank flow cell without an immobilized reactant, the signal change caused by unbound sample in the flow solution is removed and the signal change observed is only due to complex formation. An important point with the method is that the reactant concentration in the flow solution, equivalent to the unbound reactant concentration, is constant. As the reactant binds to the immobilized component, new flow solution rapidly replaces the partially depleted solution and there is no change in concentration in the flow solution such that reaction occurs at a constant concentration of the solution reactant(s). The unbound concentration of the immobilized component decreases with binding and the concentration of the complex increases by an equivalent amount. This leads to a change in SPR angle that is directly proportional to the concentration of the complex.

In order to determine the equilibrium and rate constants from the interaction in Equation 4.1, we must know all concentrations as a function of time. As described above, the solution reactant concentration is the same as its concentration in the flow solution. The immobilized reactant concentration, typically the nucleic acid, and the concentration of complex as a function of time are obtained from the changes in SPR angle, and will be described later. It should be noted that in the SPR method the light beam strikes the back side or opposite side of the sensor chip from the reaction matrix (Figure 4.2) and it is thus possible to use cloudy or scattering solutions, including whole cells in biosensor-SPR experiments. In summary, SPR biosensors are devices that monitor binding events on a surface matrix through the use of coupled light–plasmon changes for the transduction of a molecular interaction into an accurately monitored output signal.

4.2.5 SAMPLE INJECTION: DATA COLLECTION AND SENSORGRAMS

With a clean, well-functioning instrument, a stable baseline is obtained with buffer flowing over the sensor chip surface. At this point, the surface can be activated to immobilize the nucleic acid reagent in the reaction to be monitored (Section 4.2.3). After immobilization, the surface is washed to remove unbound reagent and buffer flow is again started. Once baseline stabilization is again obtained, the other reaction component can be injected (Figure 4.1). The diagram in Figure 4.3 is an illustration of the injection process: (i) initial buffer flow gives a reference baseline and sample is injected; (ii) the association is monitored until the desired amount of sample is injected—a steady-state plateau will be obtained with a relatively fast association rate and a large enough injected concentration; (iii) at this point, buffer flow is again started and dissociation of the complex can be followed; and (iv) in some cases complete dissociation is obtained in a time period of seconds to minutes while in others the dissociation is quite slow and a regeneration reagent is needed (Section 4.2.6). At the end of the regeneration, buffer flow is again started and when a stable baseline is obtained, another sample concentration can be injected.

Research-level instruments are automated to conduct this entire procedure as well as data collection with computer control. Each injection provides a reference baseline and a sensorgram for binding of the compound in the flow solution to the immobilized macromolecule. The data is processed with baseline correction and blank flow cell subtraction. The corrected set of sensorgrams can then be evaluated for reaction kinetics and equilibrium constants, as described in Section 4.2.7.

4.2.6 SURFACE REGENERATION

Following the completion of sample injection, buffer flows over the sensor chip and causes some dissociation of the bound complex, but this step can be slow depending on the dissociation rate constant. To take full advantage of the SPR instrument, it is important to develop a robust surface regeneration system so that the chip surfaces may be used at several concentrations in a reasonable time period [10]. The regeneration reagents must be determined empirically because the combination of physical interactions responsible for complex stabilization is often unknown. Good regeneration conditions should remove the bound compound completely from the surface without significantly removing or damaging the immobilized reagent. Some general regeneration solutions are listed in Ref. 3 and can serve as a place to start when evaluating the regeneration methods.

In studies with small-molecule complexes with DNA or RNA hairpin duplexes, 10 mM glycine/HCl (pH 2.5) is an efficient regeneration solution. This reagent unfolds the complex and removes the bound small molecule from the immobilized nucleic acid biosensor surface. Regeneration conditions must be harsh enough to break the complex and remove the bound reagent but mild enough to keep the nucleic acid strand intact. It is highly recommended to start with the mildest conditions and short surface contact times since regeneration solutions can cause undesired effects on the nucleic acid or immobilization matrix. To optimize regeneration, a solution mixture approach can be used to find the best regeneration conditions [12]. The main idea is to target several binding interactions simultaneously by mixing different chemicals. By using several chemicals in one cocktail, a complex can be disrupted at less harsh conditions to preserve the biosensor surface.

In some cases, regeneration may not be necessary due to rapid and complete dissociation complex from the sensor surface. However, in other cases, the surface may only be able to withstand a relatively small number of regenerations before it is disrupted. To resolve this problem, a different approach is necessary and more than one binding pulse can be used without surface regeneration [13,14]. This single-cycle kinetics method bypasses the need for surface regeneration between injections, and targets that are difficult to regenerate, or which have limited stability can be evaluated. In this method, the binding compound and dissociation buffer are injected in the standard manner but instead of waiting for complete dissociation or using a regeneration reagent, another sample is injected and this process continues until the nucleic acid-binding sites are saturated for a full-binding analysis. This procedure creates a more complicated data set for analysis since each injection does not have a reference baseline. Fortunately, the software to evaluate kinetics in systems without regeneration is now available [14].

4.2.7 DATA PROCESSING

4.2.7.1 From Raw Data to Final Sensorgrams

A typical-binding experiment would contain a series of compound injections of different concentrations over a sensor surface that is immobilized with a target nucleic acid of interest. The design of Biacore systems allows the evaluation of reactions in four flow cells simultaneously; however, in most cases one of the flow channels is used as a control or reference surface for subtraction. This is essential with the SPR detection method since injection of any solution with dissolved compounds will enter the matrix and cause a bulk refractive index change, relative to buffer alone, that is the same in all flow cells and can thus be subtracted. A typical step-by-step method for processing raw data to a final sensorgram for a small molecule interacting with a nucleic acid is described in this section. The steps described here are a guide to obtain high-quality final sensorgrams with all the essential information needed for further reaction kinetics and equilibrium-binding analysis. The data processing procedure described here yields sensorgrams of high quality (Figure 4.4). Many of the steps illustrated have been automated in the software supplied with most research instruments but a careful analysis of the steps from new data to final sensorgrams is always advisable.

4.2.7.2 Axis Calibration

Collection of a binding sensorgram, as illustrated in Figure 4.1, should begin with a baseline stabilization step (flow of the running buffer over the sensor surface); a stable baseline is an essential requirement at the beginning of each experiment. Injection of the compound for a sufficient time period provides the association and the steady-state steps. Compound injection is followed by resuming the flow of running buffer over the surface to monitor the complex dissociation phase. Finally, the surface regeneration step, which constitutes the injection of a regeneration reagent for a short period of time and followed by a series of buffer injections, again provides a stable baseline. All the pertinent interaction information is contained within the association and the disso-

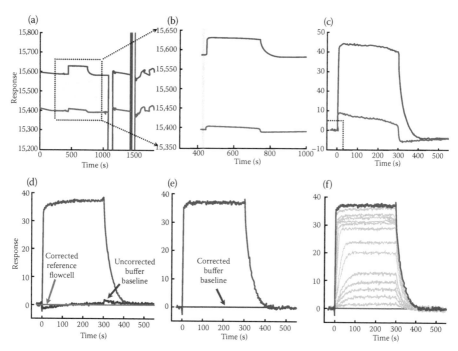

FIGURE 4.4   Sequential data processing steps involved in converting the original SPR data into final sensorgrams. (a) Raw sensorgrams for the injection of a small molecule over immobilized DNA (top trace) and the reference flow cell (bottom trace). (b) The expanded region from (a) for a more clear view of the association and dissociation steps of the cycle. (c) Zeroing of the response and the time-axes prior to the sample injection to obtain a detailed view of the binding event occurring on the flow cells. (d) Subtraction of the response of the reference flow cell from the immobilized-DNA surface. (e) Correction of the response from the blank buffer injections. (f) Overlay of the reference and buffer corrected sensorgrams for a series of concentrations of the ligand.

ciation steps of the cycle, and this information has to be accurately extracted for further processing. Figure 4.4a shows raw sensorgrams collected for a minor-groove-binding agent injected over a sensor surface immobilized with a hairpin-DNA (top trace) and also data from a blank reference surface on a flow cell with no DNA (bottom trace). This compound forms a strong complex with the hairpin-DNA sequence.

A quick visual inspection of the raw data in Figure 4.4a readily reveals the association and dissociation regions contained within a single cycle (highlighted by the dotted box). All the extraneous data within the cycle are first excluded by deleting those regions; however, precautions must be taken not to exclude regions from the cycle in which valuable information might be lost. This is particularly important for compounds that have very slow kinetics where a significantly longer dissociation time must be employed. Manually zooming into the region of interest is useful before deleting any regions of the cycle.

Figure 4.4b shows the highlighted region of Figure 4.4a after the removal of non-essential parts of the cycle. At this point, the association and dissociation steps can be readily observed; however, the response change due to the compound binding to the immobilized DNA is not yet clear. Zeroing the baselines on the response axis by choosing a small region of a few seconds (highlighted by the yellow region in Figure 4.4b in both the sample and blank sensorgrams) prior to the injection of the compound and overlaying the response for the sample and buffer injection to give Figure 4.4c provides clearer view of binding. A rough estimate for the change in the response associated with the compound binding to the immobilized nucleic acid (~45 RU in Figure 4.4c) can now be made. At this stage, the injection time associated with each step of the cycle on the time axis can also be zeroed. This is generally achieved by adjusting the beginning of the compound injection time to zero in all sensorgrams. A general sensorgram obtained after removing all the extraneous data and axes calibration is shown in Figure 4.4c.

4.2.7.3 Reference Cell Correction

In most cases, the reference cell surface is left blank on the sensor chip; however, nonbinding oligonucleotides can be immobilized for the use as a control surface. In the example in Section 4.2.7, the reference flow cell is a dextran–streptavidin blank without DNA. Flow of the compound solution over the reference surface results in a response change either due to compound adsorbing to the surface of the reference cell or due to the unbound compound in the flow solution. The response change associated with these phenomena must be subtracted from the other flow cells. Figure 4.4d shows a binding sensorgram with the elimination of background by subtracting the reference flow cell signal.

4.2.7.4 Buffer Correction

Although the sensorgram in Figure 4.4d is devoid of all the background from the reference cell, subtle systematic deviations in responses are generally present in the compound injections as well as the blank buffer injections across the individual flow cells with DNA. These effects are due to small differences in the geometry of flow cells on a sensor chip and can be removed by subtracting an average of several buffer injections (Figure 4.4d, green trace) from all compound injections over each individual flow cell surface [10]. The final sensorgrams result in elimination of any baseline distortion from each flow cell as shown in Figure 4.4e. The reference correction and the buffer correction steps are collectively termed *double referencing* [10] and result in sensorgrams that are free from the artifacts associated with the different steps of data collection. The final sensorgrams at different concentrations (Figure 4.4f) are now ready for either steady-state or kinetics analysis. If the sensorgrams do not look similar to Figure 4.4f, it suggests some problem with the flow system, or the reagents used, or extensive sticking of a component to the surface. The following sections will describe the fitting and analysis of the processed sensorgrams.

4.2.7.5 Stoichiometry (*n*)

One of the most useful features of the SPR method is that the stoichiometry, *n*, which for macromolecular interactions such as DNA with proteins or small molecules is the maximum number of molecules specifically bound to the macromolecule, comes directly from the experimental results. In many methods, the stoichiometry is evaluated indirectly, along with equilibrium-binding constants, by optimum fitting of a model to a well-determined-binding curve. The binding curve should cover a large section of the possible binding range from zero to completely bound. Direct determination of the n value in SPR obviously reduces the fitting dimensionality and improves fit quality and reliability.

In a biosensor-SPR experiment, the maximum response observed, $(RU_{max})_{obs}$, in an experiment is the response value in the steady-state region at a concentration of binding molecules sufficient to saturate all binding sites on the immobilized macromolecule, for example, the blue curve in Figure 4.4f. This is best determined in an experiment by finding the concentration of injected molecules in the flow solution that does not lead to a further increase in steady-state RU when the concentration is increased. It is also possible to predict a theoretical value for the maximum RU per molecule, $(RU_{max})_{pred}$, bound to a macromolecular site (note this is per bound molecule). This predicted value depends on several features of the immobilized and flow solution interaction components. For a small molecule, S, binding to a DNA site, NA (or any other immobilized reagent), in Equation 4.1 for example, the $(RU_{max})_{pred}$ depends on the amount of immobilized DNA, RU_{NA}; the molecular weights of the immobilized DNA, M_{NA}, and of the compound in the flow solution, M_S; and the derivative of refractive index, n', with concentration, c, for DNA $(dn'/dc)_{NA}$ and compound, $(dn'/dc)_S$ [15]:

$$\left(RU_{max}\right)_{pred} = RU_{NA} \times \left(\frac{M_S}{M_{NA}}\right) \times \frac{(dn'/dc)_S}{(dn'/dc)_{NA}} \quad (4.2)$$

The observed steady-state response, RU_{obs}, at each concentration of binding molecule also depends on the fraction of the DNA sites with compound bound, f_{bound}, and the stoichiometry, n:

$$RU_{obs} = \left(RU_{max}\right)_{pred} \times n \times f_{bound} \quad (4.3)$$

Thus, dividing the observed steady-state response RU_{obs} by the calculated response $(RU_{max})_{pred}$ yields a stoichiometry-normalized-binding isotherm, and at the maximum where f_{bound} is 1, the stoichiometry for the interaction is determined directly without the need for fitting:

$$n = \frac{\left(RU_{max}\right)_{obs}}{\left(RU_{max}\right)_{pred}} \quad \text{at } f_{bound} = 1 \quad (4.4)$$

As will be described in Sections 4.2.9 and 4.2.10, RU_{obs} in Equation 4.3 can be plotted versus the compound concentration in the flow solution, the unbound concentration, to allow determination of K, which depends on f_{bound}.

There are some features of this method of determining stoichiometry, which can lead to errors without appropriate care. From the equations above, it can be seen that accurate calculation of $(RU_{max})_{pred}$, and, as a result, stoichiometry, depend on an accurate determination of RU_{NA}. RU_{NA} is determined, for example, by carefully monitoring the RU increase as DNA is captured on a streptavidin (or other) surface during preparation of the biosensor surface (Section 4.2.3). Two possible problems with this method are that (i) some of the DNA may just absorb temporarily to the surface and dissociate in subsequent surface washes and (ii) over a period of use some of the DNA may degrade or be lost from the surface. The first problem can be minimized by carefully monitoring the buffer baseline of the immobilized DNA surface between wash cycles after immobilization. If the baseline decreases significantly, it generally means DNA loss that requires correction of RU_{NA}. A correction for the second problem can be obtained by injecting a high concentration of a well-behaved and/or well-characterized-binding molecule with a known stoichiometry, n_{std}, over the prepared surface and determining the maximum response, $(RU_{max})_{std}$, in addition to $(RU_{max})_{pred}$. If it is suspected that some DNA may have been lost after several experiments, the standard can again be passed over the surface and the new $(RU_{max})_{std}$ compared with the original to give a correction to the RU_{NA} value, if necessary.

If neither of the above two methods proves satisfactory for determining $(RU_{max})_{pred}$, then Equation 4.4 can be rewritten with $(RU_{max})_{std}$ in place of $(RU_{max})_{obs}$:

$$\left(RU_{max}\right)_{pred} = \frac{\left(RU_{max}\right)_{std}}{n_{std}} \quad \text{at } f_{bound} = 1 \tag{4.5}$$

and the value of $(RU_{max})_{std}$ is determined by fitting experimental-binding curves as described in Sections 4.2.8 and 4.2.9. This obviously requires another fit parameter and will thus decrease the reliability of the fit, but there may be no choice. With a well-determined-binding curve, the K, n, and $(RU_{max})_{std}$ can be determined quite accurately. Another possible problem with Equation 4.2 is in the refractive index gradient, (dn'/dc), values. These values are very similar with DNA, proteins, and a number of other compounds such that the (dn'/dc) ratio essentially cancels in Equation 4.2 [15,16]. These values are generally not known for small molecules of interest and the (dn'/dc) value may be very different from DNA, proteins, or other immobilized-binding reagents [16]. In this case, it is again not possible to reliably calculate $(RU_{max})_{pred}$ and the best choice is to use Equation 4.5 to determine $(RU_{max})_{std}$ as part of the fitting procedure.

There are some features of Equation 4.4 that make stoichiometry determination more accurate. Except in unusual cases, the stoichiometry of a reaction is a whole number that is generally low so that values very different from whole numbers can be rejected. If $(RU_{max})_{pred}$ is determined to be 22, for example, then a value of $(RU_{max})_{obs}$ in Equation 4.4 between 18 and 26 would clearly mean $n = 1$ while a value between 40 and 48 would mean $n = 2$ and so forth. At high n values, the differences are reduced and the method is less reliable, but this is, of course, true for all methods as they try to sort out values for high n. High n values are not a problem in the design of most biosensor experiments and n values of 1 or 2 are by far the most common. In summary, in most cases it is possible in biosensor-SPR experiments to determine the interaction stoichiometry independent of fitting a binding curve and this significantly improves the accuracy of the method. In cases where this cannot be done, the stoichiometry can be obtained by fitting the binding curve.

4.2.8 REACTION KINETICS AND BINDING AFFINITIES WITH MASS TRANSFER

Because of the varied properties of compounds that bind to nucleic acids, it is frequently difficult to find suitable methods to quantitatively monitor the interactions of a variety of compounds. If the binding reaction is followed by biosensor-SPR methods: (i) the requirement to determine the diverse properties of compounds that would be required in other techniques is removed since the method responds directly to the mass of bound compound, and (ii) direct determination of association and dissociation rates, the reaction rates in both directions, can be monitored in real time as the reaction occurs. Both the association and dissociation phases of the sensorgram can be fit to an appropriate binding model for several sensorgrams at different concentrations with global fitting, where all data are used to get the best fit to the entire experiment [10]. Fitting in this way allows the most accurate determination of the kinetics as well as the equilibrium constant from the ratio of kinetic rate constants. It is also possible to determine the binding constants independently of rate constants by fitting the steady-state response (Section 4.2.9).

Following the collection of a series of sensorgrams as a function of concentration, the next step is choosing an appropriate model to fit the processed results. This is basically the same procedure as with any data set involving a reversible interaction. Different interaction models are provided with the software of most SPR instruments and Myszka and coworkers have done extensive software development for processing and fitting biosensor-SPR results [17]. Most equilibria of interest are explained by standard one- or two-site interaction models for small molecules binding to nucleic acids [9] and standard equations for such fits are well known [18]. More complex models can be entered into most fitting programs for the analysis of the data, but as the number of variables that must be determined from the data increases, the error limits begin to increase to a point where the fit loses significance. The problem is primarily a result of random error in the experiment and variable correlation, which essentially means that the same quality fit, within experimental error, can be obtained by reducing the value of one variable and raising the value of another (or of several). Equilibria that appear complex may also be simple but seem complex due to impure materials and/or errors in experimental design [9,10].

Since the reaction occurs on a surface, there is an additional factor that must be considered in biosensor experimental design and in evaluating kinetic constants from biosensor-SPR methods. The unbound reactant is injected over the flow cell surface and must be transported from the bulk solution into the matrix, which is called mass transport, before it can react with the immobilized component. The key assumption in accurate determination of kinetic constants by this method is that the concentration of the free reactant in the matrix quickly reaches a constant value that is the same concentration as in the flow solution. For this condition to hold, mass transfer must occur significantly faster than the binding reaction. If the reagent is consumed in the binding reaction significantly faster than it can be resupplied from the flowing solution by mass transfer, then the matrix concentration will fall below the flow solution concentration and a mass transfer effect is obtained [19]. In this case, the observed k_a is less than the actual k_a without mass transfer. An illustration of the mass transfer effect and how it affects observed reaction rates is shown in Figure 4.5. Small spherical molecules typically diffuse faster and are transported rapidly while larger, asymmetric molecules diffuse slower and thus require a longer time period for transport. The mass transfer rate also depends on the flow rate at which the unbound reagent is injected over the surface

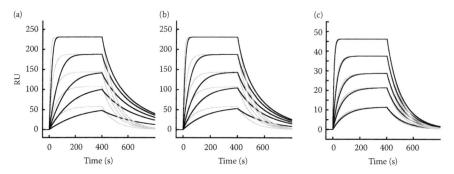

FIGURE 4.5 Simulated sensorgrams for the case of no mass transfer (gray) and mass transfer (black) under different experimental conditions. (a) A low flow rate and high immobilization level give mass transfer-dominated sensorgrams. (b) Increasing the flow rate lowers the mass transfer effect. (c) High flow rates and low immobilization levels remove most of the mass transfer effect.

since this is the way the surface is resupplied with reagent that diffuses into the matrix. All of these factors can be combined into a mass transfer constant, k_t, and the complete process for a 1:1 reaction to take place in the matrix is defined by an expanded Equation 4.1:

$$S_{bulk} \underset{k_t}{\overset{k_t}{\rightleftharpoons}} S_{matrix} + NA \underset{k_d}{\overset{k_a}{\rightleftharpoons}} C \tag{4.6}$$

where k_a and k_d are the standard association and dissociation rate constants for a reaction, and S_{bulk} and S_{matrix} are the concentrations of the free reagent in the bulk and matrix phases. The mass transfer rate constant is the same in the forward and reverse directions [19].

Although estimates of k_t are possible based on molecular properties, flow cell characteristics, and flow rate, the entire equation above can be used to fit the results in biosensor experiments when there is only a moderate effect of mass transfer on the reaction rates. In this case, k_t is determined from the fit along with k_a and k_d [19]. If the mass transfer effect is so dominant that the rates of binding are largely controlled by the transfer step in the above process, then it will not be possible to accurately determine k_a and k_d without changing the method of data collection.

The general effects of mass transfer and how it is affected by experimental variables are illustrated in Figure 4.5, which is a simulation of a binding reaction under different conditions with the BIAsimulation 2.1 software (GE Healthcare Biosciences, Sweden). The k_t value used in the simulation is not important since it will change with different instruments and flow cell designs and the same is true of the k_a and k_d values since they change with the reaction conditions. The relative changes in mass transport with changes in the experimental conditions are the important considerations in this comparison. In the left panel, the flow rate is 20 μL/min and the maximum RU of immobilized DNA is 250. As can be seen, significant mass transfer is obtained in this system and the observed kinetic constants with mass transfer are artificially reduced by about 2/3 relative to an equivalent, but ideal, system with no mass transfer.

In the center panel, the flow rate is increased to 100 μL/min and the mass transfer effect is reduced to about 50%. In the right panel, the flow rate remains 100 μL/min while the immobilized reagent is reduced by a factor of 5 (RU_{max} of 50) and the mass transfer effect is reduced to about 20%. Fitting these results with the mass transfer constant from Equation 4.6 should provide accurate values for k_a and k_d but it would be much more difficult in the experiment shown in the left panel [19]. These simulations point to the strong effect of the amount of immobilized reagent on mass transfer and of the need to keep the immobilized amount as small as possible if accurate kinetics are desired. For low-molecular-weight-binding reagents, however, the immobilized amount must be large enough to give a satisfactory signal on complex formation. The flow rate in an experiment affects the data collection time period, but the highest possible flow rate that is consistent with the data collection time needed should be used. The data collection time needs to be long enough for at least half of the association or dissociation to occur in order to be able to accurately fit the curves and determine k_a and k_d.

If within the experimental constraints described above it is not possible to find a set of conditions that will reduce mass transfer to an acceptable level, no values for k_a and k_d can be determined in the experiment. As can be seen in the simulations in Figure 4.5, however, even when mass transfer is significant, the reactions eventually reach the same steady-state plateau as without mass transfer. In this case, the K (but not k) can be determined by using the steady-state method (Section 4.2.9). Even if it is not possible to reach a steady state in the experiment for most of the injected concentrations, Karlsson [19] has shown that the ratio of the mass transfer influenced k_a and k_d and provides an accurate K value that should be the same as obtained with steady-state methods. An example of fitting the kinetics of an interaction that involves some mass transfer is given in Section 4.3.1.

4.2.9   Steady State (K)

As described above, SPR results can provide both kinetic (Section 4.2.8) and, when it is available, steady-state quantitative characterization of interactions. When the association phase collection time is long enough for a particular reaction, a steady-state plateau can be reached in the association step such that the rate of binding of the injected molecule equals the rate of dissociation of the complex and no change of signal with time is observed (e.g., see the highest concentration sensorgrams in Figure 4.6). When the association rate is too fast to be analyzed by kinetic fitting, such that the reaction basically occurs during filling of the flow cell with reactant, steady-state analysis is the only option to determine K.

In order to determine the binding constant from these results, the steady-state RU values from the experiment are plotted against free compound concentration (from the flow solution concentration). The results can then be fitted to an appropriate interaction model to determine the equilibrium-binding constant, K (e.g., the DB921-AATT complex in Section 4.3.1). Since, as shown in Section 4.2.8, mass transfer does not affect the steady-state RU, possible mass transfer effects and more complex fitting models are unnecessary. A second significant advantage of steady-state analysis, particularly for low-molecular-weight systems, is that there is no special requirement

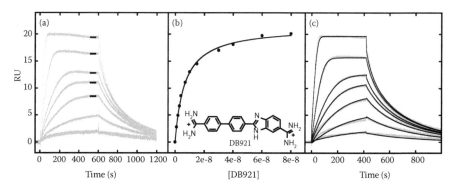

FIGURE 4.6 Sensorgrams and binding curve for the DB921 interaction with an AATT site in the DNA minor groove. (a) Sensorgrams for increasing concentrations of DB921 (bottom to top, gray lines) with best fit lines in the steady-state region (black, note that the bottom two, lowest concentration injections, do not reach a steady state). (b) The averaged RU from (a) are plotted against the concentration of DB921 in the flow solution (unbound concentration) to obtain K from the best fit (solid curve). Note that points from additional sensorgrams are included in this plot to improve the fit. (c) Sensorgrams at a higher flow rate (gray) and best fit lines (black) to a kinetic model [19]. The DB921 concentrations are 1, 3, 6, 12, 30, and 65 nM from bottom to top.

for a low-density biosensor surface to study nucleic acid–small molecule interactions. A disadvantage of steady-state analysis is that it cannot be used for cases where the reaction does not reach a steady state within the experimental conditions and a kinetic fit is the only possibility. In many cases, however, this limitation can be overcome. Since the association reaction must at least have a stoichiometry of 1, for a 1:1 complex, the association must be at least bimolecular and the rate of association will increase with injected concentration and at some point, except for the very slowest reactions, will reach a steady state. All sensorgrams above this point can be used in a steady-state fit. It is also possible to use some lower concentration points by extrapolating the association region of the sensorgrams to the steady-state time limit. Although this must be done with caution, it can be quite accurate if RU_{max} is known and the curves are close to the steady-state limit as in Section 4.3.1.

4.2.10 Multisite Interactions and Cooperativity

Experiments for interactions that involve two or more sites with or without cooperativity are conducted just as with single-site interactions, except that it is generally useful to have more data points (additional concentration injections) than with single-site complexes to clearly define the binding curves for the more complex processes. As noted above, one of the major advantages of the SPR detection method is that comparison of predicted response, $(RU_{max})_{pred}$, and observed response, $(RU_{max})_{obs}$, allows the determination of the total number of binding sites on an immobilized nucleic acid independently of fitting the binding curve (Section 4.2.7). Independently fixing the stoichiometry for a complex allows more accurate determination of K values for multiple-site interactions.

Generally, in the preparation of a nucleic acid biosensor surface for a multisite interaction, it is possible to, at least partially, simplify the system so that the number of sites on the immobilized nucleic acid is reduced to a smaller number, in many cases only one or two. Relatively short DNA hairpin or hybridized duplexes, modular quadruplexes, or other structures have been designed and immobilized to reduce the binding complexity. In the same manner, quite complex RNA structures are frequently composed of folded, relatively independent modules that can be isolated and immobilized on a sensor surface for individual study [20,21]. The most common multisite complexes studied for both protein and small-molecule interactions with DNA to this time have been two-site complexes, and we will focus on two-site complexes in this section.

In analogy with Equation 4.1, we can write for a two-site interaction:

$$S + NA_1 \xrightleftharpoons[k_{d1}]{k_{a1}} C_1 \quad K_1 \tag{4.7}$$

$$S + NA_2 \xrightleftharpoons[k_{d2}]{k_{a2}} C_2 \quad K_2 \tag{4.8}$$

where S is a small molecule (or protein, etc.) that is injected over an immobilized nucleic acid surface that has two binding sites, NA_1 and NA_2, on each nucleic acid molecule. The sites can be identical or different and can be noninteracting or can interact with positive or negative cooperativity. Cooperativity can be very important to the function of biological systems. The analysis can obviously be extended to $n > 2$.

The basic desired information about the reactions in the above equations includes the macroscopic equilibrium constants K, the association k_a, and dissociation k_d kinetic constants. The macroscopic equilibrium constants are determined with Equation 4.9 from the observed RU at each free compound injection point, C_f:

$$\frac{RU}{RU_{max}} = \frac{K_1 \cdot C_f + 2 \cdot K_1 \cdot K_2 \cdot C_f^2}{1 + K_1 \cdot C_f + K_1 \cdot K_2 \cdot C_f^2} \tag{4.9}$$

where C_f is the unbound reagent concentration in the flow solution and is constant for each injection. If $K_2 = 0$, a single-site binding equation is obtained that gives the K value for Equation 4.1 [5]. With equivalent, noninteracting sites, $K_1 = 4*K_2$, in Equation 4.9, a purely statistical value [18]. If the sites in a two-site complex are identical but interact when compounds are bound, positive, $K_1 \ll 4*K_2$, or negative, $K_1 \gg 4*K_2$, cooperativity in the interaction can be determined directly by evaluating the macroscopic constants [3].

All of the kinetic rate constants, k_a and k_d, in the above reaction equations, are also important and are needed for a full characterization of the interacting system if they can be determined. It should also be noted that the reactions above have conditional microscopic equilibrium constants that provide much additional information [18]. The problem, however, is that from binding results alone it becomes quite difficult to accurately determine all of the microscopic constants, even the four microscopic constants in a 2:1 complex, unless results from other techniques are also available.

Cooperativity in biomolecular interactions is difficult to establish in any quantitative manner with most techniques for monitor binding, and biosensor-SPR methods are among the best for this type of analysis as shown below (Section 4.3.2).

4.3   APPLICATIONS AND CASE STUDIES

4.3.1   SINGLE-SITE BINDING: 1:1 INTERACTIONS

Small-molecule complexes with DNA that incorporate a linking water are rare and DB921, synthesized by Boykin and coworkers [22,23], which binds in the DNA minor groove, provides a unique and well-defined system for the analysis of water-mediated binding in the context of a DNA complex (Figure 4.6). The DB921 benzimidazole-biphenyl system has terminal amidines with an unusual linear conformation without the appropriate radius of curvature to match the DNA minor groove shape. A crystallographic structure shows DB921 bound at the AATT site with a water-mediated interaction between the phenylamidine of DB921 and DNA [23] that can serve to complete the curvature of the bound system and yields a very energetically stable complex. Biosensor-SPR experiments to obtain quantitative binding affinity and stoichiometry information on the complex were performed with a four-channel Biacore T100 optical biosensor system (GE Healthcare Biosciences, Sweden). A 5′-biotin-labeled AATT DNA sequence was immobilized onto streptavidin-coated sensor chips (Biacore SA). Steady-state-binding analysis was performed with multiple injections of different compound concentrations over the immobilized DNA surface at a flow rate of 25 µL/min. Sensorgrams for the interaction are shown in Figure 4.6a and RU values from the steady-state region of SPR sensorgrams are plotted against C_{free} in Figure 4.6b to obtain the equilibrium-binding constant ($K = 1.5 \times 10^8$ M^{-1}; fitted with Equation 4.9 with $K_2 = 0$). Binding results from the SPR experiments were fit with a single-site interaction model well within experimental error. In this figure, the steady-state was not reached at the lowest injected concentrations and the steady-state values were determined by extrapolation to longer times.

A kinetic analysis was also performed with multiple injections of different DB921 concentrations over the immobilized DNA surface at a flow rate of 50 µL/min (Figure 4.6c). The binding results were evaluated with global fitting to a mass transport kinetic-binding model with one-site (Section 4.2.8). The association ($k_{\text{a}} = 2.0 \times 10^6$ M^{-1} s^{-1}) and dissociation ($k_{\text{d}} = 0.014$ s^{-1}) rate constants and the derived equilibrium-binding constant ($K = 1.4 \times 10^8$ M^{-1}), as a ratio of the association and dissociation rate constants, were determined from the global best fit (Figure 4.6c). As can be seen from these results, the binding affinity derived from kinetic models is in good agreement with the K obtained from steady-state fitting. The kinetic fitting results for DB921 binding with AATT meet the criteria for an acceptable kinetic model (Section 4.2.8) [19] since the rates are not dominated by mass transport.

4.3.2   TWO-SITE INTERACTION–COOPERATIVITY: DIMER FORMATION

A number of minor-groove-binding agents interact with the groove in a cooperative manner to form a stacked dimer complex; dimer formation can be quite strong and selective for specific DNA sequences [24–26]. Since the SPR-biosensor method directly establishes the stoichiometry of complex formation (Section 4.2.7), it is quite effective in defining the affinity and level of cooperativity in the interaction. A number of heterocyclic cations have recently been found to form cooperative complex and the furan-benzimidazole, DB915 (Figure 4.7), is an example that forms a 2:1 stacked complex in the minor groove of a CGTTAACG sequence in a 5′-biotin hairpin duplex [27].

A streptavidin-coated sensor chip (Biacore SA) was used in the experiment to immobilize the TTAA hairpin duplex and multiple injections of different DB915 concentrations were injected over the surface (Figure 4.7a). The sensorgrams increase in response as the compound concentration is increased and, as can be seen in Figure 4.7a, reach a steady-state plateau. The steady-state RU values were obtained by equilibrium fitting (Section 4.2.9) and the observed RU_{max} value clearly indicated a 2:1 complex formation (Section 4.2.7). The RU values are plotted against free compound concentration in Figure 4.7b and global fitting with a two-site model using Equation 4.9 gave values of $K_1 = 2.3 \times 10^6$ M^{-1} and $K_2 = 3.5 \times 10^7$ M^{-1} with a very strong indication of positive cooperativity for the dimer formation. The maximum RU value from the fit is 57 and the predicted value for two DB915 molecules bound to TTAA is 60 (Section 4.2.7) as described above. The RU versus C_{free} plot in Figure 4.7b has a sigmoidal shape and the fact that K_2 is higher than K_1 illustrates that DB915 binds as a cooperative dimer. This is

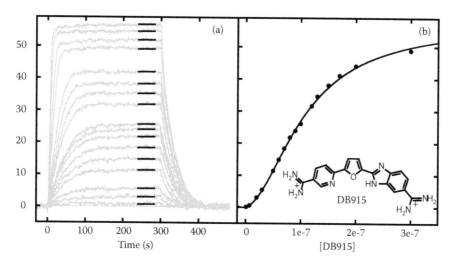

FIGURE 4.7 Sensorgrams (gray) for the interaction of DB915 with a TTAA sequence. The sensorgrams (a) were collected in 0.1 M NaCl, 10 mM cacodylate buffer, 1 mM EDTA at pH 6.25. The individual sensorgrams represent responses at different DB915 concentrations; the concentrations are from 1 nM (bottom) to 300 nM (top). DB915 solutions were injected at a flow rate of 25 μL/min over a period of 5 min followed by a 5 min dissociation period. RU values in the steady-state region were obtained by linear fitting (black lines). (b) The steady-state RU are plotted versus the DB915 concentration in the flow solution and the plot is fitted with a two-site model Equation 4.9. The sigmoidal nature of the plot clearly indicates strong, positive cooperativity for the binding of DB915 to the TTAA sequence. Many more sensorgrams (concentration injections) are needed to fit this more complex, cooperative interaction than for the simple 1:1 interaction shown in Figure 4.6.

an example of how SPR technology can be used to study the binding affinity, stoichiometry, as well as cooperativity of biomolecular interactions in a single experiment.

4.3.3 DNA HYBRIDIZATION

DNA hybridization is one of the most important methods in the detection and characterization of microorganisms, mutation detection, and a number of other biotechnology applications. It is a key tool for the diagnosis of several diseases, including some types of cancer related to inherited mutations. The high sensitivity and specificity of DNA hybridization techniques make them powerful tools for DNA diagnosis [28,29]. The biosensor-SPR method is a powerful alternative to conventional DNA hybridization screening methods because it can provide a detailed quantitative analysis of the DNA interaction in real time without the need for labeling [30,31].

In addition to its use in biotechnology applications, DNA hybridization can also be used to prepare biosensor surfaces for use in quantitative SPR-binding studies. In this method, instead of immobilizing a fold-back single-strand DNA sequence, a noncomplementary single strand is immobilized by using standard methods (Section 4.2.3). This strand is then hybridized to form a duplex by flow of the complementary strand. In order for this to be useful for SPR, the duplex must have a high T_m and withstand solution flow throughout a standard SPR experiment. This type of procedure can also be used to characterize the kinetics and affinity for the formation and dissociation of a duplex, including the effects of base pair mismatches, backbone modifications, and other similar features of interest.

4.3.4 G-QUADRUPLEXES

One of the areas of considerable interest in the field of targeting regulatory nucleic acid elements is designing and developing small molecules that can selectively target noncanonical DNA conformations such as G-quadruplexes [32–34]. The remarkable structural diversity exhibited by quadruplex-forming sequences throughout the genome, coupled with the elucidation of their potential role in important biological processes, provides an excellent platform to study these structures using biosensor-SPR techniques [35,36]. There is a growing body of literature documenting small molecule–quadruplex interactions using SPR techniques that demonstrates the accuracy and reliability of this method for quadruplex studies [34,37,38]. Additionally, the technique has found applications in probing folding kinetics of single-stranded oligonucleotides into quadruplexes, ligand-induced conformational transitions of quadruplex structures, and in monitoring quadruplex–protein interactions [39–42].

Azacyanines are a class of planar, bis-benzimidazole analogs that were designed as ion channel activators [43,44] and can serve as an example for the discovery of quadruplex-binding agents. Although they were shown to have high affinity for single-stranded poly(A) sequences [45], the large fused ring systems of azacyanines are too large to intercalate in a Watson–Crick duplex but could potentially stack at the terminal G-quartets of a quadruplex [46]. The introduction of exocyclic groups along the aromatic core of the compound was predicted by Dr. Nicholas Hud and coworkers to produce greater ligand diversity for discrimination among different types of G-quadruplexes. NMR structural studies of aza-3 (which lacks any exocyclic groups) with a modified human telomeric sequence, Tel24 d[TTGGG(TTAGGG)$_3$A], showed the compound preferentially stacking at the 3′-terminal of the quadruplex with a secondary weak affinity for the 5′-terminal tetrad. SPR results for the binding of these compounds to quadruplexes show an excellent correlation with the results obtained from NMR structural studies and fluorescence-binding experiments [46].

The interaction of aza-5, which contains a methoxy exocyclic group, with the Tel24 quadruplex and a hairpin-DNA duplex sequence is illustrated here with SPR methods (Figure 4.8). The biosensor-SPR experiments were performed with a four-channel Biacore 2000 optical biosensor system (Biacore, Inc.) and streptavidin-coated sensor chips. All DNA quadruplex and hairpin-DNA sequences were used as folded single strands to prevent dissociation in the SPR flow system. A series of different aza-5 concentrations were injected onto the chip (flow rate of 50 µL/min, 5–10 min) until a constant steady-state response was obtained followed by a dissociation period. At the end of the dissociation period, the chip surface was regenerated (20 s injection of 10 mM glycine solution, pH 2.0) and this was followed with multiple buffer injections to obtain a stable baseline for the next round of compound injections. The sensorgrams for the interaction of aza-5 with the quadruplex Tel24 and hairpin-DNA are shown in Figure 4.8b and b, respectively.

The instrument response in the steady-state region is proportional to the amount of bound drug and was determined by linear averaging over a 40–60 s time span (indicated with black bars in Figure 4.8a and b). The steady-state response values of Tel24 and hairpin-DNA (Figure 4.8c) were plotted as a function of unbound compound concentration in the flow system. To obtain the binding constants, the steady-state responses were evaluated as described in Section 4.2.9.

SPR shows an initial strong binding for the quadruplex-DNA with an affinity of 2.9×10^6 M^{-1}, followed by secondary binding that is almost 10-fold weaker.

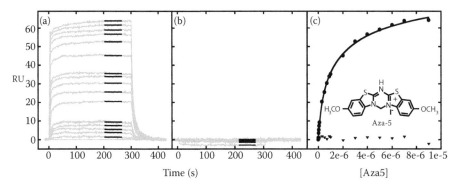

FIGURE 4.8 SPR sensorgrams for binding of Aza5 to a quadruplex-forming sequence, Tel24 (a), and a hairpin-DNA (b) in HEPES buffer containing 80 mM KCl at 25°C. The quadruplex curves range in Aza5 concentration from 100 nM to 10 µM. Steady-state-binding plots (c) are shown along with best-fit lines to obtain the binding constants (Equation 4.9); circles are for the quadruplex and triangles are for the duplex. The concentration values in (c) are for Aza5 in the flow solution.

This suggests that the two molecules could be potentially interacting at different ends of the quadruplex structure; this was in excellent correlation with the NMR studies. The response value per site obtained using this model (RU = 38) also agrees well with the predicted maximum response per bound compound, which is 40. A quite weak, nonspecific binding could also be detected at higher concentrations. Remarkably, no significant aza-5 interaction with hairpin-DNA sequences was detected under the SPR conditions used (Figure 4.8b) and the extremely low steady-state response data for the duplex could not be effectively fit to a binding curve (Figure 4.8c). This example clearly shows that complex quadruplex interactions can be evaluated with high accuracy by using biosensor-SPR techniques.

In spite of the success of SPR studies with quadruplexes, precautions must be taken with quadruplex-forming DNA sequences because of their structural plasticity. The immobilization of the quadruplex-DNA on to the sensor chip surface is essentially the same as duplex-DNA immobilization. Even though the nature of the quadruplex analysis is not dependent on the type of sensor chip, the streptavidin–dextran gold surface sensor chips have been the preferred choice of most researchers and biotinyl-ated quadruplex-forming oligonucleotides are commercially available for immobili-zation on the streptavidin surface. With the abundance of duplex-DNA in the genome, the binding affinity and selectivity of small molecules for quadruplex structures over duplex sequences is always in question, and it is imperative to employ the control duplex sequences for evaluation.

The nature of cations in the buffer can directly affect the stability and the folding of quadruplex sequences and it is useful to have the quadruplex sequence folded into the preferred conformation prior to immobilization. An alternative is to employ quadruplex destabilizing agents such as lithium in the immobilization buffer and after immobilization, to treat the surface with a buffer containing a stabilizing cation such as potassium to allow folding into the preferred conformation on the surface. This is a case where a known quadruplex-binding molecule can be used to check for correct folding on the surface.

The exposed terminal tetrads of quadruplex units are the most preferred-binding sites for small molecules. Therefore quadruplex-interactive compounds are typically cationic and composed of large aromatic ring systems to mimic the surface of a quar-tet. However, solubility of compounds with large ring systems is a problem that is gener-ally observed and extends to quadruplex-binding compounds also. SPR requires the delivered compound to be devoid of any particulate matter to prevent blockage of the narrow channels of the microfluidics. To assist with solubility, organic solvents such as dimethyl sulfoxide (DMSO) can be used in very small quantities to enhance the solubil-ity of the compounds. However, care must be taken not to use organic solvents in excess due to the possible damage to the microfluidics. In order to cancel out the refractive index change caused by the added solvent, it must be included in exactly the same amount in the injected buffer that is used to establish the baseline (Section 4.2.7).

4.3.5   PROTEIN–DNA INTERACTIONS

Sequence-specific DNA recognition by proteins modulates many cellular pro-cesses such as replication, recombination, and transcription, and understanding their interactions is fundamentally important for deciphering the molecular mechanisms involved [47–49]. Many of the techniques used to monitor protein–DNA binding do not provide direct information related to binding affinity or kinetics of protein–DNA interactions. Real-time detection of interactions is also an added plus for the analysis of complex macromolecular interactions of this type and this makes biosensor-SPR methods to monitor these reactions especially attractive.

Several protein–DNA interactions have been successfully investigated using SPR and some selected examples will be presented here. The mammalian high-mobility-group transcriptional factor HMGA2 (high-mobility group AT-hook 2) is an AT-hook protein that targets the DNA minor groove in AT sequences [50,51]. The protein has important functions in disease processes from cancer to obesity. For HMGA2-binding investigations, DNA sequences with sites containing two AT-target sequences, the minimum required for strong binding of HMGA2, were selected. Biosensor-SPR methods have been used to monitor HMGA2 binding to the target sites on immobilized DNA and binding constants from $3–5 \times 10^7$ M^{-1} on different AT-binding sequences were found [52]. The design of small synthetic molecules that can be used to affect gene expression is an area of active interest for development of agents in therapeutic and biotechnology applications. Many compounds that target the minor groove in AT sequences in DNA are well characterized and are promising reagents for use as modulators of protein–DNA complexes, especially minor-groove-binding proteins like HMGA2. In order to discover HMGA2–DNA-binding inhibitors, a biosensor-SPR screening assay, using a competition-binding method, was designed for inhibition of the HMGA2–DNA interaction by small molecules [52].

The well-characterized minor-groove-binding polyamide netropsin (Figure 4.9) was used to develop and test the assay. The compound has two binding sites in the protein–DNA interaction sequence, and this provides an advantage for the inhibition of HMGA2 binding. A series of samples with a fixed concentration of HMGA2 and varying concentrations of netropsin were injected over the immobilized DNA surface (Figure 4.9). The observed response for competition binding at the AT sites is the sum of the bound masses from HMGA2 and netropsin. Netropsin competition with HMGA2 at the same sites will decrease the response signal because the molecular weight of netropsin is much smaller than that of HMGA2. Figure 4.9 shows a typical SPR-binding sensorgram for netropsin competition with HMGA2 binding to DNA. The large decrease in SPR signal on adding netropsin indicates direct competition with HMGA2 for binding to the DNA sites. The top and bottom curves in Figure 4.9 represent 100 nM HMGA2 alone and 1 μM netropsin alone, respectively. The middle curves represent 100 nM HMGA2 protein in the presence of increasing concentrations of netropsin (top to bottom). As netropsin competes with the higher-molecular-weight HMGA2 for DNA-binding sites, the bound mass on the surface is reduced and a decreased RU response is easily observed.

Figure 4.9 also illustrates competition curves of unusual shape in the association part of competition-binding sensorgrams. The shape can be understood in terms of the quite different kinetics and molecular weights of the protein and small molecule. In the case of HMGA2 alone, the steady state is reached rapidly and the sensorgram is normal. In the presence of HMGA with netropsin, the observed response is high at the beginning of injection due to rapid HMGA binding and then decreases as netropsin

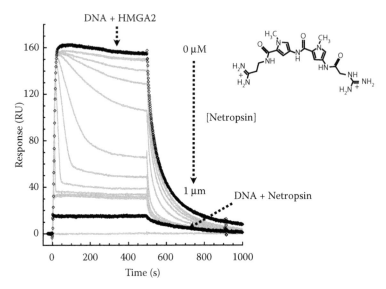

FIGURE 4.9 SPR sensorgrams for HMGA2 and netropsin competing for binding sites at a minor groove AT-binding sequence. The top curve (dark with diamond symbols) represents HMGA2 binding without competitor, and the bottom curve (also with diamond symbols) represents netropsin binding without HMGA2. The gray curves are SPR competition sensorgrams for binding of a mixture with a fixed concentration of HMGA2 (100 nM) and increasing concentrations of netropsin (from 0 to 1 µM) to the AT DNA. (Adapted from Miao, Y., Cui, T., Leng, F., et al. *Anal. Biochem.* **374**, 7–15, 2008.)

binds more slowly to the sites and displaces HMGA2. Netropsin is an excellent inhibitor and inhibits the protein–DNA complex at an IC_{50} of approximately 6 nM. The SPR-based assay thus provides an excellent platform for discovery of HMGA2 inhibitors.

A quite different example can illustrate the application of the SPR technique to investigate major-groove-binding transcription factors, which are extremely important for control of gene expression. The *lac* repressor is a DNA-binding protein that acts by inhibiting gene expression involved in lactose metabolism in bacteria [53]. The protein binds as a tetramer with strong affinity to its operator, a 20-mer DNA sequence [53]. Initial binding kinetics studies of *lac* repressor to the operator DNA sequence were evaluated by Bondenson and coworkers with gel-mobility shift assays and the resulting binding affinity (2.4×10^8 M^{-1}) was further tested with SPR experiments [54]. The biotinylated lac operator DNA and its complementary strand were immobilized on a streptavidin-coated CM5 sensor chip. Different concentrations of the lac repressor ranging from 3.4 to 13.4 nM were injected onto the surface for a sufficient period until the steady state was reached, and this was followed by a longer dissociation period to completely remove the repressor from the surface. The resulting sensorgrams were used to extract the rate constants for the binding and dissociation. The rate constants obtained using SPR ($k_a = 2 \times 10^6$ M^{-1} s^{-1}; $k_d = 3 \times 10^{-4}$ s^{-1}), however, did not correspond to the values determined using filter-binding technique ($k_a = 1 \times 10^9$ M^{-1} s^{-1}; $k_d = 2 \times 10^{-2}$ s^{-1}) for the same system [55]. Furthermore, the equilibrium-binding constant obtained through SPR (5.1×10^9 M^{-1}) was 20-fold higher than the value obtained from the gel-mobility shift assay. Interestingly, the dissociation phase of the sensorgrams exhibited a biphasic behavior with a fast dissociation event, followed by a slow dissociation event (Figure 2 in Ref. 54). The slower dissociation phase was attributed to the rebinding of the repressor protein to the free receptor on the surface due to the longer dissociation times. The authors argued that if the rebinding event was indeed occurring during the dissociation phase, the obtained rate constant values would be generally underestimated.

A modified SPR technique was recently conducted with the same protein–DNA by Myszka and coworkers to specifically address the effect of rebinding during dissociation and obtain better estimates for the dissociation rates [56]. A series of varying concentrations of DNA containing the *lac* repressor-binding site were included as an additional component during the dissociation phase to prevent the rebinding of the repressor to the immobilized receptor. This additional dissociation factor was integrated into the rate equations to account for a more accurate limiting behavior and precise measurement of rate constants. A new value of 7.5×10^{-2} s^{-1} obtained for the protein dissociation rate was significantly higher than the previously underestimated value using SPR. The addition of the extra component during dissociation phase increased the rate of dissociation by removing the rebinding event. This example is an important case study to show how SPR can be successfully employed to investigate the binding events by altering individual components of a reaction and studying the effects.

4.4 CONCLUSIONS

SPR agreement with solution methods: Since biosensor-SPR analysis of molecular interactions occurs on a surface, the question arises as to how the results from this method agree with comparative studies in solution. Fortunately, quite a number of comparative studies of different systems have been done in different laboratories to help answer this question. An early study with nucleic acids by Boger and coworkers [57] investigated the DNA binding of the bisintercalator, sandramycin, and a series of chromophore analogs. They compared binding constant measurements of the analogs by fluorescence quenching and SPR detection. In all the cases, the agreement between solution and biosensor results was quite good. In addition, they noted that the SPR technique allowed the extension of the binding studies to analogs that failed to provide an effective fluorescence signal. They also found that the SPR method with this system was experimentally more reliable and general than with fluorescence-quenching measurements. With SPR detection, they were able to determine association and dissociation rate constants for the compounds that they could not do with fluorescence. Where comparative results could be obtained, the base sequence selectivities and the overall affinities for the SPR and fluorescence methods were in excellent agreement.

Our laboratory has conducted comparative solution fluorescence and biosensor-SPR studies on the interaction of minor-groove-binding compounds with DNA. Heterocyclic diamidines, related to and including the clinically tested antiparasitic compound furamidine, were evaluated for minor-groove binding to an AATT site in which they had been crystallized [58]. The crystal structures clearly showed the compounds binding in the deep, narrow minor groove in the AATT sequence. Both fluorescence and SPR measurements showed that the compounds bind strongly as a 1:1 complex to the AATT site with K values near 1×10^7 M^{-1}. As with the sandramycin results, the SPR experiments were able to provide considerable information that was not possible with fluorescence measurements, including results for compounds with weak fluorescence as well as kinetics information on minor groove complex formation. These studies have been extremely useful in understanding the molecular features that lead to strong minor-groove-binding and biological activity. In addition, SPR methods have played a key role in the discovery of linear agents that bind to the minor groove with a linking water molecule (Section 4.3.1) or that bind to the groove as dimers with expanded sequence recognition capability (Section 4.3.2).

Quite a number of comparative-binding studies have also been conducted with protein complexes. The most detailed and thorough comparative protein-binding studies of biosensor-SPR with solution methods have been conducted by Myszka and coworkers [59–61]. As an example, they investigated the binding of small-molecule inhibitors to carbonic anhydrase by SPR, isothermal titration calorimetry, and stopped-flow kinetics methods [59]. They found that when the experiments were carried out accurately, the K values, thermodynamic and kinetic constants determined by the biosensor method, were in excellent agreement with those determined in solution. Rich and Myszka also published a detailed yearly analysis of biosensor-SPR literature reports with excellent critical analysis and comments on how to obtain high-quality biosensor results that should agree well with solution studies (e.g., [9]). The same group has also conducted multilab, blind studies of biosensor-SPR methods versus those from solution experiments (e.g., [61]). In all cases, they were able to clearly demonstrate that those labs that conducted the biosensor experiments properly obtained results in excellent agreement with results from solution methods.

Another protein system, protein kinases, plays a central role in many physiologically and clinically important cellular signaling events and is at the center of many drug development efforts. The p38α protein kinase has been used as a model system to study the comparative binding of small-molecule p38α inhibitors by biosensor-SPR methods and solution methods [62]. Both the equilibrium and kinetic constants were found to be in excellent agreement by the biosensor and solution methods. The biosensor method thus provides an efficient way to directly and reproducibly evaluate potential protein kinase inhibitors as well as those for other important protein systems.

Even though the SPR method has most often been used with relatively large proteins and nucleic acids, it has also been useful in determining stability constants for noncovalent interactions of small-molecule complexes. Results were compared, for example, for solution and biosensor-SPR methods for cyclodextrin host complex formation with small organic guests [63]. High-quality, reproducible SPR-binding data could be obtained with these small systems. In addition, where comparative results could be obtained, the magnitudes and trends in the stability constants were consistent between values for solution and SPR studies. These results clearly show that SPR is suitable for study of the interactions among small molecules.

The conclusion from these and other comparative solution and biosensor-SPR results is clearly that, for experiments that are conducted properly, the results from biosensor and solution studies agree quite well. When there is disagreement, it generally appears that proper care was not taken with some features of the experiment such as corrections for, or inclusion of, mass transport and/or proper immobilization methods to ensure that the molecule on the surface has the same activity as in solution [9]. The biosensor-SPR method thus provides results that agree with those in solution and it provides both equilibrium and kinetic information in real time without the need for labels.

ACKNOWLEDGMENTS

We gratefully acknowledge the support for biosensor-SPR studies on DNA complexes, for a number of years, by NIH (NIH AI064200) and for biosensor-SPR equipment by the Georgia Research Alliance. We would like to recognize our many coworkers and collaborators who have helped make our biosensor-SPR work more interesting and successful. Most of them are coauthors of the references in this chapter. We specifically acknowledge Professor David Boykin and coworkers for a long and very enjoyable collaboration on DNA–small molecule design, synthesis, and biophysical analysis. Dr. Boykin and coworkers also synthesized DB921 and DB915, which are shown in figures in this chapter.

REFERENCES

1. Davis, T. M. and Wilson, W. D. 2001. Surface plasmon resonance biosensor analysis of RNA-small molecule interactions. In *Methods Enzymol*, eds. J. B. Chaires and M. J. Waring, Vol. **340**, pp. 22–51. San Diego: Academic Press.
2. Homola, J. Surface plasmon resonance sensors for detection of chemical and biological species. *Chem. Rev.* **108**, 462–493, 2008.
3. Liu, Y. and Wilson, W. D. 2010. Quantitative analysis of small molecule-nucleic acid interactions with a biosensor surface and surface plasmon resonance detection. In *Drug-DNA Interaction Protocols*, ed. K. R. Fox, Vol. **613**, pp. 1–23. New Jersey: Humana Press.
4. Malmqvist, M. Surface plasmon resonance for detection and measurement of antibody-antigen affinity and kinetics. *Curr. Opin. Immunol.* **5**, 282–286, 1993.
5. Nguyen, B., Tanious, F. A., and Wilson, W. D. Biosensor-surface plasmon resonance: Quantitative analysis of small molecule-nucleic acid interactions. *Methods.* **42**, 150–161, 2007.
6. Piliarik, M., Vaisocherová, H., and Homola, J. 2009. Surface plasmon resonance biosensing. In *Biosensors and Biodetection*, ed. A. Rasooly and K. E. Herold, pp. 65–88. New Jersey: Humana Press.
7. Rich, R. L. and Myszka, D. G. Advances in surface plasmon resonance biosensor analysis. *Curr. Opin. Biotechnol.* **11**, 54–61, 2000.
8. Schuck, P. Reliable determination of binding affinity and kinetics using surface plasmon resonance biosensors. *Curr. Opin. Biotechnol.* **8**, 498–502, 1997.
9. Rich, R. L. and Myszka, D. G. Grading the commercial optical biosensor literature-class of 2008: 'The mighty binders'. *J. Mol. Recognit.* **23**, 1–64, 2010.
10. Myszka, D. G. Improving biosensor analysis. *J. Mol. Recognit.* **12**, 279–284, 1999.
11. Löfås, S. and McWhirter, A. 2006. The art of immobilization for SPR sensors. In *Surface Plasmon Resonance Based Sensors*, ed. J. Homola, pp. 117–151. Berlin: Springer.
12. Andersson, K., Hämäläinen, M., and Malmqvist, M. Identification and optimization of regeneration conditions for affinity-based biosensor assays. A multivariate cocktail approach. *Anal. Chem.* **71**, 2475–2481, 1999.

13. Karlsson, R., Katsamba, P. S., Nordin, H., et al. Analyzing a kinetic titration series using affinity biosensors. *Anal. Biochem.* **349**, 136–147, 2006.
14. Rich, R. L., Quinn, J. G., Morton, T., et al. Biosensor-based fragment screening using faststep injections. *Anal. Biochem.* **407**, 270–277, 2010.
15. Davis, T. M. and Wilson, W. D. Determination of the refractive index increments of small molecules for correction of surface plasmon resonance data. *Anal. Biochem.* **284**, 348–353, 2000.
16. Di Primo, C. and Lebars, I. Determination of refractive index increment ratios for protein-nucleic acid complexes by surface plasmon resonance. *Anal. Biochem.* **368**, 148–155, 2007.
17. Myszka, D. G. 2010. SCRUBBER-2. Center for biomolecular interaction analysis. http://www.cores.utah.edu/interaction/scrubber.html.
18. Holde, K. V., Johnson, C., and Ho, P. S. 2006. Chemical equilibra involving macromolecules. In *Principles of Physical Biochemistry*, eds. K. V. Holde, C. Johnson, and P. S. Ho, pp. 605–659. Upper Saddle River, NJ: Pearson Prentice Hall.
19. Karlsson, R. Affinity analysis of non-steady-state data obtained under mass transport limited conditions using BIAcore technology. *J. Mol. Recognit.* **12**, 285–292, 1999.
20. Apostolico, A., Ciriello, G., Guerra, C., et al. Finding 3D motifs in ribosomal RNA structures. *Nucleic Acids Res.* **37**, e29, 2009.
21. Rios, A. C. and Tor, Y. Model systems: How chemical biologists study RNA. *Curr. Opin. Chem. Biol.* **13**, 660–668, 2009.
22. Liu, Y., Kumar, A., Depauw, S. et al. Water-mediated binding of agents that target the DNA minor groove. *J. Am. Chem. Soc.* **133**, 10171–10183, 2011.
23. Miao, Y., Lee, M. P. H., Parkinson, G. N., et al. Out-of-shape DNA minor groove binders: Induced fit interactions of heterocyclic dications with the DNA minor groove. *Biochemistry.* **44**, 14701–14708, 2005.
24. Munde, M., Ismail, M. A., Arafa, R., et al. Design of DNA minor groove binding diamidines that recognize GC base pair sequences: A dimeric-hinge interaction motif. *J. Am. Chem. Soc.* **129**, 13732–13743, 2007.
25. Pelton, J. G. and Wemmer, D. E. Structural characterization of a 2:1 distamycin A.d(CGCAAATTGGC) complex by two-dimensional NMR. *Proc. Natl. Acad. Sci. U.S.A.* **86**, 5723–5727, 1989.
26. Wang, L., Bailly, C., Kumar, A., et al. Specific molecular recognition of mixed nucleic acid sequences: An aromatic dication that binds in the DNA minor groove as a dimer. *Proc. Natl. Acad. Sci. U.S.A.* **97**, 12–16, 2000.
27. Munde, M., Kumar, A., Nhili, R., et al. DNA minor groove induced dimerization of heterocyclic cations: Compound structure, binding affinity, and specificity for a TTAA site. *J. Mol. Biol.* **402**, 847–864, 2010.
28. Wang, J. Towards genoelectronics: Electrochemical biosensing of DNA hybridization. *Chem. Eur. J.* **5**, 1681–1685, 1999.
29. Wilson, P. K., Jiang, T., Minunni, M. E., et al. A novel optical biosensor format for the detection of clinically relevant TP53 mutations. *Biosens. Bioelectron.* **20**, 2310–2313, 2005.
30. Carrascosa, L., Calle, A. and Lechuga, L. Label-free detection of DNA mutations by SPR: Application to the early detection of inherited breast cancer. *Anal. Bioanal. Chem.* **393**, 1173–1182, 2009.
31. Lucarelli, F., Tombelli, S., Minunni, M., et al. Electrochemical and piezoelectric DNA biosensors for hybridisation detection. *Anal. Chim. Acta.* **609**, 139–159, 2008.
32. Balasubramanian, S. and Neidle, S. G-quadruplex nucleic acids as therapeutic targets. *Curr. Opin. Chem. Biol.* **13**, 345–353, 2009.

33. Hurley, L. H., Wheelhouse, R. T., Sun, D., et al. G-quadruplexes as targets for drug design. *Pharmacol. Ther.* **85**, 141–158, 2000.
34. White, E. W., Tanious, F., Ismail, M. A., et al. Structure-specific recognition of quadruplex DNA by organic cations: Influence of shape, substituents and charge. *Biophys. Chem.* **126**, 140–153, 2007.
35. Dai, J., Carver, M., and Yang, D. Polymorphism of human telomeric quadruplex structures. *Biochimie.* **90**, 1172–1183, 2008.
36. Phan, A. T., Kuryavyi, V., and Patel, D. J. DNA architecture: From G to Z. *Curr. Opin. Struct. Biol.* **16**, 288–298, 2006.
37. Dash, J., Shirude, P. S., Hsu, S. D., et al. Diarylethynyl amides that recognize the parallel conformation of genomic promoter DNA G-quadruplexes. *J. Am. Chem. Soc.* **130**, 15950–15956, 2008.
38. Rezler, E. M., Seenisamy, J., Bashyam, S., et al. Telomestatin and diseleno sapphyrin bind selectively to two different forms of the human telomeric G-quadruplex structure. *J. Am. Chem. Soc.* **127**, 9439–9447, 2005.
39. Halder, K. and Chowdhury, S. Kinetic resolution of bimolecular hybridization versus intramolecular folding in nucleic acids by surface plasmon resonance: Application to G-quadruplex/duplex competition in human c-myc promoter. *Nucleic Acids Res.* **33**, 4466–4474, 2005.
40. Ladame, S., Schouten, J. A., Roldan, J., et al. Exploring the recognition of quadruplex DNA by an engineered Cys2-His2 zinc finger protein. *Biochemistry,* **45**, 1393–1399, 2006.
41. Zeng, Z. X., Zhao, Y., Hao, Y. H., et al. Tetraplex formation of surface-immobilized human telomere sequence probed by surface plasmon resonance using single-stranded DNA binding protein. *J. Mol. Recognit.* **18**, 267–271, 2005.
42. Zhao, Y., Kan, Z. Y., Zeng, Z. X., et al. Determining the folding and unfolding rate constants of nucleic acids by biosensor. Application to telomere G-quadruplex. *J. Am. Chem. Soc.* **126**, 13255–13264, 2004.
43. Galietta, L. J. V., Springsteel, M. F., Eda, M., et al. Novel CFTR chloride channel activators identified by screening of combinatorial libraries based on flavone and benzoquinolizinium lead compounds. *J. Biol. Chem.* **276**, 19723–19728, 2001.
44. Huang, K. S., Haddadin, M. J., Olmstead, M. M., et al. Synthesis and reactions of some heterocyclic azacyanines 1. *J. Org. Chem.* **66**, 1310–1315, 2001.
45. Çetinkol, Ö. P. and Hud, N. V. Molecular recognition of poly(A) by small ligands: An alternative method of analysis reveals nanomolar, cooperative and shape-selective binding. *Nucleic Acids Res.* **37**, 611–621, 2009.
46. Persil Çetinkol, Ö., Engelhart, A. E., Nanjunda, R. K., et al. Submicromolar, selective G-quadruplex ligands from one pot: Thermodynamic and structural studies of human telomeric DNA binding by azacyanines. *ChemBioChem.* **9**, 1889–1892, 2008.
47. Garvie, C. W. and Wolberger, C. Recognition of specific DNA sequences. *Mol. Cell.* **8**, 937–946, 2001.
48. Luscombe, N. M., Austin, S. E., Berman, H. M., et al. An overview of the structures of protein-DNA complexes. *Genome Biol.* **1**, 1–37, 2000.
49. Rohs, R., Jin, X., West, S. M., et al. Origins of specificity in protein-DNA recognition. *Annu. Rev. Biochem.* **79**, 233–269, 2010.
50. Dragan, A. I., Liggins, J. R., Crane-Robinson, C., et al. The energetics of specific binding of AT-hooks from HMGA1 to target DNA. *J. Mol. Biol.* **327**, 393–411, 2003.
51. Reeves, R. Molecular biology of HMGA proteins: Hubs of nuclear function. *Gene.* **277**, 63–81, 2001.
52. Miao, Y., Cui, T., Leng, F., et al. Inhibition of high-mobility-group A2 protein binding to DNA by netropsin: A biosensor-surface plasmon resonance assay. *Anal. Biochem.* **374**, 7–15, 2008.

53. Lewis, M. The lac repressor. *Comptes Rendus Biologies.* **328**, 521–548, 2005.
54. Bondeson, K., Frostellkarlsson, A., Fagerstam, L., et al. Lactose repressor-operator DNA interactions: Kinetic analysis by a surface plasmon resonance biosensor. *Anal. Biochem.* **214**, 245–251, 1993.
55. Goeddel, D. V., Yansura, D. G., and Caruthers, M. H. How lac repressor recognizes lac operator. *Proc. Natl. Acad. Sci. U.S.A.* **75**, 3578–3582, 1978.
56. He, X., Coombs, D., Myszka, D., et al. A theoretical and experimental study of competition between solution and surface receptors for ligand in a Biacore flow cell. *Bull. Math. Biol.* **68**, 1125–1150, 2006.
57. Boger, D. L. and Saionz, K. W. DNA binding properties of key sandramycin analogues: Systematic examination of the intercalation chromophore. *Biorg. Med. Chem.* **7**, 315–321, 1999.
58. Laughton, C. A., Tanious, F., Nunn, C. M., et al. A crystallographic and spectroscopic study of the complex between d(CGCGAATTCGCG)$_2$ and 2,5-Bis(4-guanylphenyl) furan, an analogue of berenil. Structural origins of enhanced DNA-binding affinity. *Biochemistry.* **35**, 5655–5661, 1996.
59. Day, Y. S., Baird, C. L., Rich, R. L., et al. Direct comparison of binding equilibrium, thermodynamic, and rate constants determined by surface- and solution-based biophysical methods. *Protein Sci.* **11**, 1017–1025, 2002.
60. Myszka, D. G. 2000. Kinetic, equilibrium, and thermodynamic analysis of macromolecular interactions with BIACORE. In *Methods Enzymol.*, eds. G. K. Ackers and Michael L. Johnson, Vol. **323**, pp. 325–332. San Diego: Academic Press.
61. Navratilova, I., Papalia, G. A., Rich, R. L., et al. Thermodynamic benchmark study using Biacore technology. *Anal. Biochem.* **364**, 67–77, 2007.
62. Casper, D., Bukhtiyarova, M. and Springman, E. B. A Biacore biosensor method for detailed kinetic binding analysis of small molecule inhibitors of p38[alpha] mitogen-activated protein kinase. *Anal. Biochem.* **325**, 126–136, 2004.
63. Brown, S. E., Easton, C. J. and Kelly, J. B. Surface plasmon resonance to determine apparent stability constants for the binding of cyclodextrins to small immobilized guests. *J. Inclusion Phenom. Macrocycl. Chem.* **46**, 167–173, 2003.

5 Studying Aptamer/Ligand Interactions Using Fluorescence Correlation Spectroscopy

Eileen Magbanua, Katja Eydeler, Volker Alexander Lenski, Arne Werner, and Ulrich Hahn

CONTENTS

Fluorescence correlation spectroscopy (FCS) can be used to study nucleic acid/ligand interactions in solution. Measuring diffusion times of fluorescent moieties at the single-molecule level gives valuable insights into the formation of complexes. The advantages of this method include the infinitesimal measuring volumes required (femtoliter range) and the possibility to study interactions on cell surfaces.

5.1 INTRODUCTION

FCS is an optical method based on a confocal microscope [1,2]. Statistical processing of fluorescence signals, arising from fluorescing particles in solution, gives insights into the diffusional behavior of fluorescing molecules. Diffusion times of fluorescing molecules depend on molecule size and shape. If this fluorescing molecule interacts with a molecule species of significantly larger molecular weight, its diffusion time changes. Thus, monitoring diffusion allows studying molecular interaction.

Additionally, in case of binding, singlet-to-triplet transitions may occur and can also be determined by FCS.

Aptamers are short single-stranded nucleic acids, which are selected by a procedure called "systematic evolution of ligands by exponential enrichment" (SELEX) [3]. Since their invention two decades ago in the labs of Szostak and Gold [3–5], aptamers have been selected for numerous kinds of ligands, ranging from small molecules to peptides, proteins, and whole cells [6–9]. For example, aptamers with specificity for fluorophores have been selected [7]. Studying aptamer/ligand interactions via FCS requires fluorescence labeling of at least one binding partner. This could be achieved by labeling the aptamer, which allows using it as a probe to detect the target molecule. An alternative of this approach is to label aptamers with fluorophores that only fluoresce upon binding of the ligand [10]. Photoinduced electron transfer FCS (PET FCS) [11] allows probing conformational changes through aptamer binding using a reporter fluorophore. Alternatively, the ligand itself could be fluorescently labeled.

When planning the labeling strategy, it has to be considered that the difference in mass between free and bound state should be at least eightfold to obtain a significant signal. Therefore, labeling and tracking the smaller-binding partner is often advantageous.

5.2 BASICS OF FLUORESCENCE CORRELATION SPECTROSCOPY

In 1972, Magde et al. [1] introduced FCS. This technique monitors fluorescent particles passing through a well-defined femtoliter-scale detection volume. Rigler et al. improved the method in the 1990s by combining FCS with a confocal setup that allowed detection at the single-molecule level and led to an increased signal-to-noise ratio [12–15]. The setup was as follows: A laser beam is directed onto the microscope objective via several lenses and a dichroic mirror, resulting in a focused detection volume as small as 0.25–1 fL, approximately the volume of a single bacterial cell. The molecules in the focused volume are excited for fluorescence emission. The signal is collected by the same objective and passes the dichroic and the emission filter. Light not originating from the focus is blocked by a pinhole. The fluorescence signal reaches the detector, which is commonly an avalanche photo diode or a photomultiplier tube. Depending on the dye system, different lasers (e.g., argon, argon–krypton, or helium–neon lasers) could be used for excitation. The wavelengths typically range from 405 to 633 nm.

Diffusion of the particles into and out of the volume causes fluctuations in the fluorescence intensity, and these are analyzed using an autocorrelation function (Figure 5.1). This is the (temporal) correlation $G(\tau)$ of a time series with itself, shifted by time τ, as a function of τ:

$$G(\tau) = \frac{<\delta I(t)\delta I(t+\tau)>}{<I(t)>^2} \qquad (5.1)$$

where $I(t+\tau)$ is the fluorescence intensity obtained from the detection volume at delay time τ. In other words, the self-similarity of a signal is analyzed in relation to the signal after the lag time τ. The intensity-normalized form is commonly used because in this case the correlation at $\tau = 0$, $G(0)$, is linked to the average number of particles in the focal volume. At the inflection point of the correlation function, the residence time τ_{Diff} of the fluorescing particle in the detection volume is found. Since for globular molecules $\tau_{Diff} \sim \sqrt[3]{M}$ different diffusion species could be distinguished from each other if their masses differ by a factor of eight or more. Therefore, changes in mobility due to binding processes could be detected easily if a fluorescing particle is bound by a significantly larger interaction partner. In a typical measurement, the fluorescence signal is autocorrelated for 10–120 s. In most cases, the autocorrelation data are fitted with Levenberg–Marquardt nonlinear least-square algorithm.

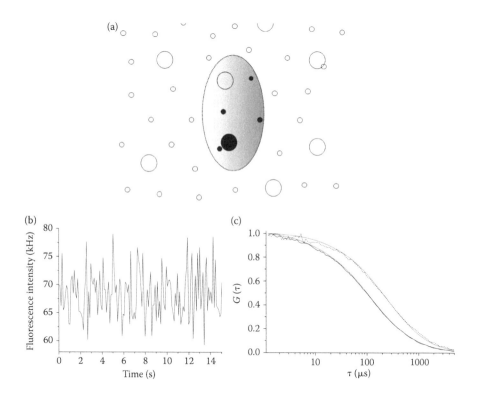

FIGURE 5.1 Principle of FCS. The confocal volume is defined by a focused laser beam (a). Labeled molecules (small circles) and unlabeled molecules (big circles) diffuse through the confocal volume and thereby labeled molecules are excited to fluorescence. Small uncomplexed-labeled molecules are diffusing quicker than those interacting with bigger-binding partners. Over a time period of several seconds, the overall fluorescence intensity in the confocal volume is monitored with nanosecond resolution (b). The decay of the time-dependent autocorrelation function of the fluorescence signal gives the weighted average of the diffusion time (c). The amplitude-normalized autocorrelation functions of two mobility species of different molecular weights (solid or dashed line) and the results of the fits according to Equation 5.2 (dotted line) are shown.

Assuming free three-dimensional diffusion and considering triplet-excited state, the data are fit according to

$$G(\tau) = 1 + \frac{1}{N} \cdot \left(\frac{1 - T + T \cdot e^{-\frac{\tau}{\tau_T}}}{1 - T} \right) \cdot \left\{ \sum_{i=1}^{n} \frac{f_i}{\left(1 + \frac{\tau}{\tau_{\text{Diff},i}} \right) \sqrt{1 + \frac{\tau}{\tau_{\text{Diff},i}} S^{-2}}} \right\} \tag{5.2}$$

The structure parameter S describes the ratio of the radial and axial distances between the maximum and $1/e^2$ laser intensities in the detection volume. T is used for describing triplet fraction and τ_T for triplet relaxation time, which is the time the molecule needs for relaxing from triplet state back to singlet state. If more than one diffusion species is present, i describes the number of this species and f_i its fraction. As mentioned above, the intensity-normalized form $G(0)$ gives information about the number of particles N in the detection volume. Using standard fluorophores with a known diffusion coefficient, the detection volume could be determined using

$$D = \frac{\omega_1^2}{4\tau_{\text{Diff}}} \tag{5.3}$$

where ω_1 represents half of the short axis of the volume element. With ω_2 corresponding to half of the long axis of the volume, the effective volume is

$$V_{\text{eff}} = \omega_1^2 \cdot 2\pi \cdot \omega_2 = 2\pi \cdot S \cdot \omega_1^3 \tag{5.4}$$

With knowledge of the detection volume, the concentration of the analyte could be calculated as

$$G(0) = \frac{1}{\langle N \rangle} = \frac{1}{V_{\text{eff}} \langle C \rangle} \tag{5.5}$$

Other models of diffusion, including anomalous diffusion, polydisperse diffusion, and diffusion with flow or chemical relaxation, are described in detail elsewhere [16–21].

To analyze aptamer target interaction by FCS, at least one component has to be fluorescent. This fluorescence can either be natural or one component must be labeled. Nucleic acids modified with fluorescence dyes could either be purchased or prepared in the lab using methods such as standard N-hydroxysuccinimide (NHS) coupling [22,23] or click chemistry [24]. For RNA aptamers, nucleotides like guanosine 5'-O-(α thiomonophosphate) (GMPS) or guanosine 5'-O-(γ thiotriphosphate) (GTPS) could be introduced during in vitro transcription, allowing subsequent labeling with a maleimide derivative of the desired fluorophore [25–27].

However, not all standard dyes are suitable for FCS measurements. Because of the laser power in the focus, a high quantum yield together with a great photostability is preferable. Owing to fast photobleaching, fluorescein, for example, would not be a good choice. A high extinction coefficient and a low tendency to singlet-to-triplet transition are also advantageous [28].

5.3 CASE STUDIES

In this chapter, we discuss case studies of successful FCS titration experiments. First, we show the binding properties of a fluorescently labeled antibiotic to its corresponding RNA aptamer. The second scenario describes how the binding properties of a fluorophore-binding aptamer to sulforhodamine B were analyzed. In a third case study, receptor-binding aptamers were fluorescently labeled to map receptor densities on the surfaces of live cells.

5.3.1 INTERACTIONS OF RHODAMINE-LABELED MOENOMYCIN A WITH AN RNA APTAMER

Moenomycin is an antibiotic that inhibits bacterial cell wall synthesis by binding to transglycosylases that catalyze the formation of the carbohydrate chains of peptidoglycan [37]. Schürer et al. selected $2'$-NH_2-modified RNA aptamers-binding moenomycin A [38]. For fluorescence labeling of the ligand moenomycin A (Figure 5.2, 1a), first, a carbamoyl derivative was generated by reaction with butylamine in methanol (Figure 5.2, 1b) [39]. This carbamoyl derivative reacted with the fluorophore tetramethylrhodamine isothiocyanate, yielding the fluorescently labeled moenomycin derivative (Figure 5.2, 2) [40]. Attachment of fluorophore to moenomycin A did not influence aptamer binding.

To investigate the interaction of aptamer and moenomycin, the fluorescently labeled moenomycin derivative concentration was kept constant (10 nM) and the aptamer amount was varied. Resulting differences in diffusion times were monitored by FCS. Using the two-component model, the fraction of formed complex could be calculated based on the diffusion time of the free fluorescent moenomycin derivative (0.075 ms). In Figure 5.3, the percentage of aptamer–moenomycin complex was plotted against increasing aptamer concentrations revealing a dissociation constant K_d of 437 nM. Comparable results ($K_d = 320$ nM) were obtained via affinity chromatography.

TABLE 5.1
Common FCS Dyes

Fluorescent Dye	D (10^{-10} m^2 s^{-1})	Excitation (nm)
Alexa Fluor 488	1.96 [29], 4.35 [30]	488
Alexa Fluor 546	3.41 [30]	543
Alexa Fluor 633	3.00 [31]	633
Atto 655-maleimide	4.07 [32]	663
Atto 655-carboxylic acid	4.26 [32]	663
Carboxyfluorescein	3.20	488
Cy3	2.80	543
Cy5	2.50 [33], 3.70 [34]	633
$2',7'$-Difluorofluorescein (Oregon Green 488)	4.10 [32]	498
Rhodamine 6G	2.80 [35], 4.26 [30]	514
Rhodamine 110	4.30 [36]	488
Tetramethylrhodamine (TMR)	2.60	543

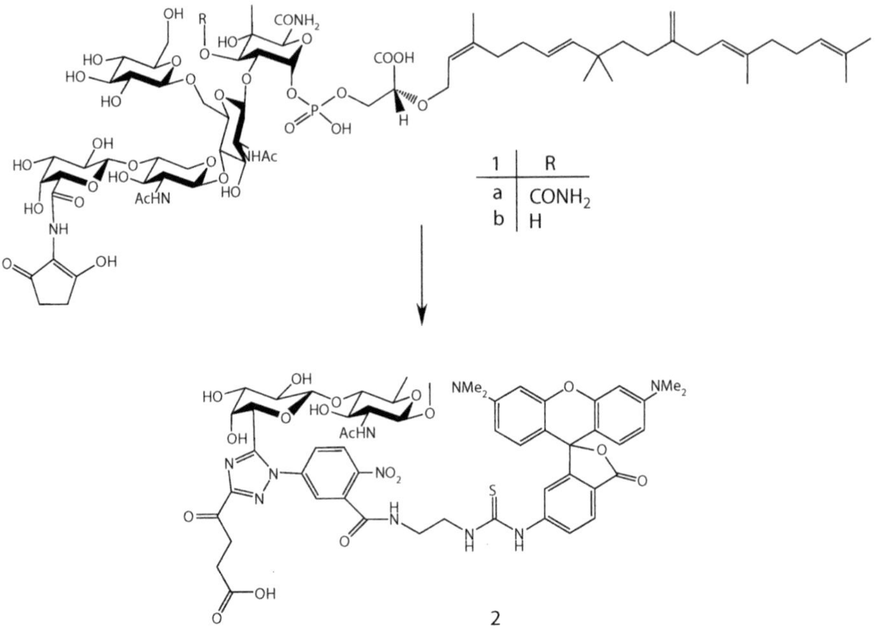

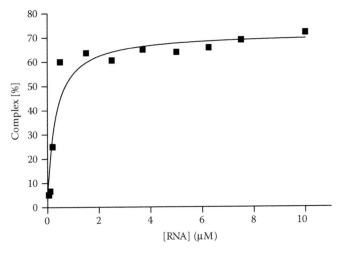

FIGURE 5.2 Tetramethylrhodamine-labeled moenomycin A. Moenomycin A (1a) was decarbamoylated (1b) by reaction with butylamine in methanol and labeled with the fluorophore tetramethylrhodamine isothiocyanate yielding the fluorescent derivative (2) [38,40].

FIGURE 5.3 Monitoring interaction of tetramethylrhodamine-labeled moenomycin A with its corresponding aptamer via FCS [40].

RNA	Length	Schematic overview		
SRB2m	54 nt		SRB2m	
SRB2m Monomer	134 nt	Noncoding RNA	SRB2m	
SRB2m Dimer	193 nt	Noncoding RNA	SRB2m	SRB2m
EGFP SRB2m	2254 nt	EGFP	SRB2m	Noncoding RNA

FIGURE 5.4 Sulforhodamine B-binding aptamer SRB2m and fusion constructs with different noncoding RNA stretches and EGFP-coding mRNA. RNAs were obtained by *in vitro* T7 transcription.

5.3.2 FLUOROPHORE-BINDING APTAMERS AS A TOOL FOR RNA VISUALIZATION

FCS is also suitable to detect fluorescing molecules in living cells. To image a selected mRNA, a fluorophore-binding aptamer was inserted into its corresponding 3′-untranslated region (UTR). The mRNA aptamer chimera could be detected upon fluorophore binding.

The mentioned strategy was investigated by Eydeler et al. [41] in the case of the aptamer SRB2m. SRB2m binds the fluorophore sulforhodamine B with high specificity and affinity (310 ± 60 nM) [7]. Different fusion constructs consisting of the aptamer SRB2m and RNA (Figure 5.4) were produced by *in vitro* transcription and analyzed by FCS. Secondary structure predictions of elongated SRB2m by *m*-fold [42] indicated that the aptamer still folded into its original structure. The binding constants of individual SRB2m aptamer, fusion constructs of SRB2m with different noncoding RNAs, and a dimer arrangement of the SRB2m to the fluorophore sulforhodamine B were derived through FCS titration experiments using a model for one diffusion species. The diffusion time of sulforhodamine B ($\tau_{Diff} = 30 \pm 1.5$ μs) increased up to $\tau_{Diff} > 300$ μs after addition of RNA, suggesting the binding of fluorophore to aptamer. Subsequently, the sulforhodamine B aptamer was inserted into the 3′-UTR of enhanced green fluorescent protein (EGFP) mRNA. Again, titration experiments showed increasing diffusion times of the fluorescing molecule, indicating complex formation between fluorophore and aptamer in conjunction with flanking RNA stretches of up to 2 kb. Figures 5.5 and 5.6 depict the corresponding data. Additionally, it could be shown that even in the context of total RNA extract, the fluorophore sulforhodamine B bound specifically to the RNA aptamer SRB2m.

5.3.3 DETECTION OF RECEPTORS ON LIVING CELLS

Fluorescently labeled high-affinity aptamers binding to specific receptors are one of the few approaches to quantify membrane receptor density [43,44].

Chen et al. [45] used FCS to map the density of the human protein tyrosine kinase-7 receptor (PTK7) presented on leukemia (CCRF-CEM) and HeLa cells. For this purpose, fluorescently labeled aptamers were selected and chosen for receptor recognition. To demonstrate the binding of the aptamer to membrane receptors, the laser beam was focused on the cell membrane. Receptor-binding aptamers stuck to the cell surface and thus diffused significantly slower. Membrane receptor-bound aptamers were best described by a two dimensional diffusion model, whereas free aptamers not connected

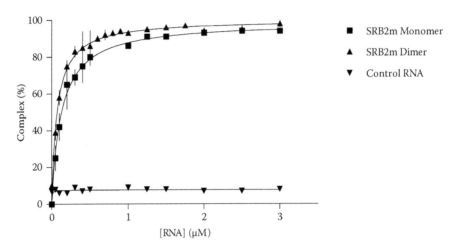

FIGURE 5.5 Binding of sulforhodamine B to SRB2m constructs was analyzed by FCS. Percentage of complex formation was plotted against concentration.

to the cell surface were described by a three-dimensional diffusion model. Specific binding of aptamers could be demonstrated by investigating different cell types and negative control cells lacking PTK7 on their cell surfaces. To ensure inhibition of receptor-mediated endocytosis, experiments were performed at 4°C.

Thus, the receptor density for different cell surfaces could be obtained by monitoring aptamer receptor binding. The laser beam was focused onto the cell membrane. The circular area covered by the focus volume could be estimated to be $\pi \cdot \omega_1^2$. Within this covered area, the number of specific receptors that occupied this area could be interpreted from the amount of specifically bound aptamers under saturation conditions. The number N of membrane-bound labeled aptamers was obtained by varying the

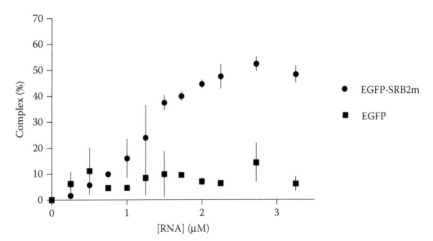

FIGURE 5.6 Binding of sulforhodamine B to EGFP-SRB2m RNA was analyzed by FCS. Percentage of complex formation was plotted against concentration.

amounts of totally labeled aptamer added. The density of receptors on the cell surface could be calculated by dividing the number of membrane-bound aptamers in the confocal volume by the circular area covered by the focus volume.

5.4 CONCLUSIONS

FCS is a versatile technique and has a wide application range. Single molecules can be studied in solution and immobilization of the binding partners is not necessary. Binding affinities of two single particles can be measured. Interactions between the antibiotic moenomycin A and its corresponding aptamer have been studied with FCS. A fluorophore-binding aptamer like SRB2m could be fused to an mRNA of interest to facilitate its visualization.

In a confocal microscope, it is possible to define a region of interest, in which FCS can be conducted. Densities of the PTK7 receptor in the membrane of leukemia and HeLa cells were monitored. These case studies emphasize the wide application range of FCS for studying aptamer/ligand interactions.

5.5 PROSPECTS AND OUTLOOK

Laser and detector techniques have significantly improved since the first FCS experiments. More stable lasers achieve more constant excitation and therefore more consistent data. State-of-the-art computer hardware allows fast time-resolved detection of fluorescence signals. Assuming progress in this technological field, there will be even faster and even more sensitive detectors available in the future. This will allow more sensitive measurements at lower concentrations of labeled molecules and at the same time a higher spatial resolution. Pulsed lasers can be used for time-resolved FCS setups like fluorescence lifetime FCS where fluorescence signals are filtered by fluorescence lifetime information before correlation. Modifications of standard FCS setups will become more widespread. Absolute diffusion coefficients can be determined by scanning FCS [30,46] or two-focus FCS [47]. Two-photon FCS [48] offers a better signal-to-noise ratio, and last but not least in fluorescence cross-correlation spectroscopy (FCCS), both binding partners are labeled individually. FCCS overcomes the requirement in FCS that the free and complexed forms of the labeled molecule have to differ in mass by a factor of eight.

ACKNOWLEDGMENTS

We are grateful to Andrea Rentmeister for providing helpful advices.

REFERENCES

1. Magde, D., E.L. Elson, and W.W. Webb. Thermodynamic fluctuations in a reacting system: Measurement by fluorescence correlation spectroscopy. *Phys. Rev. Lett.* **29,** 705–8, 1972.
2. Riegler, R. and Elson, E. *Fluorescence Correlation Spectroscopy: Theory and Applications.* Ann Arbor, MI: Springer Series in Chemical Physics, 2001.

3. Ellington, A.D. and J.W. Szostak. *In vitro* selection of RNA molecules that bind specific ligands. *Nature.* **346** (6287), 818–22, 1990.

4. Ellington, A.D. and J.W. Szostak. Selection *in vitro* of single-stranded DNA molecules that fold into specific ligand-binding structures. *Nature.* **355** (6363), 850–2, 1992.

5. Tuerk, C. and L. Gold. Systematic evolution of ligands by exponential enrichment: RNA ligands to bacteriophage T4 DNA polymerase. *Science.* **249** (4968), 505–10, 1990.

6. Raddatz, M.-S.L., et al. Enrichment of cell-targeting and population-specific aptamers by fluorescence-activated cell sorting. *Angew. Chem. Int. Ed. Engl.* **47** (28), 5190–3, 2008.

7. Holeman, L.A., et al. Isolation and characterization of fluorophore-binding RNA aptamers. *Folding Design.* **3** (6), 423–31, 1998.

8. Bock, L.C., et al. Selection of single-stranded DNA molecules that bind and inhibit human thrombin. *Nature.* **355** (6360), 564–6, 1992.

9. Famulok, M. Oligonucleotide aptamers that recognize small molecules. *Curr. Opin. Struct. Biol.* **9** (3), 324–9, 1999.

10. Nutiu, R. and Y. Li. Aptamers with fluorescence-signaling properties. *Methods.* **37** (1), 16–25, 2005.

11. Schuttpelz, M., et al. Changes in conformational dynamics of mRNA upon AtGRP7 binding studied by fluorescence correlation spectroscopy. *J. Am. Chem. Soc.* **130** (29), 9507–13, 2008.

12. Eigen, M. and R. Rigler. Sorting single molecules: Application to diagnostics and evolutionary biotechnology. *Proc. Natl. Acad. Sci. USA.* **91** (13), 5740–7, 1994.

13. Rigler, R., J. Widengren, and P. Kask. Fluorescence correlation spectroscopy with high count rate and low background: Analysis of translational diffusion. *Eur. Biophys. J.* **22** (3), 169–75, 1993.

14. Rigler, R. Ultrasensitive detection of single molecules by fluorescence correlation spectroscopy. *Bioscience.* **3**, 180–3, 1990.

15. Kinjo, M. and R. Rigler. Ultrasensitive hybridization analysis using fluorescence correlation spectroscopy. *Nucleic Acids Res.* **23** (10), 1795–9, 1995.

16. Wu, J. and K.M. Berland. Propagators and time-dependent diffusion coefficients for anomalous diffusion. *Biophys. J.* **95** (4), 2049–52, 2008.

17. Banks, D.S. and C. Fradin. Anomalous diffusion of proteins due to molecular crowding. *Biophys. J.* **89** (5), 2960–71, 2005.

18. Sengupta, P., K. Garai, J. Balaji, N. Periasamy, and S. Maiti. Measuring size distribution in highly heterogeneous systems with fluorescence correlation spectroscopy. *Biophys. J.* **84** (3), 1977–84, 2003.

19. Kohler, R.H., P. Schwille, W.W. Webb, and M.R. Hanson. Active protein transport through plastid tubules: Velocity quantified by fluorescence correlation spectroscopy. *J. Cell Sci.* **113** (Pt 22), 3921–30, 2000.

20. Gräslund, A.R., R. Rigler, and J. Widengren. *Single Molecule Spectroscopy in Chemistry, Physics and Biology.* Heidelberg: Springer, 2010.

21. Rigler, R. *Fluorescence Correlation Spectroscopy, Theory and Applications.* Ann Arbor, MI: Springer, 2000.

22. Blank, M., T. Weinschenk, M. Priemer, and H. Schluesener. Systematic evolution of a DNA aptamer binding to rat brain tumor microvessels. Selective targeting of endothelial regulatory protein pigpen. *J. Biol. Chem.* **276** (19), 16464–8, 2001.

23. Ulrich, H., A.H. Martins, and J.B. Pesquero. RNA and DNA aptamers in cytomics analysis. *Cytometry A.* **59** (2), 220–31, 2004.

24. Kolb, H.C., M.G. Finn, and K.B. Sharpless. Click chemistry: Diverse chemical function from a few good reactions. *Angew. Chem. Int. Ed. Engl.* **40** (11), 2004–21, 2001.

25. Qin, P.Z. and A.M. Pyle. Site-specific labeling of RNA with fluorophores and other structural probes. *Methods.* **18** (1), 60–70, 1999.

26. Sengle, G., A. Jenne, P.S. Arora, B. Seelig, J.S. Nowick, A. Jaschke, and M. Famulok. Synthesis, incorporation efficiency, and stability of disulfide bridged functional groups at RNA 5′-ends. *Bioorg. Med. Chem.* **8** (6), 1317–29, 2000.
27. Eisenfuhr, A., P.S. Arora, G. Sengle, L.R. Takaoka, J.S. Nowick, and M. Famulok. A ribozyme with michaelase activity: Synthesis of the substrate precursors. *Bioorg. Med. Chem.* **11** (2), 235–49, 2003.
28. Krichevsky, O. and G. Bonnet Fluorescence correlation spectroscopy: The technique and its applications. *Rep. Prog. Phys.* **65** (2), 2002.
29. Pristinski, D., V. Kozlovskaya, and S.A. Sukhishvili. Fluorescence correlation spectroscopy studies of diffusion of a weak polyelectrolyte in aqueous solutions. *J. Chem. Phys.* **122** (1), 14907, 2005.
30. Petrasek, Z. and P. Schwille. Precise measurement of diffusion coefficients using scanning fluorescence correlation spectroscopy. *Biophys. J.* **94** (4), 1437–48, 2008.
31. van den Bogaart, G., V. Krasnikov, and B. Poolman. Dual-color fluorescence-burst analysis to probe protein efflux through the mechanosensitive channel MscL. *Biophys. J.* **92** (4), 1233–40, 2007.
32. Müller, C.B., A. Loman, V. Pacheco, F. Koberling, D. Willbold, W. Richtering, and J. Enderlein. Precise measurement of diffusion by multi-color dual-focus fluorescence correlation spectroscopy. *Europhys. Lett.* **83** (46001), 2008.
33. Widengren, J. and P. Schwille. Characterization of photoinduced isomerization and back-isomerization of the cyanine dye Cy5 by fluorescence correlation spectroscopy. *J. Phys. Chem. A.* **104**, 6416–28, 2000.
34. Loman, A., T. Dertinger, F. Koberling, and J. Enderlein. Comparison of optical saturation effects in conventional and dual-focus fluorescence correlation spectroscopy. *Chem. Phys. Lett.* **459**, 18–21, 2008.
35. Magde, D., E.L. Elson, and W.W. Webb. Fluorescence correlation spectroscopy. II. An experimental realization. *Biopolymers.* **13** (1), 29–61, 1974.
36. Gendron, P.O., F. Avaltroni, and K.J. Wilkinson. Diffusion coefficients of several rhodamine derivatives as determined by pulsed field gradient-nuclear magnetic resonance and fluorescence correlation spectroscopy. *J. Fluoresc.* **18** (6), 1093–101, 2008.
37. van Heijenoort, J. Formation of the glycan chains in the synthesis of bacterial peptidoglycan. *Glycobiology.* **11** (3), 25R–36R, 2001.
38. Schurer, H., K. Stembera, D. Knoll, G. Mayer, M. Blind, H.H. Forster, M. Famulok, P. Welzel, and U. Hahn. Aptamers that bind to the antibiotic moenomycin A. *Bioorg. Med. Chem.* **9** (10), 2557–63, 2001.
39. Vogel, S., A. Buchynskyy, K. Stembera, K. Richter, L. Hennig, D. Muller, P. Welzel, F. Maquin, C. Bonhomme, and M. Lampilas. Some selective reactions of moenomycin A. *Bioorg. Med. Chem. Lett.* **10** (17), 1963–5, 2000.
40. Schurer, H., A. Buchynskyy, K. Korn, M. Famulok, P. Welzel, and U. Hahn. Fluorescence correlation spectroscopy as a new method for the investigation of aptamer/target interactions. *Biol. Chem.* **382** (3), 479–81, 2001.
41. Eydeler, K., E. Magbanua, A. Werner, P. Ziegelmueller, and U. Hahn, Fluorophore binding aptamers as a tool for RNA visualization. *Biophys. J.* **96** (9), 3703–7, 2009.
42. Zuker, M. Mfold web server for nucleic acid folding and hybridization prediction. *Nucleic Acids Res.* **31** (13), 3406–15, 2003.
43. Colabufo, N.A., F. Berardi, R. Calo, M. Leopoldo, R. Perrone, and V. Tortorella. Determination of dopamine D(4) receptor density in rat striatum using PB12 as a probe. *Eur. J. Pharmacol.* **427** (1), 1–5, 2001.
44. Bice, A.N. and B.R. Zeeberg. Quantification of human caudate d2 dopamine receptor density with positron emission tomography: A review of the model. *IEEE Trans. Med. Imaging.* **6** (3), 244–9, 1987.

45. Chen, Y., A.C. Munteanu, Y.F. Huang, J. Phillips, Z. Zhu, M. Mavros, and W.H. Tan. Mapping receptor density on live cells by using fluorescence correlation spectroscopy. *Chemistry*. **15** (21), 5327–36, 2009.

46. Ries, J. and P. Schwille. Studying slow membrane dynamics with continuous wave scanning fluorescence correlation spectroscopy. *Biophys. J.* **91** (5), 1915–24, 2006.

47. Dertinger, T., A. Loman, B. Ewers, C.B. Mueller, B. Kramer, and J. Enderlein. The optics and performance of dual-focus fluorescence correlation spectroscopy. *Opt. Express*. **16** (19), 14353–68, 2008.

48. Petrášek, Z. and P. Schwille. Photobleaching in two-photon scanning fluorescence correlation spectroscopy. *ChemPhysChem*. **9** (1), 147–58, 2008.

6 Studying Nucleic Acid– Drug Interactions at the Single-Molecule Level Using Optical Tweezers

Thayaparan Paramanathan, Micah J. McCauley, and Mark C. Williams

CONTENTS

In this chapter, we quantify ligand–DNA interactions and characterize various DNA-binding modes. Optical tweezers measurements are used as a DNA-binding assay for drugs and potential drug candidates. Changes in the measured DNA force versus extension reveal the DNA-binding mode and affinity of these molecules as they bind to the DNA.

6.1 INTRODUCTION

The first optical tweezers, a single-beam gradient force optical trap, was developed in 1986 [1] to trap micron-sized particles using the force from the radiation pressure of a continuous laser [2,3]. Optical tweezers entered the field of biology in 1987 when they were used to trap viruses and bacteria [4,5]. Since then, optical tweezers have been used as a powerful tool to study biological systems at the single-molecule level. In 1996, optical tweezers were introduced to stretch deoxyribonucleic acid (DNA) [6]. A variety of optical tweezers experiments have advanced the understanding of the elastic and thermodynamic properties of nucleic acids and their interactions with proteins and small molecules [7–9], including the interactions of drugs and potential drugs designed specifically to interact with DNA.

Optical tweezers can be used to isolate a single DNA or RNA molecule by attaching the ends of the molecule to optically trapped dielectric beads or cylinders. The force experienced by the molecule is measured as a function of the end-to-end distance, yielding the elastic properties of the nucleic acid. These properties change measurably when DNA is packaged, bent or looped during interactions with other molecules. DNA-stretching experiments performed in the presence of small molecules have shown that by measuring the changes of DNA elastic properties, various binding modes can be characterized. Compared to bulk experiments, which also study these interactions, single-molecule techniques often require less sample while still yielding precise, quantitative results, and can also be performed under solution conditions not available to bulk experiments. Recent experiments have measured the equilibrium binding properties and binding kinetics of small molecules that interact with DNA as a function of force and extrapolate to obtain these properties in the absence of force [10]. These extrapolated results are in excellent agreement with the bulk experiment whenever results from bulk experiments are available. Thus, it is clear that optical tweezers can be used to characterize the binding properties of drugs with nucleic acids.

6.2 BACKGROUND AND BASICS

6.2.1 TRAPPING AND STRETCHING SINGLE MOLECULES USING OPTICAL TWEEZERS

An optical trap consists of a focused laser beam that exerts a dipole force on a dielectric object. The force is due to the interaction between the dipole induced by the electric field from the laser and the electric field itself. The force ($F_{gradient}$) is proportional to the gradient of the intensity of the laser field E

$$F_{gradient} = 1/2\alpha\nabla E^2 \qquad (6.1)$$

where α is the polarizability of the material. As the diameter of trapped particle (typically several μm in most cases) is larger than the laser wavelength, a simple explanation is provided by ray optics. The particle will experience two types of force. The scattering force ($F_{scattering}$) pushes the particle in the direction of laser propagation. The gradient force ($F_{gradient}$) acts as a restoring force, pulling the particle toward the laser focus.

Consider two rays from the edge of the trapping laser as it is brought to focus (Figure 6.1). Incident photons are scattered in random directions. This will induce a momentum change to the bead, and the net momentum change pushes the bead in the direction of the beam. Now, consider a particle that is slightly displaced from the center of the beam (Figure 6.1 inset). The refraction of the incident beams now generates a strong restoring force, pulling the bead in the opposite direction of the displacement. There are now two components to the force. The scattering force pushes the particle in the direction of laser propagation and the gradient force pulls the particle toward the center of the beam. Typically, for micron-sized beads, the scattering force is much smaller than the gradient force. Furthermore, in a dual-beam optical tweezers setup, there are two counterpropagating laser beams focused tightly to trap the particle. In this case, scattering forces from the counterpropagating beams oppose each other, resulting in very little net-scattering force. Therefore, the gradient force primarily acts as the net-trapping force.

When a trapped particle is displaced from the trap center by a displacement x, the trap may be approximated as a linear elastic spring, exerting a force F that is proportional to the displacement

$$\vec{F} = -k\vec{x} \tag{6.2}$$

The proportionality constant k is determined by applying a known force on the trapped bead and measuring the displacement. The estimated trap stiffness is then used to calculate the force exerted on the bead when it is displaced by a known amount. A further correction considers the change in equilibrium position of the

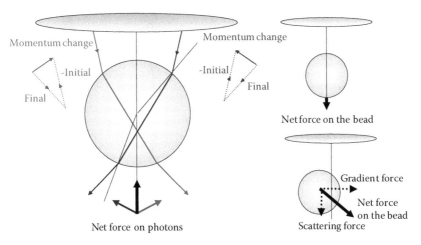

FIGURE 6.1   (See color insert.) Ray optics description of bead trapping with single-beam optical tweezers. A bead is perturbed slightly from the laser focus. (The perturbation is toward the microscope objective in this figure.) Individual rays (red and blue lines) will experience a net change in momentum, due to refraction. The bead will experience an equal and opposite force (as shown in the lowest inset). This restoring force will be proportional to the displacement. The total force will consist of two components, a gradient force and a scattering force (the scattering force is not shown), though in a dual counterpropagating beam trap, the scattering forces will cancel.

trapped bead at various forces. This change in position is proportional to the trap strength, or stiffness, and is used to refine the extension measurement.

In an optical tweezers experiment, a single DNA molecule is trapped by chemically attaching it to polystyrene beads at one or both ends. A typical chemical attachment labels DNA with biotin while the beads are coated with streptavidin. There are several specific trapping configurations available. A bead may be held by the optical trap while the other end may be held either on a bead on the end of a micropipette tip (Figure 6.2a) or by another optically trapped bead in the dual trap design (Figure 6.2b) or a bead held or attached directly to a glass surface (Figure 6.2c and d). While moving the bead or the glass surface at one end by a known extension, the corresponding displacement of the bead in the trap is measured. The force exerted on the bead in the trap is determined using the known trap calibration (Equation 6.2).

The DNA stretching can be divided into two main categories. Stretching torsionally constrained DNA and stretching torsionally unconstrained DNA, depending on the DNA–bead attachment geometry. Most small molecule–DNA interaction studies are done by stretching torsionally unconstrained DNA; therefore, we will consider the stretching properties of the latter type of experiment. Torsionally unconstrained DNA can be obtained by chemically attaching the 3′ ends of opposite strands of the DNA to the bead or glass surface, which allows the DNA to freely rotate around its axis while stretching.

A typical DNA-stretching curve (black solid circles in Figure 6.3a) obtained with counterpropagating dual-beam optical tweezers with torsionally unconstrained

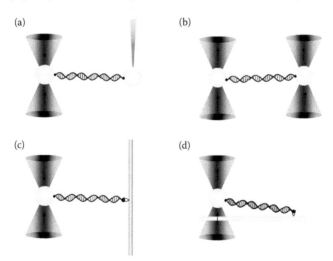

FIGURE 6.2 **(See color insert.)** Different configurations used in optical tweezers to trap and manipulate single molecules. The black arrows indicate the direction of propagation of the laser light. (a) A dual counterpropagating beam optical trap holds a bead attached to one end of the single molecule and the bead at the other end is held by a micropipette. (b) Both beads attached at ends of the single molecule are held by unique traps created by two beams, or a time-shared single beam. (c) One end of the single molecule is attached to a glass slide and the other end attached to a bead that is held by a dual-beam optical trap. (d) One end of the single molecule attached to the glass slide and the other end attached to the bead held by single-beam optical trap.

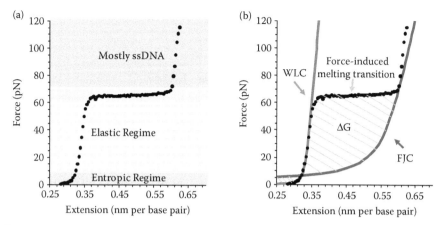

FIGURE 6.3 **(See color insert.)** Force measured as a function of DNA extension during a typical-stretching experiment with optical tweezers. (a) DNA-stretching data (black circles) exhibits four distinct regimes. At low extensions, the force change is very small upon extension across the entropic regime (violet-shaded area). Further stretching shows a rapid increase in force where the DNA backbone responds, in an elastic regime (light blue-shaded area). This is followed by a transition plateau around 65 pN (light green-shaded area). Finally, at the end of the transition, the force again increases rapidly (light red-shaded area), where two single-stranded DNA held by GC-rich regions are stretched. (b) The worm-like chain model fit (blue solid curve) to the double-stranded DNA and the freely jointed chain model (red solid curve) of ssDNA clearly indicates that the transition at 65 pN is a melting transition from dsDNA to ssDNA (see discussion in the text). The shaded area between the DNA-stretching curve and ssDNA model yields the melting free energy. Experimental conditions: 100 mM Na+, 10 mM HEPES, pH 7.5, 20°C.

lambda phage DNA (λ DNA) of length 16.5 μm (48,500 bp) is shown in Figure 6.3a. Initially, as the DNA is stretched at low extensions, there is little change in the force, in a region known as the entropic-stretching regime (data in violet shaded area in Figure 6.3a). Once the DNA is stretched to its normal contour length, the sugar phosphate backbone starts to respond to further stretching as the DNA behaves like an elastic polymer in this region and the force increases rapidly as the DNA is extended. This regime is known as the elastic regime (data in light blue-shaded area in Figure 6.3a). Once the force reaches ~65 pN, a clear transition is observed (data in light green-shaded area in Figure 6.3a) where the force remains nearly constant as the extension is almost doubled (increasing by a factor of ~1.7). Initially, this overstretching transition at 65 pN was believed to preserve the base pairing while most of the base stacking was broken by slanting the bases into a structure resembling a skewed ladder, a model termed S-DNA [6,11]. However, later experiments confirmed that the base pairs progressively open during a force-induced melting transition [12–16]. At the end of the melting transition, most of the double-stranded DNA (dsDNA) is converted into single-stranded DNA (ssDNA) although a few GC-rich regions hold the two strands together. Further stretching beyond the transition exhibits distinct elastic properties again (data in light red-shaded area in Figure

6.3a). Relaxing back slowly recovers the original elasticity of dsDNA indicating that this is a reversible process.

6.2.2 POLYMER MODELS AND MELTING FREE ENERGY

Polymer models of dsDNA (Figure 6.3b) effectively characterize DNA elasticity. The area confined between the experimental data (including melting) and the ssDNA polymer model (green-striped area in Figure 6.3b) yields the melting free energy associated with converting dsDNA into ssDNA [12]. The worm-like chain (WLC) model used to describe polymers like dsDNA assumes a smooth distribution of bending angles. For a polymer of length L, the tangent vector $\vec{t}(S)$ at position $\vec{r}(S)$ at distance S along the chain (Figure 6.4a) can be written as

$$\vec{t}(S) = \frac{d\vec{r}(S)}{dS} \tag{6.3}$$

The orientation correlation function for WLC model decays exponentially [17]:

$$\langle \vec{t}(0) \cdot \vec{t}(S) \rangle = \exp\left(-\frac{S}{P}\right) \tag{6.4}$$

P is known as the persistence length, which reflects the bend stiffness of the polymer. The WLC model describes dsDNA in terms of observed length b_{ds} of an elastic polymer under the influence of tension F [18–22]. Though no exact solutions to this model are known, an approximate solution is appropriate for high forces:

$$b_{ds} = B_{ds}\left[1 - \frac{1}{2}\left(\frac{k_B T}{P_{ds} F}\right)^{1/2} + \frac{F}{S_{ds}}\right] \tag{6.5}$$

where P_{ds} is the persistence length, B_{ds} is the contour length of dsDNA, usually expressed as the overall length, L, divided by the number of base pairs. A stretch

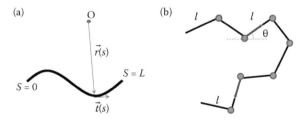

FIGURE 6.4 **(See color insert.)** Polymer models of nucleic acid flexibility. (a) The worm-like chain model describes polymers with smooth continuous bends. (b) The freely jointed chain model describes the polymer as a collection of independent monomers with varying bond angles connected together by hinges.

modulus S_{ds} is added to account for backbone extensibility. Finally, k_B is the Boltzmann's constant and T is the temperature.

The freely jointed chain (FJC) model describes a collection of independent monomers with varying bond angles connected together by free hinges (Figure 6.4b). This model is used to describe more flexible polymers like ssDNA [23]:

$$b_{ss} = B_{ss}\left[\coth\left(\frac{2P_{ss}F}{k_BT}\right) - \frac{k_BT}{2P_{ss}F}\right]\left[1 + \frac{F}{S_{ss}}\right]$$ (6.6)

where P_{ss} is the persistence length, B_{ss} is the end-to-end or contour length of ssDNA, and a stretch modulus S_{ss} is added to account for backbone extensibility.

6.2.3 SMALL-MOLECULE INTERACTIONS WITH DNA

Interactions of small molecules with DNA play an essential role in cellular processes. Small molecules can also interfere with transcription or replication processes during cell division. Thus, the understanding of their interaction with nucleic acids is essential in designing drugs for challenging disease. These small molecules can interact with DNA either through nonreversible covalent binding or through reversible noncovalent-binding interactions. Single-molecule experiments, including optical and magnetic tweezers, as well as atomic force microscopy (AFM), have explored the latter category. The results can be divided into three major binding modes: intercalation, ssDNA binding, and groove binding (Figure 6.5). Small molecules reversibly bind to DNA through one or a combination of these modes. Each of these three binding modes may be identified by unique features within a single-molecule experiment. These features may also quantify the effectiveness of drug binding. Ligands that bind through a combination of these modes, or via a sequence of modes, require careful consideration.

6.3 CASE STUDIES

6.3.1 INTERCALATORS

The majority of small molecules that are used in clinical applications to target DNA are intercalators. Thus, the properties of intercalators play an important role in rational drug design. Proflavin, used for psoriasis and herpes virus infections [24,25], psoralen, used for skin diseases [24], daunomycin [26,27], and actinomycin D [28], used in cancer chemotherapy, are a few examples of intercalators used for different therapeutic applications. Intercalators generally incorporate a flat aromatic structure, which stacks between the adjacent base pairs of the dsDNA to stabilize the dsDNA structure, possibly stalling replication.

Ruthenium polypyridyl complexes have a flat aromatic moiety that can intercalate and though they are not currently used as a drug, they have recently entered clinical trials for intestinal tumors [29]. Since $[Ru(phen)_3]^{2+}$ (phen = 1,10 phenanthroline) was introduced [30], Ru complexes have played an important role in designing and testing small molecules that bind to DNA. To increase the binding affinity, several modifications have been made to these ruthenium polypyridyl complexes, where one

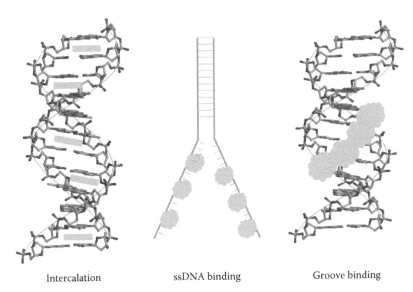

Intercalation   ssDNA binding   Groove binding

FIGURE 6.5 (See color insert.) Reversible-binding modes of small molecules. Intercalation, where the flat aromatic ring of the drug (green) stacks between the DNA base pairs (left), single-stranded DNA binding where ligands (green) bind to the exposed bases or the stands of the DNA (middle), and groove binding where the drug (green) interacts with DNA by fitting in either the major or minor groove of the DNA helical structure (right). DNA structure obtained from PDB file 1BNA from Ref. 70.

or more phen moieties were replaced by different groups. These drugs, along with classical intercalators like ethidium and DNA-intercalating dyes like oxazole yellow (YO), have been examined with optical tweezers experiments to quantify the effects of intercalation on DNA at the single-molecule level.

6.3.1.1 Qualitative Effects with Stretching Experiments

While single-molecule force spectroscopy can characterize intercalation, these molecules have also been utilized to investigate the overstretching transition and other effects observed during single-molecule DNA stretching. Ethidium, a well-studied intercalator, was used [11] at saturated concentrations (~25 μM) and showed lengthening of the DNA upon binding to ethidium with no hysteresis upon relaxation, indicating reversibility on the time scale of DNA-stretching experiments. It was also apparent that the melting transition disappeared at this concentration. This early discovery was interpreted as an indication that DNA overstretching was not force-induced melting.

In general, intercalator binding to dsDNA lengthens dsDNA while an increase in melting force indicates stabilization of dsDNA structure (Figure 6.6a). Not only the optical tweezers studies of intercalators, ethidium [10,27,31,32], Ru complexes [10,33], daunomycin [27], and YO [27] but also other force spectroscopic studies of intercalators, ethidium [24,26], proflavin [24,34], and psoralen [24] exhibit this effect. More detailed studies, which measure force at varying concentrations [10,32,33],

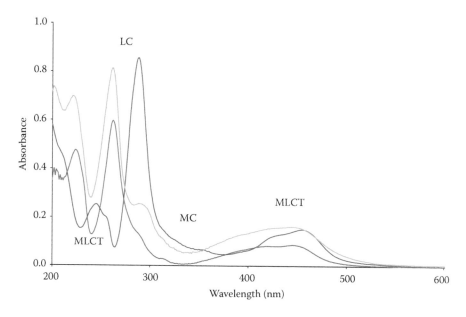

FIGURE 1.11 UV–visible absorption spectra of $[Ru(bpy)_3]^{2+}$ (blue), $[Ru(phen)_3]^{2+}$ (red), and $[Ru(phen)_2(dpq)]^{2+}$ (green) in water (band assignments for $[Ru(bpy)_3]^{2+}$ from Campagna et al.[78]).

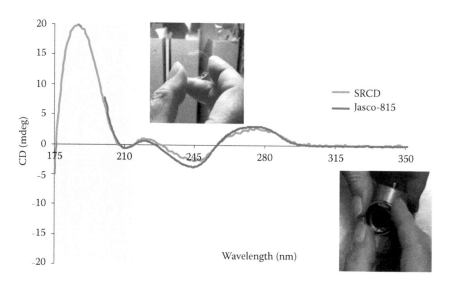

FIGURE 1.15 The SRCD spectra of ct-DNA is shown as the solid green line (indicates B-type DNA conformation) and the CD spectra as the solid red line showing the increase in wavelength range. Insets show the cells and the cell holder required.

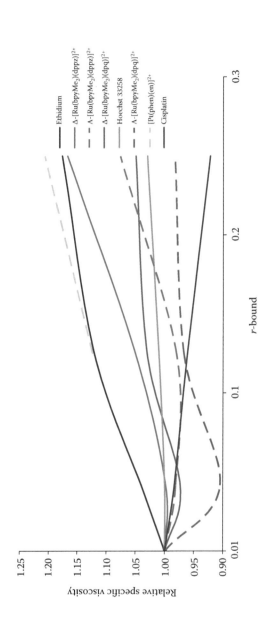

FIGURE 1.21 A plot of r-bound (ratio of bound drugs per DNA nucleotide) versus $(\eta/\eta_0)^{1/3}$. An increase in viscosity is seen with intercalators like ethidium bromide, $[Pt(phen)(en)]^{2+}$,[106] a negligible effect on viscosity with the addition of the groove binder Hoechst 33258, whereas a decrease in viscosity is seen with the addition of cisplatin. Subtle difference in enantiomer binding, as the ratio increases, can also be observed. Δ-$[Ru(bpyMe_2)_2dppz]^{2+}$ increases viscosity and as such binds by intercalation; however, Λ-$[Ru(bpyMe_2)_2dppz]^{2+}$ and Δ-$[Ru(bpyMe_2)_2dpq]^{2+}$ produce a reduction in viscosity initially, due to partial intercalation, as the concentration increases, and so does the viscosity indicating that full insertion is taking place. Λ-$[Ru(bpyMe_2)_2dpq]^{2+}$ only partial intercalations even at higher concentrations.[107]

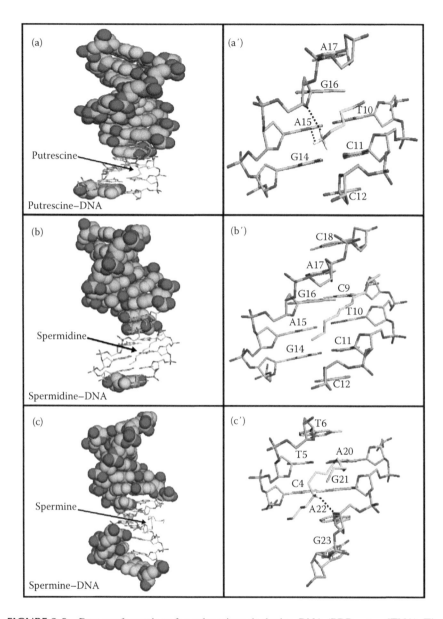

FIGURE 2.8 Best conformations for polyamines docked to DNA (PDB entry 6TNA). The polyamines are shown in green color. (a) shows DNA in sphere-filling model with the putrescine-binding site in sticks and (a′) shows putrescine-binding sites represented in sticks with the corresponding base residues. (b) shows DNA in sphere-filling model with the spermidine-binding site in sticks and (b′) shows spermidine-binding sites represented in sticks. (c) shows DNA in sphere-filling model with the spermine-binding sites represented in sticks and (c′) shows the binding sites represented in sticks.

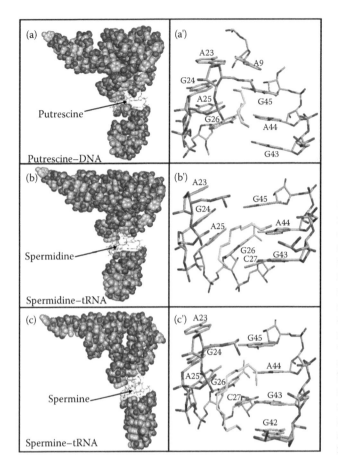

FIGURE 2.9 Best conformations for polyamines docked to tRNA (PDB entry 6TNA). The polyamines are shown in green color. (a) shows tRNA in sphere-filling model with the putrescine-binding site in sticks and (a′) shows putrescine-binding sites represented in sticks with the corresponding base residues. (b) shows tRNA in sphere-filling model with the spermidine-binding site in sticks and (b′) shows spermidine-binding sites represented in sticks. (c) shows tRNA in sphere filling model with the spermine-binding sites represented in sticks and (c′) shows the binding sites represented in sticks.

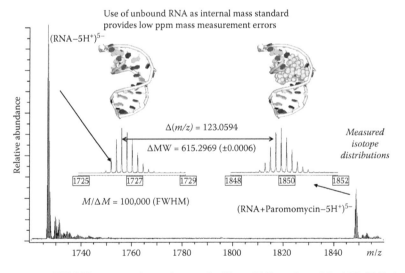

FIGURE 3.1 ESI-MS spectrum for a mixture of a 27mer RNA analog of the 16S rRNA A site with paromomycin. The signal from the [M−5H]⁵⁻ ion of the free RNA appears at m/z 1746, while the signal from the [M−5H]⁵⁻ ion of RNA bound to paromomycin appears at m/z 1849. The accurate measurement of the difference in mass using an FT-ICR mass spectrometer allows the identity of the ligand to be determined from the limited set of elemental compositions consistent with the measured mass. For paromomycin, the measured mass of 615.2969 + 0.0006 Da is only consistent with one chemically plausible elemental composition.

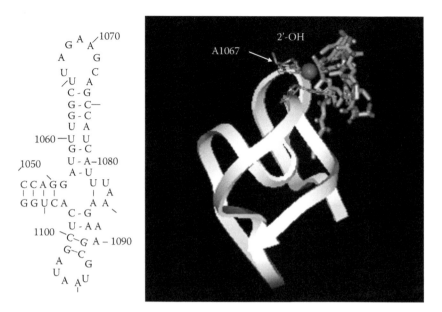

FIGURE 3.10 RNA sequence of the 58mer RNA analog for the L11-binding site in the 23S rRNA and a model structure with bound thiostrepton. The large thiostrepton molecule makes multiple surface contacts with the RNA.

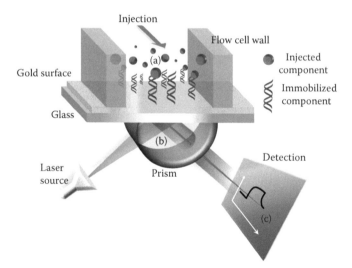

FIGURE 4.2 Schematic representation of biomolecular interactions observed in a flow cell and SPR angle change with some of the critical components labeled. The flow of a compound (a) over the immobilized nucleic acid (blue) on the biosensor surface will result in a change in the refractive index with a change in the light beam–gold metal interaction and a change in the SPR angle (b) that can be monitored in real time by an array detector (c). The sensorgram overlaid on the array detector is simply to illustrate how the measured SPR angle change is converted into a time-dependent output signal.

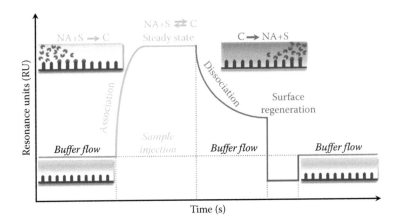

FIGURE 4.3 Diagram of the steps and observed signal for buffer and compound injections over a sensor chip surface with immobilized DNA. These are typical responses observed in each cycle of a biomolecular interaction experiment. The immobilized NA is actually in a 3D hydrophilic matrix. In the sensorgram, the SPR angle change from Figure 4.2 has been converted to resonance units (RU) where a 1000 RU response is equivalent to binding of 1 ng/mm² for most biomolecules. In the 3D matrix of most sensor chips, this would be equivalent to approximately 6 mg/mL amount of bound material.

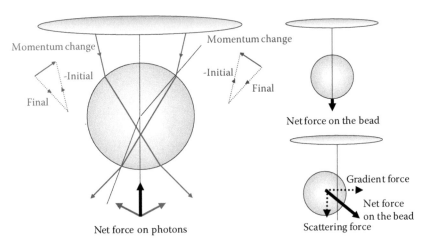

FIGURE 6.1 Ray optics description of bead trapping with single-beam optical tweezers. A bead is perturbed slightly from the laser focus. (The perturbation is toward the microscope objective in this figure.) Individual rays (red and blue lines) will experience a net change in momentum, due to refraction. The bead will experience an equal and opposite force (as shown in the lowest inset). This restoring force will be proportional to the displacement. The total force will consist of two components, a gradient force and a scattering force (the scattering force is not shown), though in a dual counterpropagating beam trap, the scattering forces will cancel.

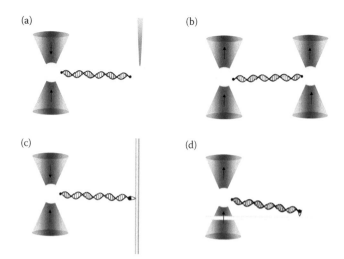

FIGURE 6.2 Different configurations used in optical tweezers to trap and manipulate single molecules. The black arrows indicate the direction of propagation of the laser light. (a) A dual counterpropagating beam optical trap holds a bead attached to one end of the single molecule and the bead at the other end is held by a micropipette. (b) Both beads attached at ends of the single molecule are held by unique traps created by two beams, or a time-shared single beam. (c) One end of the single molecule is attached to a glass slide and the other end attached to a bead that is held by a dual-beam optical trap. (d) One end of the single molecule attached to the glass slide and the other end attached to the bead held by single-beam optical trap.

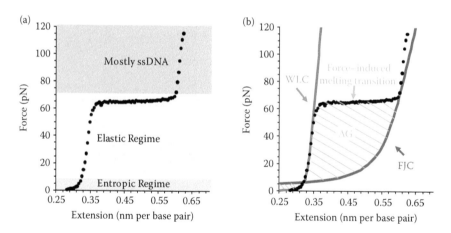

FIGURE 6.3 Force measured as a function of DNA extension during a typical-stretching experiment with optical tweezers. (a) DNA-stretching data (black circles) exhibits four distinct regimes. At low extensions, the force change is very small upon extension across the entropic regime (violet-shaded area). Further stretching shows a rapid increase in force where the DNA backbone responds, in an elastic regime (light blue-shaded area). This is followed by a transition plateau around 65 pN (light green-shaded area). Finally, at the end of the transition, the force again increases rapidly (light red-shaded area), where two single-stranded DNA held by GC-rich regions are stretched. (b) The worm-like chain model fit (blue solid curve) to the double-stranded DNA and the freely jointed chain model (red solid curve) of ssDNA clearly indicates that the transition at 65 pN is a melting transition from dsDNA to ssDNA (see discussion in the text). The shaded area between the DNA-stretching curve and ssDNA model yields the melting free energy. Experimental conditions: 100 mM Na$^+$, 10 mM HEPES, pH 7.5, 20°C.

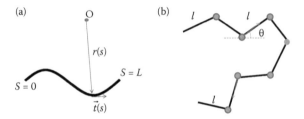

FIGURE 6.4 Polymer models of nucleic acid flexibility. (a) The worm-like chain model describes polymers with smooth continuous bends. (b) The freely jointed chain model describes the polymer as a collection of independent monomers with varying bond angles connected together by hinges.

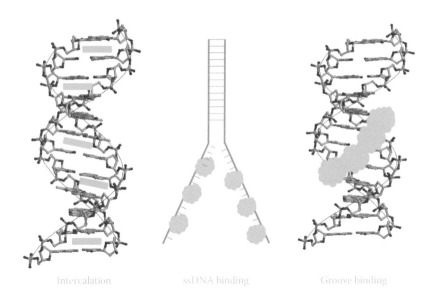

FIGURE 6.5 Reversible-binding modes of small molecules. Intercalation, where the flat aromatic ring of the drug (green) stacks between the DNA base pairs (left), single-stranded DNA binding where ligands (green) bind to the exposed bases or the stands of the DNA (middle), and groove binding where the drug (green) interacts with DNA by fitting in either the major or minor groove of the DNA helical structure (right). DNA structure obtained from PDB file 1BNA from Ref. 70.

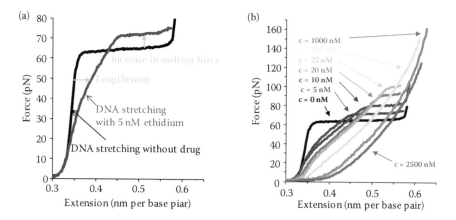

FIGURE 6.6 Effects of intercalation on DNA-stretching curves. (a) DNA-stretching curve in the presence (purple solid line) and absence (black solid line) of classical intercalator ethidium clearly indicates the lengthening of DNA and increase in melting force upon binding to ethidium. (b) DNA-stretching curves in the presence of different ethidium concentrations show the lengthening of DNA and increase in melting force increasing with ethidium concentration. The melting transition plateau shortens gradually and vanishes at a critical concentration between 25 and 125 nM. Experimental conditions: 100 mM Na$^+$, 10 mM HEPES, pH = 7.5, 20°C. (Adapted from Vladescu, I. D. et al. *Phys Rev Lett* **95**, 4pp., 2005.)

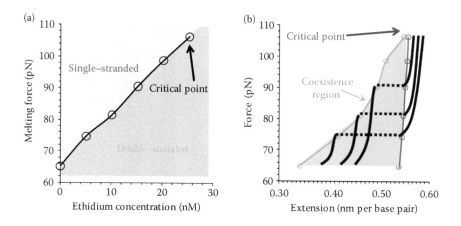

FIGURE 6.7 Phase transition and critical point observed in the presence of intercalators. (a) Melting force obtained as a function of ethidium concentration separates the two DNA phases, dsDNA and ssDNA (dsDNA-shaded area in light green and ssDNA-shaded area in light red). (b) The phase diagram shows that beyond a critical ethidium concentration these phases cannot be distinguished by stretching experiments. (Adapted from Vladescu, I. D. et al. *Phys Rev Lett* **95**, 4pp., 2005.)

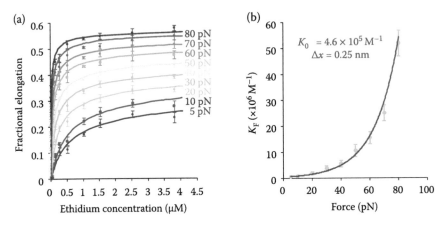

FIGURE 6.8 Quantifying intercalation to obtain binding constants in the absence of force. (a) McGhee–von Hippel fits (6.10) at different forces (colored solid lines) for the fractional elongations obtained as a function of ethidium concentration yields the binding constant and binding site size (values not shown in the figure) at each force. (b) Binding constants obtained from the fits in (a) (green circles) agrees well with the exponential behavior (6.12) as predicted (solid red line). The results are extrapolated to zero force to obtain the binding constant (K_0) and lengthening upon single intercalation (Δx) (values shown as inset in the plot) in the absence of force. Experimental conditions: 100 mM Na^+, 10 mM HEPES, pH = 7.5, 20°C. (Adapted from Vladescu, I. D. et al. *Nat Methods* **4**, 517–522, 2007.)

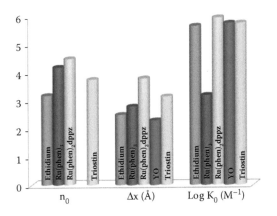

FIGURE 6.9 Comparison of binding parameters of different intercalators obtained using the method introduced by Vladescu et al. Binding site size n_0 in base pairs, binding constant K_0 in M^{-1} in the absence of force, and lengthening upon a single intercalation event Δx in Å estimated for intercalators, ethidium (blue), $Ru(phen)_3^{2+}$ (brown), $Ru(phen)_2dppz^{2+}$ (green), YO (purple), and bis-intercalator triostin (orange).

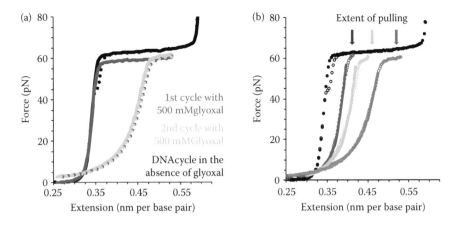

FIGURE 6.10 Exploring the force-induced melting transition with glyoxal binding. (a) The DNA was stretched in the presence of 500 mM glyoxal (solid blue line), which followed exactly the same path as the stretching of DNA in the absence of glyoxal (black solid line) except for the fact that the melting transition is lowered a little. The relaxation curve (blue broken line) obtained after holding at a fixed extension (blue arrow) for nearly 30 min exhibits significant hysteresis supporting the glyoxal binding to exposed ssDNA. The second stretch (green solid line) follows the previous relaxation curve (blue broken line), which suggests that modification is permanent. (b) Relaxation data (blue, green, and red open circles) obtained by progressively stretching the DNA along the transition plateau in the presence of 500 mM glyoxal and holding at different fixed extensions along the plateau (indicated by blue, green, and red arrows) for nearly 30 min fits very well to the predicted linear combination of the dsDNA and ssDNA models (blue, green, and red solid lines). (Adapted from Shokri, L. et al. *Biophys J* **95**, 1248–1255, 2008.)

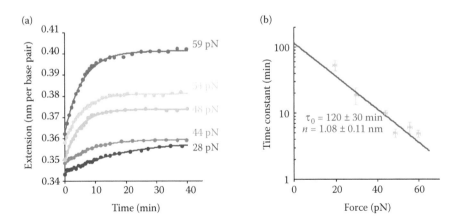

FIGURE 6.11 Exploring threading intercalation with optical tweezers. (a) Extension measurements as a function of time obtained at constant forces of 28 pN (purple circles), 44 pN (blue circles), 48 pN (green circles), 54 pN (orange circles), and 59 pN (red circles) in the presence of threading intercalator ΔΔ-[μ-bidppz(phen)$_4$Ru$_2$]$^{4+}$ and single exponent fits (solid lines in corresponding colors). (b) Characteristic time constants (green circles) obtained from the fits in panel (a) fit well to the predicted exponential dependence (6.17) on force (red solid line). The fit suggests that only one base pair must melt in order for this molecule to thread through the DNA bases and extrapolation yields the time constant at zero force shown as an inset in the plot. (Adapted from Paramanathan, T. et al. *J Am Chem Soc* **130**, 3752–3753, 2008.)

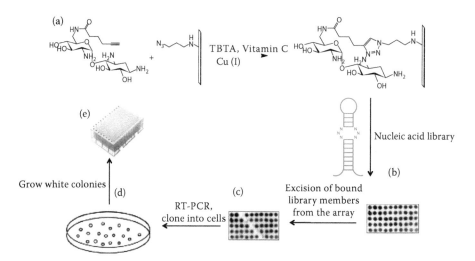

FIGURE 9.1 Schematic of the microarray-based selection process. (a) Immobilization of small molecules on an activated agarose microarray via a Huisgen dipolar cycloaddition reaction. (b) Image of the slide after hybridization with radioactively labeled RNA. (c) Image of the same slide after mechanical removal of bound RNA. (d) Schematic of an agar plate after transformation with a vector containing selected members of the library. (e) Colonies that are white from blue/white screening are grown in medium and sequenced to deconvolute the selected members of the library.

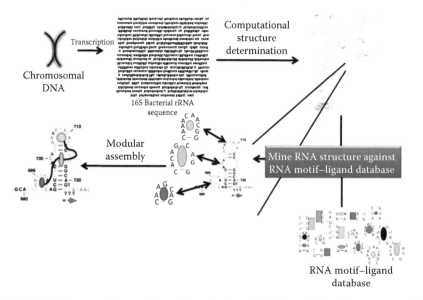

FIGURE 9.2 The overall approach to using a database of RNA motif–ligand partners to rationally design small molecules targeting RNA. First, a target RNA sequence is subjected to structure determination (free energy minimization, phylogenic comparison, or selective 2′-hydroxyl acylation analyzed by primer extension (SHAPE)). Next, the secondary structures within the target RNA and their relative orientation are extracted and mined against the RNA motif–ligand database. Once overlaps between the database and the secondary structures are found, modular assembly routes are used to target multiple motifs in the RNA simultaneously.

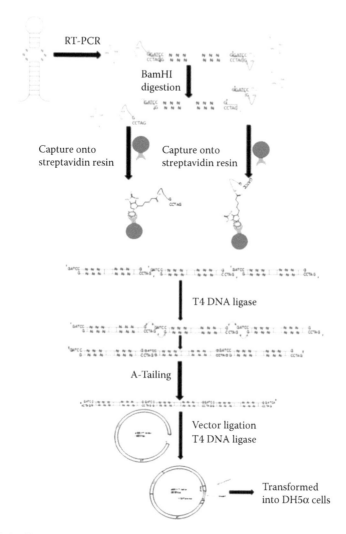

FIGURE 9.4 The schematic of a concatenation approach that is used to ligate multiple copies of the selected members of the RNA motif library together. Key features of the method are the installation of two BamHI restriction sites into the RT-PCR product and the use of biotinylated primers. After RT-PCR amplification, the cDNAs are digested with BamHI and purified using streptavidin resin. The products are ligated together using T4 DNA ligase.

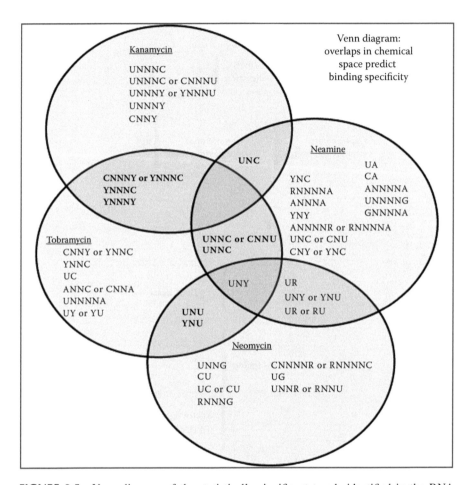

FIGURE 9.5 Venn diagram of the statistically significant trends identified in the RNA sequence space from a 6-nucleotide hairpin library selected for four aminoglycoside derivatives. Overlapping trends are shown in bold. The most statistically significant trend for the kanamycin A derivative is 5′UNNNC3′; for the tobramycin derivative, 5′UNNC3′; for the neamine derivative, 5′UNC3′; and for the neomcyin B derivative, 5′UNNG3′. (Reproduced from Paul, D. J., Seedhouse, S. J. and Disney, M. D. *Nucleic Acids Res.* 37, 5894–5907, 2009. With permission.)

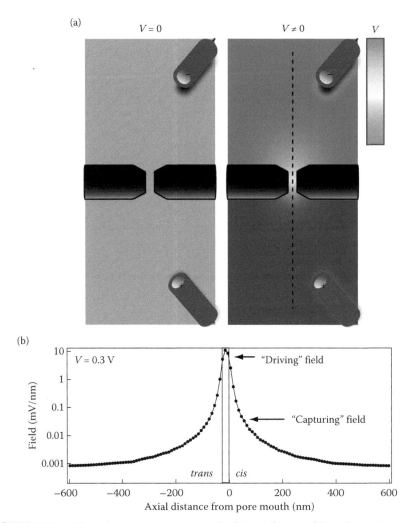

FIGURE 12.2 Voltage drop across a nanopore. (a) Schematic map of the voltage drop across a membrane with a nanoscale pore when a voltage is applied across the pore using a pair of electrodes that are far away from the pore (depicted for high electrolyte concentration). Application of voltage results in ion flow toward the membrane, where ion flow is then impeded by the membrane and restricted to flow through the nanopore. (b) Electric field at various axial positions from the mouth of a 4 nm hourglass-shaped pore fabricated in a 25 nm thick silicon nitride membrane. (Adapted from Wanunu, M. et al. *Nat. Nanotechnol.* **2010,** 5 (2), 160–165.) The axis of the nanopore is defined by the dashed line in (a). For the simulation, 1 M KCl buffer was used as the electrolyte and a 300 mV voltage was applied at the two ends of the simulation box (see Ref. 35 for more details). The semilog plot shows that the field is very strongly concentrated at the pore (membrane height outlined by gray area in plot), with a very steep decay that is symmetric about the membrane center. Although weak, the interaction of the "capturing" field with the highly charged nucleic acid molecules captures nucleic acids toward the pore, where they are threaded through the pore by the strong "driving" field.

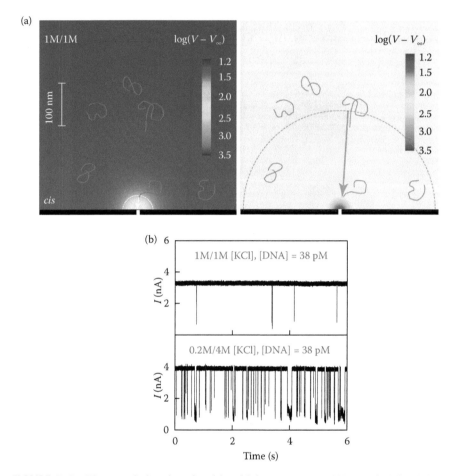

FIGURE 12.3 Electrostatic focusing of nucleic acids into a nanopore: (a) Numerical simulations of the electrical potential in the *cis* side of a 4 nm nanopore for both symmetric (left) and asymmetric (right) KCl concentrations, where the *cis/trans* KCl concentrations are indicated in each image (simulation taken from Ref. 35). The pore mouth is positioned at the bottom center of each image (*trans* chamber not shown). The potential outside the pore increases with salt asymmetry, thereby funneling molecules to be captured. (b) Experimental time traces for a 38 pM 8000 bp solution under 1 M/1 M (left, rate = 0.4 s⁻¹) and 0.2 M/4 M (left, rate = 11.20 s⁻¹) *cis/trans* KCl concentrations. The 30-fold enhancement in the capture rate is a result of the electrical field outside the pore. (From Wanunu, M. et al. *Nat. Nanotechnol.* **2010,** 5 (2), 160–165. With permission.)

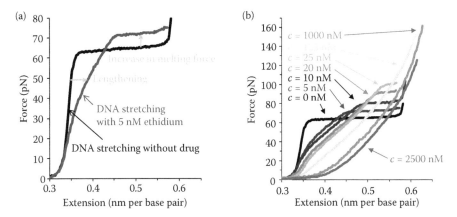

FIGURE 6.6 **(See color insert.)** Effects of intercalation on DNA-stretching curves. (a) DNA-stretching curve in the presence (purple solid line) and absence (black solid line) of classical intercalator ethidium clearly indicates the lengthening of DNA and increase in melting force upon binding to ethidium. (b) DNA-stretching curves in the presence of different ethidium concentrations show the lengthening of DNA and increase in melting force increasing with ethidium concentration. The melting transition plateau shortens gradually and vanishes at a critical concentration between 25 and 125 nM. Experimental conditions: 100 mM Na^+, 10 mM HEPES, pH = 7.5, 20°C. (Adapted from Vladescu, I. D. et al. *Phys Rev Lett* **95**, 4pp., 2005.)

explained the disappearance of the overstretching plateau as a simple thermodynamic effect due to the use of very high-ligand concentrations. While low concentrations of intercalator stabilize dsDNA and increase the DNA overstretching force, the concomitant lengthening of dsDNA also causes the melting plateau to shorten (Figure 6.6b). At some critical concentration (between 25 and 125 nM for ethidium in Figure 6.6b), the lengthening of dsDNA upon drug binding makes it impossible to distinguish the dsDNA–drug complex from ssDNA and the melting transition disappears. The concentration-dependent effects of intercalators on DNA overstretching are fully consistent with the interpretation of DNA overstretching as a force-induced melting transition.

We may consider the phase conversion of dsDNA to ssDNA by noting the similarity to the conversion of liquid to the gas phase in the classical thermodynamic case. The melting force dependence on drug concentration (Figure 6.7a) corresponds to the relationship of pressure and melting temperature. Mapping the force–extension-ethidium concentration phase diagram (Figure 6.7b) shows clear similarities with the classical pressure-volume-temperature (PVT) phase diagrams of gaseous systems, where force is analogous to pressure, extension corresponds to volume in a one-dimensional system, and concentration is analogous to temperature [32].

6.3.1.2   Measuring Persistence and Contour Lengths

The drug–DNA complex stretching curves obtained with optical tweezers were fitted to the WLC model (Equation 6.5) at very low force ($F > 2$ pN) [35], low force ($F > 15$ pN) [31], and high force ($F > 40$ pN) [27] limits to obtain the contour and persistence length of the complex. All intercalators studied under different conditions

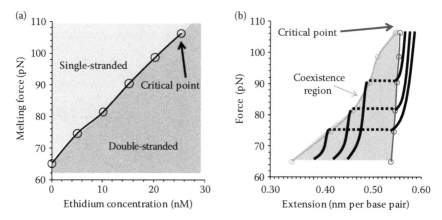

FIGURE 6.7 **(See color insert.)** Phase transition and critical point observed in the presence of intercalators. (a) Melting force obtained as a function of ethidium concentration separates the two DNA phases, dsDNA and ssDNA (dsDNA-shaded area in light green and ssDNA-shaded area in light red). (b) The phase diagram shows that beyond a critical ethidium concentration these phases cannot be distinguished by stretching experiments. (Adapted from Vladescu, I. D. et al. *Phys Rev Lett* **95**, 4pp., 2005.)

exhibited increases in contour length compared to dsDNA in the absence of the drug. Persistence lengths obtained in very low force measurements ($F > 2$ pN) [35] showed a surprising initial increase of the persistence length with increasing concentration although P_{ds} dropped back down to the value for dsDNA (~45 nm) after a critical ligand concentration. Other studies done with various intercalators at saturation [27,31,36] show decreases in the measured DNA persistence length upon binding. WLC model fittings at saturated concentrations [10] of these drugs exhibited an almost 70% decrease of persistence length and almost 25% smaller-elastic modulus compared to dsDNA.

6.3.1.3 Estimation of Melting Free Energy Change

The melting free energy of dsDNA in the presence of intercalators ΔG(drug–DNA) can be estimated by integrating the area between force–extension curve of the drug–DNA complex and drug-free ssDNA [32]:

$$\Delta G(\text{drug} - \text{DNA}) = \int_0^{F_m} \left(x_{ss}(F) - x_{\text{drug−DNA}}(F) \right) dF \tag{6.7}$$

Subtracting the melting free energy of dsDNA in the absence of the drug (ΔG) yields the melting free energy increase due to the drug intercalation. This free energy increase is a measure of the stabilization of the dsDNA structure upon intercalator binding. Ligand concentration dependence measurements of the melting free energy increase for ethidium obtained from optical tweezers stretching experiments [32] show that the melting free energy increases with concentration. The results agree well with estimated free energy changes from thermal-melting experiments [37].

6.3.1.4 Fitting to McGhee–von Hippel Isotherm to Obtain Binding Parameters

More careful quantitative studies consider the fractional ligand binding as a function of the force applied by optical tweezers. These fractional elongations were fitted to the (noncooperative) McGhee–von Hippel isotherm to obtain the binding constant and individual-binding site size [10,33]. The fractional ligand-binding γ is given by the ratio between the lengthening observed due to the binding of intercalator at a particular concentration (δb) compared to saturation (δb_{max}):

$$\gamma = \frac{\delta b}{\delta b_{max}} = \frac{b_{ds}(C) - b_{ds}(0)}{b_{ds}(C_{sat}) - b_{ds}(0)} \tag{6.8}$$

Now, $b_{ds}(C)$ is the extension observed for the drug–DNA complex at drug concentration C, $b_{ds}(C_{sat})$ the extension of the complex at saturated drug concentration, and $b_{ds}(0)$ the extension of dsDNA in the absence of drug. These extensions also vary with force. The McGhee–von Hippel site exclusion-binding isotherm relates the fractional binding (γ) and concentration (C) at a particular force. Fits to this equation yield the binding constant (K) and binding site size (n):

$$\gamma = KnC \frac{(1-\gamma)^n}{\left(1 - \gamma + \frac{\gamma}{n}\right)^{n-1}} \tag{6.9}$$

Initial studies using this method on ethidium [32] gave a binding constant of $K = 10^7$ M^{-1} and binding site size $n = 2$ at low force ($F \leq 10$ pN), which was in excellent agreement with bulk experiments. Although the high force ($F \geq 20$ pN) measurement of the binding constant, $K = 1.5 \times 10^7$ M^{-1}, was within an appropriate range, the binding site size $n = 1$ contradicted the site exclusion theory that was established in bulk experiments. According to this principle, the maximum saturated binding of an intercalator to the DNA lattice is to every other base pair due to sugar puckering of the backbone, giving a minimum value of $n = 2$. However, apparently increasing tension in the DNA removes this constraint and enables binding of ethidium to nearest-neighbor base pairs.

This method was extended to study three ruthenium polypyridyl complexes: [Ru(phen)$_3$]$^{2+}$, [Ru(phen)$_2$dppz]$^{2+}$, and [Ru(bpy)$_3$]$^{2+}$ [33]. Fits to concentration dependence experiments yielded a binding constant of $K = (3.2 + 0.1) \times 10^6$ M^{-1} for [Ru(phen)$_2$dppz]$^{2+}$, compared to $K = (8.8 + 0.3) \times 10^3$ M^{-1} for [Ru(phen)$_3$]$^{2+}$. Both molecules were estimated to have binding site size $n = 3 \pm 0.5$ base pairs, while [Ru(bpy)$_3$]$^{2+}$ did not show any significant low force intercalation (up to nearly mM concentrations).

6.3.1.5 Quantifying the Force Dependence to Obtain Zero Force Intercalation

A new method to obtain the binding parameters in the absence of the force was introduced by Vladescu et al. [10], in which the drug–DNA complexes were stretched at different drug concentrations. The fractional ligand binding per binding site ($\upsilon_F = \gamma_F/n_F$) at a particular force was calculated and then fitted to a modified site exclusion isotherm (Figure 6.8a)

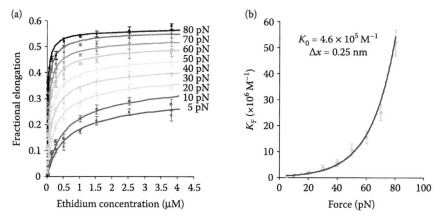

FIGURE 6.8 **(See color insert.)** Quantifying intercalation to obtain binding constants in the absence of force. (a) McGhee–von Hippel fits (6.10) at different forces (colored solid lines) for the fractional elongations obtained as a function of ethidium concentration yields the binding constant and binding site size (values not shown in the figure) at each force. (b) Binding constants obtained from the fits in (a) (green circles) agrees well with the exponential behavior (6.12) as predicted (solid red line). The results are extrapolated to zero force to obtain the binding constant (K_0) and lengthening upon single intercalation (Δx) (values shown as inset in the plot) in the absence of force. Experimental conditions: 100 mM Na$^+$, 10 mM HEPES, pH = 7.5, 20°C. (Adapted from Vladescu, I. D. et al. *Nat Methods* **4**, 517–522, 2007.)

$$\upsilon_F = K_F C \frac{\left(1 - n_F \upsilon_F\right)^{n_F}}{\left(1 - n_F \upsilon_F + \upsilon_F\right)^{n_F - 1}} \tag{6.10}$$

The binding constant (K_F) and binding site size (n_F) obtained from force-specific fits may be extrapolated to obtain the values at zero force. The drug–DNA complex is extended by Δx during a single intercalation event in the presence of an external force F, as work done is on the system. The work $F\Delta x$ reduces the free energy of the drug intercalation in the presence of force $\Delta G(F)$ by a factor $F\Delta x$ from the free energy in the absence of force $\Delta G(0)$:

$$\Delta G(F) = \Delta G(0) - F\Delta x \tag{6.11}$$

As a result, the force dependence of binding constant can be written as

$$K_F = K_0 \exp\left(\frac{F\Delta x}{k_B T}\right) \tag{6.12}$$

Fitting the values of K_F determines the binding constant in the absence of the force (K_0). The length change upon a single intercalation event, Δx, is also given by the fit (Figure 6.8b). A complete analysis using this method determined the zero force binding constant K_0 and lengthening upon an individual intercalation event Δx for several intercalators (and refined the results stated in the section above). The results obtained for ethidium [10], [Ru(phen)$_3$]$^{2+}$ [10], [Ru(phen)$_2$dppz]$^{2+}$ [10], and YO [36] are shown in Figure 6.9. While these intercalators show similar binding

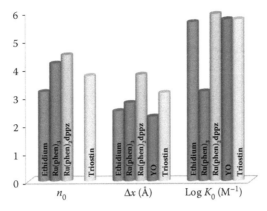

FIGURE 6.9 (**See color insert.**) Comparison of binding parameters of different intercalators obtained using the method introduced by Vladescu et al. Binding site size n_0 in base pairs, binding constant K_0 in M^{-1} in the absence of force, and lengthening upon a single intercalation event Δx in Å estimated for intercalators, ethidium (blue), Ru(phen)$_3$$^{2+}$ (brown), Ru(phen)$_2$dppz^{2+} (green), YO (purple), and bis-intercalator triostin (orange).

and induced length changes, [Ru(phen)$_3$]$^{2+}$ is distinguished by 1000-fold weaker binding.

The zero force binding site size n_0 can be estimated from fitting the drug–DNA complex-stretching curve at saturated concentration to the WLC model (Equation 6.5) by rewriting the WLC model in the following form:

$$b(F) = B(1 + \gamma_0)\left(1 - \frac{1}{2}\left(\frac{k_B T}{FP}\right)^{1/2} + \frac{F}{S}\right) \tag{6.13}$$

where γ_0 is the zero force fractional ligand binding and the remaining terms were defined above (Equation 6.5). The reciprocal of γ_0 yields the zero force binding site size n_0. The n_0 values calculated for ethidium, [Ru(phen)$_3$]$^{2+}$, and [Ru(phen)$_2$dppz]$^{2+}$ using this method [10] are also tabulated in Figure 6.9. These intercalators yield similar results of $5 > n_0 > 3$ base pairs, for binding at nearly every other base pair.

6.3.2 SINGLE-STRANDED DNA BINDING

Although many proteins bind to ssDNA [38–45], only a few small molecules and drugs are known to do so. Exceptions include the drug actinomycin D, which is reported to bind to ssDNA [46–49], and glyoxal, which binds directly to exposed bases [50–54]. While actinomycin studies at the single-molecule level have yet to be published, results with glyoxal binding are well understood. Glyoxal (C$_2$H$_2$O$_2$) is known to bind to the exposed guanine bases of DNA, with relatively slow kinetics [50]. Stretching relaxation curves of λ-DNA obtained with optical tweezers in the presence of glyoxal [55] showed lowering of melting force and huge hysteresis upon the binding of glyoxal (Figure 6.10a) similar to the results that is observed with

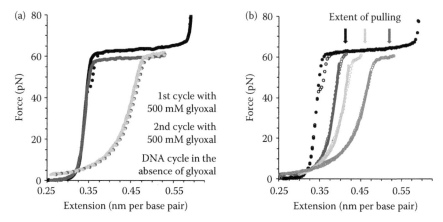

FIGURE 6.10 **(See color insert.)** Exploring the force-induced melting transition with glyoxal binding. (a) The DNA was stretched in the presence of 500 mM glyoxal (solid blue line), which followed exactly the same path as the stretching of DNA in the absence of glyoxal (black solid line) except for the fact that the melting transition is lowered a little. The relaxation curve (blue broken line) obtained after holding at a fixed extension (blue arrow) for nearly 30 min exhibits significant hysteresis supporting the glyoxal binding to exposed ssDNA. The second stretch (green solid line) follows the previous relaxation curve (blue broken line), which suggests that modification is permanent. (b) Relaxation data (blue, green, and red open circles) obtained by progressively stretching the DNA along the transition plateau in the presence of 500 mM glyoxal and holding at different fixed extensions along the plateau (indicated by blue, green, and red arrows) for nearly 30 min fits very well to the predicted linear combination of the dsDNA and ssDNA models (blue, green, and red solid lines). (Adapted from Shokri, L. et al. *Biophys J* **95**, 1248–1255, 2008.)

ssDNA-binding proteins [41–44,56], where the observed hysteresis is due to ligand binding to ssDNA. As glyoxal is known to bind only to the exposed guanine bases, it was used to explore the nature of the overstretching transition that occurs around 65 pN during dsDNA stretching. DNA in the presence of glyoxal was stretched to different extensions along this transition plateau and relaxed after holding at that extension for 30 min to allow enough time for the reaction to occur [55]. The results showed more hysteresis with progressive stretching into the plateau, suggesting that more glyoxal is bound at greater extensions. This indicates that more bases are exposed at increasing extensions supporting a model that describes overstretching as force-induced melting of dsDNA. Thus, the relaxation curves of the DNA–glyoxal complex should result from a mixture of dsDNA and ssDNA bound to glyoxal. The observed extension of this mixture at a particular force can be written as a function of the ssDNA fraction (γ_{ss}) [55]

$$b = b_{ds}(1 - \gamma_{ss}) + b_{ss}\gamma_{ss} \tag{6.14}$$

In the above equation, b_{ds} is the extension of pure dsDNA (a function of the force) that is determined from the WLC model (Equation 6.5) and b_{ss} is the extension of the ssDNA (also a function of the force) using the FJC model (Equation 6.6). The glyoxal-bound ssDNA is well approximated by this equation, as glyoxal only binds to the exposed bases, and not the backbone. The fraction of ssDNA (γ_{ss}) obtained from the fits (Equation 6.14) clearly indicates that this is a direct measure of the fractional length of the transition plateau into which the DNA was stretched and held (Figure 6.10b). This result clearly indicates that the transition that occurs in the force plateau is a force-induced melting transition.

6.3.3 GROOVE BINDERS

Small molecules that have a net positive charge are likely to interact with DNA via an electrostatic interaction, binding to the major or minor groove of the negatively charged DNA. These molecules are known as groove binders and several types of drugs are known to have this binding mode either as an intermediate or final binding state. The primary mode of electrostatic interaction is often complemented by hydrogen bonding and van der Waals interactions [26,57].

Netropsin, used as an antibiotic that also exhibits antitumor and antiviral activity, and distamycin A, an antitumor agent [58], are drugs that are believed to bind in the minor groove of DNA. Berenil, used for the effective control of trypanosome infections and known to be a chemotherapeutic agent [59], also binds to the minor groove at low concentrations. These groove binding drugs were first characterized in AFM experiments. The force required to melt dsDNA in the presence of netropsin and low concentration of berenil was increased in DNA-stretching experiments. In contrast, stretching experiments performed with poly(dG–dC) dsDNA in the presence of distamycin A [26] lowered the melting force to 50 pN. These results could not conclude whether or not minor groove binders stabilize dsDNA structure. To resolve this contradiction, λ-DNA was stretched using optical tweezers in the presence of distamycin A [27] and the results confirmed an increase in the melting force obtained previously with other minor groove binders. WLC model (Equation 6.5) fittings done for dsDNA–minor groove binder complexes at low force limit ($F < 15$ pN) in the presence of minor groove binders netropsin and low concentrations of Berenil [31] showed an increase in the persistence length and almost no change in contour length from that of dsDNA. However, the persistence length obtained for the distamycin A–DNA complex with high force limits (up to 40 pN) indicated a decrease in persistence length [27]. Thus, while minor groove binders appear to change the persistence length, this change is not well understood.

In contrast to the abundance of minor groove binders, relatively few drug types bind into the major groove of DNA. Methyl green, used in the diagnosis of trachoma [60], is a dye that has been used to probe major groove binding. AFM-stretching experiments with poly(dG–dC) dsDNA in the presence of synthetic major groove binders yielded similar results to the minor groove binders, exhibiting an increase in the melting transition force [26]. The major groove binders also decrease melting cooperativity, as melting occurred over an increasing range of forces. Optical tweezers experiments with synthetic peptide α-helix Ac–(Leu–Ala–Arg–Leu)$_3$–NH–linker (linker:1,8-diamino-3,6-dioxaoctane) [27] and another major groove binder SYBR-Green I [61] confirmed an increase in the melting transition force and noncooperative melting. Fitting the peptide–DNA complex to the WLC model yielded an increase in contour length and decrease in persistence length upon binding to an α-helix Ac–(Leu–Ala–Arg–Leu)$_3$–NH–linker (linker:1,8-diamino-3,6-dioxaoctane) [27]. The clear stabilization of dsDNA upon binding to major groove binders suggests that this property can be used to design potential drugs or diagnostic tools.

6.3.4 MULTIPLE-BINDING MODES

The drug Berenil, introduced in the previous section as a minor groove binder, was observed to bind to the minor groove at small concentrations. Increasing concentrations of Berenil indicate a new binding mode, where the drug intercalates. Psoralen is another drug that exhibits multiple-binding modes. These multiple modes may be distinguished using force spectroscopic studies. Binding of Berenil to DNA was first explored using AFM studies at the single-molecule level [24,57]. Low concentrations show some lengthening while the melting transition force increases, effects that are similar to those due to minor groove binding. As drug concentrations increase, a significant increase in the length is observed and melting becomes increasingly noncooperative. This behavior is characteristic of intercalators at saturating concentrations. This activity was quantitatively characterized in an optical tweezers experiment where T7 bacteriophage DNA (14,046 bp) was stretched in the presence of Berenil. The concentration of Berenil was gradually increased and the stretching curves obtained were fitted to the WLC model (Equation 6.5) at low force limits ($F < 15$ pN) [31]. At low drug concentrations, the persistence length (P_{ds}) nearly doubled with little change in the contour length (B_{ds}). These observations are consistent with minor groove binding. As the drug concentration is increased gradually, the persistence length (P_{ds}) decreases to a value that is less than the persistence length of pure dsDNA, while the contour length simultaneously increases. Thus, the binding mode switches at a critical concentration, changing from minor groove binding to intercalative binding.

Psoralen is used to treat certain skin diseases. This therapy is commonly known as PUVA (psoralen ultraviolet-A) therapy [62]. The process of drug binding to DNA begins with intercalation. A covalent attachment is made with a pyrimidine in one DNA strand to form a mono-adduct in the next step. When exposed to ultraviolet-A light, the drug binds with the opposite strand to create a cross-link. The cross-link can be broken with exposure to ultraviolet-B light. Thus, the drug initially binds through intercalation and then binds covalently after exposure to ultraviolet light. The binding mechanism of psoralen was explored with optical tweezers by stretching λ-DNA in the presence of psoralen and exposing it to both UV-A and UV-B light [62]. The stretching curves obtained at very low force limit ($F < 2$ pN) were fitted to the WLC model (Equation 6.5) to obtain the persistence length (P_{ds}). It was observed that the persistence length increases dramatically to almost double the persistence length of dsDNA after 35 min of UV-A exposure, indicating that the complex becomes more rigid due to the formation of cross-links between the opposite strands. The cross-links may be broken through exposure of the drug–DNA complex to UV-B light. Measuring the persistence length again indicates that it decreases as the cross-links are broken and left out as mono-adducts. Exploration of psoralen cross-links using optical tweezers proves that this technique can also be employed to study the drugs that bind irreversibly to DNA.

6.3.5 COMPLEX-BINDING MODES

In this section, we will discuss the small molecules that bind through complex modes. These modes include threading intercalators such as binuclear ruthenium complexes. YOYO and triostin are examples of another class, known as bis-intercalators. The ruthenium complexes (discussed in the intercalators section) have been systematically modified to improve their effectiveness as an anticancer drug. Potential drug candidates should exhibit high-binding affinity to DNA and a slow dissociation rate. Among those mononuclear ruthenium complexes that intercalate dsDNA, $Ru(phen)_2dppz^{2+}$ showed the highest affinity to dsDNA. Therefore, it was the first choice to make binuclear ruthenium complexes. Covalently linking two of these molecules forms Δ,Δ-[μ-didppz $Ru_2(phen)_4$]$^{4+}$ significantly altering the binding kinetics. We shall denote this molecule in short as Δ,Δ-P, where Δ denotes the right-handed coordination of ruthenium.

Initial bulk studies of these dumbbell-shaped molecules showed that they bind to the major groove of the dsDNA. Later studies reported threading through the dsDNA base pairs such that the bridging bi-dppz part is fully intercalated [63]. Bulk experiments require hours to reach this final orientation of the threaded form since the molecule must wait until a few base pairs melt due to thermal fluctuations. This necessary step provides enough space to thread the bulky propeller-shaped end through the bases. But in the case of stretching experiments with optical tweezers, the DNA is mechanically manipulated to favor these fluctuations, enhancing threading. The binding kinetics are obtained as a function of force and then extrapolated to obtain the zero force kinetics. It has been observed that during traditional stretching–relaxing experiments this threading intercalation of Δ,Δ-P does not reach equilibrium during the stretching–relaxation cycles. A new method was introduced to study the slow drug-binding kinetics, where DNA in the presence of drug was stretched rapidly and held at constant force to measure the extension until it reached equilibrium [64]. These constant force–extension measurements showed a characteristic time constant that depends upon the force (Figure 6.11a).

The slow-threading rate of Δ,Δ-P through the DNA bases (k) requires melting of n DNA base pairs and should be proportional to the probability of melting n base pairs. This rate varies exponentially:

$$k = k_c \exp\left(\frac{-n\Delta G}{k_B T}\right) \tag{6.15}$$

where k_c is the proportionality constant, ΔG is the melting free energy of a single base pair, k_B is the Boltzmann's constant, and T is the absolute temperature. In the stretching experiments, when a force F is applied to melt the base pairs, the DNA is lengthened by Δx per base pair during the conversion from dsDNA to ssDNA. Therefore, applying a force reduces the melting free energy of a single base pair ΔG by an amount of $F\Delta x$. This increases the probability of melting of n base pairs by an exponential factor ($Fn\Delta x/k_B T$) [65]

$$k = k_c \exp\left[\frac{-n(\Delta G_0 - F\Delta x)}{k_B T}\right] \tag{6.16}$$

(a)

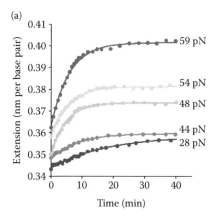

(b)

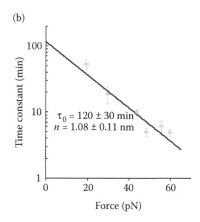

FIGURE 6.11 **(See color insert.)** Exploring threading intercalation with optical tweezers. (a) Extension measurements as a function of time obtained at constant forces of 28 pN (purple circles), 44 pN (blue circles), 48 pN (green circles), 54 pN (orange circles), and 59 pN (red circles) in the presence of threading intercalator $\Delta\Delta$-[μ-bidppz(phen)$_4$Ru$_2$]$^{4+}$ and single exponent fits (solid lines in corresponding colors). (b) Characteristic time constants (green circles) obtained from the fits in panel (a) fit well to the predicted exponential dependence (6.17) on force (red solid line). The fit suggests that only one base pair must melt in order for this molecule to thread through the DNA bases and extrapolation yields the time constant at zero force shown as an inset in the plot. (Adapted from Paramanathan, T. et al. *J Am Chem Soc* **130**, 3752–3753, 2008.)

Here, ΔG_0 is the free energy associated with the melting of single base pair in the absence of force. The threading time constants (τ), obtained from the reciprocals of the rates, indicate that the time constants fall exponentially with force:

$$\tau = \tau_0 \exp\left(\frac{-Fn\Delta x}{k_B T}\right) \tag{6.17}$$

Time constants obtained from constant force measurements fits (Figure 6.11a), plotted as function of force, are in good agreement with this predicted exponential behavior, as shown in Figure 6.11b. In the linear approximation region Δx, the length increase observed when a base pair of dsDNA is converted to ssDNA is expected to be 0.22 nm [56]. Combining this with the exponential prefactor obtained from the fitting suggests that only one base pair has to be melted to allow threading of the binuclear ruthenium complex [64]. Extrapolating the force dependence to zero force gives 120 ± 30 min as the time constant (τ_0) associated with the threading in the absence of force [64]. These experiments suggest that optical tweezers can be used to determine the slow-binding rates, which are difficult to determine in bulk experiments.

The key difference between threading intercalators and bis-intercalators is that the latter have the intercalating aromatic rings at the ends of the molecule whereas the threading molecule contains fused rings in the middle. Therefore, there is no base pair melting requirement for bis-intercalating molecules. Once an end of a bis-intercalator intercalates between bases, the other end must find an appropriate-binding site depending on the length of the linker. Two characteristic bis-intercalators have been explored with optical tweezers.

YOYO-1 is a bridged bis-intercalator that has two YO molecules that stack into two intercalating sites. It can also bind to the major groove at high concentrations, which complicates binding. YOYO-1 is nonfluorescent in solution but highly fluorescent when it binds to dsDNA [66]. The DNA-stretching experiments with optical tweezers in the presence of YOYO-1 exhibited effects similar to normal intercalators except for the presence of hysteresis during relaxation [27,36,67]. Additionally, the lengthening is strongly dependent on the pulling rate, indicating slow-binding kinetics for YOYO-1 [27,36,67]. To understand the kinetics of YOYO-1 binding, the DNA was stretched to and held at a constant length, while monitoring the retention force that decreases with time [27,36]. The time constants measured from the exponential force decay exhibit a linear relationship with the retention force up to the dsDNA-melting force [27]. Fluorescence properties of YOYO when it binds to DNA may be used to estimate the number of molecules bound by measuring the fluorescence intensity while stretching with optical tweezers [36]. As the YOYO molecules bind to DNA, both the measured fluorescence intensity and the fractional elongation calculated from stretching experiments increased linearly. Equilibrium extensions estimated from these experiments allow force-dependent-binding parameters to be calculated (as described above for intercalators) and extrapolated to obtain the binding constant $K_o = (38.75 \pm 1.28) \times 10^5$ M^{-1} and lengthening upon single intercalation of $\Delta x = 0.095 \pm 0.002$ nm in the absence of force. Finally, the elastic properties of the DNA–YOYO complex were obtained by fitting the stretching curves obtained at different pulling rates to the WLC model (Equation 6.5).

In a recent study that combines fluorescence microscopy with optical tweezers, the bis-intercalator YOYO was used to explore the dsDNA-melting transition. Since bis-intercalation can only occur in the presence of dsDNA, YOYO will only bind to DNA that is double stranded. The fluorescence obtained while stretching dsDNA suggests that as the extension increases along the transition plateau, the number of YOYO bound is reduced. This provides additional proof that the overstretching transition is indeed a melting transition where the dsDNA is melted to form ssDNA to which YOYO cannot bind [68]. This was further verified by adding ssDNA-binding protein that was observed to bind at places where fluorescence due to YOYO disappeared.

Triostin A is an antibiotic that has two covalently linked quinoxaline that are oriented to favor the bis-intercalation binding mode. Optical tweezers DNA-stretching studies in the presence of Triostin A [69] were similar to the ones obtained in the case of YOYO as explained above, except for the fact that the kinetics were still slower than the binding kinetics of YOYO. DNA was stretched rapidly to a particular force and then held at constant force to obtain an equilibrium extension in the presence of the drug (as was done for threading intercalators discussed earlier). Since the bis-intercalators contain two intercalating moieties, fractional elongation obtained from Equation 6.8 must be divided by two [69]. The binding constant (K) and binding site size (n) were estimated from ligand concentration dependence (Equation 6.9) as a function of applied force. This method, introduced by Vladescu et al. [10] for intercalators, obtains the binding parameters in the absence of force by studying the force dependence. Similar analysis gives the zero force-binding constant $K_0 = (5.8 \pm 0.3) \times 10^5$ M^{-1} and a lengthening upon single intercalation of $\Delta x = 0.316$ nm for Triostin [69]. These values are compared to the values for simple intercalators in Figure 6.9.

6.4   CONCLUSIONS

Stretching DNA in optical tweezers experiments can characterize a wide array of drug-binding modes. The elastic properties of DNA and the melting transition force differ for various types of binding of small molecules. These changes can be used to distinguish and quantify these major-binding modes.

Both intercalators and groove binders bind to the dsDNA and stabilize the double-stranded structure, while ssDNA binders can bind to the ssDNA portions that are exposed during thermal fluctuations, resulting in destabilization of the double-stranded structure. As a result, intercalators and groove binders increase the melting transition force, whereas ssDNA binders lower the melting transition force. This property can be used to distinguish intercalators and groove binders from ssDNA binders in force spectroscopic studies. When the DNA bases are exposed to the ssDNA binders during the melting transition, they bind to the exposed single strands and prevent reannealing of these strands during relaxation, introducing hysteresis in the force–extension curves upon relaxation. This is another distinct feature that distinguishes ssDNA binders from other binding modes.

Furthermore, intercalators and groove binders can be distinguished by fitting the dsDNA-stretching curve obtained in the presence of these small molecules to the WLC model, which yields the contour and persistence lengths of the drug–DNA complex. Intercalators show clear, measurable force-dependent increases in the contour length while the groove binders exhibit no change in contour length. Persistence lengths obtained from these fittings can be used to discern minor and major groove binders. Major groove binders characteristically decrease the persistence length while minor groove binders generally increase the persistence length.

Complex-binding modes like bis-intercalation and threading intercalation have unique characteristics. Bis-intercalators exhibit an increase in the contour length and melting transition similar to the intercalators. However, their slow-binding kinetics also produces hysteresis, as in the case of ssDNA binders. On the other hand, extremely slow-threading intercalators require melting to bind DNA. Therefore, they show very little of the effects of intercalation during their first stretch and then subsequently exhibit nonequilibrium stretching and relaxation curves. The kinetics of these slow-binding molecules can be explored by measuring this lengthening while holding the DNA at a constant force.

Optical tweezers have proven to be a useful method to explore interactions at the single-molecule level. These experiments may require less time and material compared to the bulk experimental techniques and can quantify the binding properties more precisely than many bulk techniques. Due to several signatures in DNA-stretching curves that indicate specific-binding modes, DNA stretching can also often distinguish multiple-binding modes. New methods obtain drug binding as function of force, determining the DNA–drug-binding properties even in the absence of force. These methods allow researchers to quantify the thermodynamics and kinetics of DNA–ligand interactions over a wide range of concentrations and binding affinity in unprecedented detail. Understanding these properties of drugs and other small molecules that interact with DNA will help develop potential drugs for challenging diseases.

ACKNOWLEDGMENTS

We thank Ioulia Rouzina for many valuable discussions concerning DNA–ligand interactions. This work was supported by NIH grant R01 GM072462 and NSF grant MCB-0744456.

REFERENCES

1. Ashkin, A., Dziedzic, J. M., Bjorkholm, J. E., and Chu, S. Observation of a single-beam gradient force optical trap for dielectric particles. *Opt Lett* **11**, 288, 1986.
2. Ashkin, A. Acceleration and trapping of particles by radiation pressure. *Phys Rev Lett* **24**, 156–159, 1970.
3. http://physics.nist.gov/News/Nobel/OtherSites/physics97.html.
4. Ashkin, A. and Dziedzic, J. M. Optical trapping and manipulation of viruses and bacteria. *Science* **235**, 1517–1520, 1987.
5. Ashkin, A., Dziedzic, J. M., and Yamane, T. Optical trapping and manipulation of single cells using infrared laser beams. *Nature* **330**, 769–771, 1987.
6. Smith, S. B., Cui, Y. J., and Bustamante, C. Overstretching B-DNA: The elastic response of individual double-stranded and single-stranded DNA molecules. *Science* **271**, 795–799, 1996.
7. McCauley, M. J. and Williams, M. C. Mechanisms of DNA binding determined in optical tweezers experiments. *Biopolymers* **85**, 154–168, 2007.
8. McCauley, M. J. and Williams, M. C. Optical tweezers experiments resolve distinct modes of DNA-protein binding. *Biopolymers* **91**, 265–282, 2009.
9. Chaurasiya, K. R., Paramanathan, T., McCauley, M. J., and Williams, M. C. Biophysical characterization of DNA binding from single molecule force measurements. *Phys Life Rev* **7**, 299–341, 2010.
10. Vladescu, I. D., McCauley, M. J., Nunez, M. E., Rouzina, I.. and Williams, M. C. Quantifying force-dependent and zero-force DNA intercalation by single-molecule stretching. *Nat Methods* **4**, 517–522, 2007.
11. Cluzel, P., Lebrun, A., Heller, C., et al. DNA: An extensible molecule. *Science* **271**, 792–794, 1996.
12. Rouzina, I. and Bloomfield, V. A. Force-induced melting of the DNA double helix-1. Thermodynamic analysis. *Biophys J* **80**, 882–893, 2001.
13. Rouzina, I. and Bloomfield, V. A. Force-induced melting of the DNA double helix. 2. Effect of solution conditions. *Biophys J* **80**, 894–900, 2001.
14. Williams, M. C., Rouzina, I., Wenner, J. R., Gorelick, R. J., Musier-Forsyth, K., and Bloomfield, V. A. Mechanism for nucleic acid chaperone activity of HIV-1 nucleocapsid protein revealed by single molecule stretching. *Proc Natl Acad Sci USA* **98**, 6121–6126, 2001.
15. Williams, M. C., Wenner, J. R., Rouzina, I., and Bloomfield, V. A. Entropy and heat capacity of DNA melting from temperature dependence of single molecule stretching. *Biophys J* **80**, 1932–1939, 2001.
16. Williams, M. C., Wenner, J. R., Rouzina, I., and Bloomfield, V. A. Effect of pH on the overstretching transition of double-stranded DNA: Evidence of force-induced DNA melting. *Biophys J* **80**, 874–881, 2001.
17. Storm, C. and Nelson, P. C. Theory of high-force DNA stretching and overstretching. *Phys Rev E Stat Nonlin Soft Matter Phys* **67**, 12pp., 2003.
18. Marko, J. F. and Siggia, E. D. Stretching DNA. *Macromolecules* **28**, 8759–8770, 1995.
19. Odijk, T. Stiff chains and filaments under tension. *Macromolecules* **28**, 7016–7018, 1995.

20. Baumann, C. G., Smith, S. B., Bloomfield, V. A. and Bustamante, C. Ionic effects on the elasticity of single DNA molecules. *Proc Natl Acad Sci USA* **94**, 6185–6190, 1997.
21. Podgornik, R., Hansen, P. L., and Parsegian, P. A. Elastic moduli renormalization in self-interacting stretchable polyelectrolytes. *J Chem Phys* **113**, 9343–9350, 2000.
22. Wenner, J. R., Williams, M. C., Rouzina, I., and Bloomfield, V. A. Salt dependence of the elasticity and overstretching transition of single DNA molecules. *Biophys J* **82**, 3160–3169, 2002.
23. Smith, S. B., Finzi, L. and Bustamante, C. Direct. mechanical measurements of the elasticity of single DNA molecules by using magnetic beads. *Science* **258**, 1122–1126, 1992.
24. Krautbauer, R., Fischerlander, S., Allen, S., and Gaub, H. E. Mechanical fingerprints of DNA drug complexes. *Single Mol* **3**, 97–103, 2002.
25. Chang, T. W., Fiumara, N., and Weinstein, L. Genital herpes: Treatment with methylene blue and light exposure. *Int J Dermatol* **14**, 69–71, 1975.
26. Eckel, R., Ros, R., Ros, A., Wilking, S. D., Sewald, N., and Anselmetti, D. Identification of binding mechanisms in single molecule-DNA complexes. *Biophys J* **85**, 1968–1973, 2003.
27. Sischka, A., Toensing, K., Eckel, R., et al. Molecular mechanisms and kinetics between DNA and DNA binding ligands. *Biophys J* **88**, 404–411, 2005.
28. Waring, M. J. DNA modification and cancer. *Annu Rev Biochem* **50**, 159–192, 1981.
29. Bergamo, A., Masi, A., Jakupec, M. A., Keppler, B. K., and Sava, G. Inhibitory effects of the ruthenium complex KP1019 in models of mammary cancer cell migration and invasion. *Met Based Drugs* **2009**, 681270, 2009.
30. Barton, J. K., Danishefsky, A., and Goldberg, J. Tris(phenanthroline)ruthenium(II): Stereoselectivity in binding to DNA. *J Am Chem Soc* **106**, 2172–2176, 1984.
31. Tessmer, I., Baumann, C. G., Skinner, G. M., et al. Mode of drug binding to DNA determined by optical tweezers force spectroscopy. *J Mod Optics* **50**, 1627–1636, 2003.
32. Vladescu, I. D., McCauley, M. J., Rouzina, I., and Williams, M. C. Mapping the phase diagram of single DNA molecule force-induced melting in the presence of ethidium. *Phys Rev Lett* **95**, 4pp., 2005.
33. Mihailovic, A., Vladescu, L., McCauley, M., et al. Exploring the interaction of ruthenium(II) polypyridyl complexes with DNA using single-molecule techniques. *Langmuir* **22**, 4699–4709, 2006.
34. Zhang, W., Barbagallo, R., Madden, C., Roberts, C. J., Woolford, A., and Allen, S. Progressing single biomolecule force spectroscopy measurements for the screening of DNA binding agents. *Nanotechnology* **16**, 2325–2333, 2005.
35. Rocha, M. S., Ferreira, M. C., and Mesquita, O. N. Transition on the entropic elasticity of DNA induced by intercalating molecules. *J Chem Phys* **127**, 7pp., 2007.
36. Murade, C. U., Subramaniam, V., Otto, C., and Bennink, M.L. Interaction of oxazole yellow dyes with DNA studied with hybrid optical tweezers and fluorescence microscopy. *Biophys J* **97**, 835–843, 2009.
37. Karapetyan, A. T., Permogorov, V. I., Frank-Kamenetskii, M. D., and Lazurkin, Y. S. Thermodynamic studies of DNA-dye complexes. *Mol Biol* **6**, 703–708, 1972.
38. Shokri, L., Marintcheva, B., Eldib, M., Hanke, A., Rouzina, I., and Williams, M. C. Kinetics and thermodynamics of salt-dependent T7 gene 2.5 protein binding to single- and double-stranded DNA. *Nucleic Acids Res* **36**, 5668–5677, 2008.
39. McCauley, M. J., Shokri, L., Sefcikova, J., Venclovas, C., Beuning, P. J., and Williams, M. C. Distinct double- and single-stranded DNA binding of *E. coli* replicative DNA polymerase III alpha subunit. *ACS Chem Biol* **3**, 577–587, 2008.
40. Shokri, L., Marintcheva, B., Richardson, C. C., Rouzina, I., and Williams, M. C. Single molecule force spectroscopy of salt-dependent bacteriophage T7 gene 2.5 protein binding to single-stranded DNA. *J Biol Chem* **281**, 38689–38696, 2006.

41. Cruceanu, M., Urbaneja, M. A., Hixson, C. V., et al. Nucleic acid binding and chaperone properties of HIV-1 Gag and nucleocapsid proteins. *Nucleic Acids Res* **34**, 593–605, 2006.

42. Cruceanu, M., Stephen, A. G., Beuning, P. J., Gorelick, R. J., Fisher, R. J., and Williams, M. C. Single DNA molecule stretching measures the activity of chemicals that target the HIV-1 nucleocapsid protein. *Anal Biochem* **358**, 159–170, 2006.

43. Cruceanu, M., Gorelick, R. J., Musier-Forsyth, K., Rouzina, I., and Williams, M. C. Rapid kinetics of protein-nucleic acid interaction is a major component of HIV-1 nucleo-capsid protein's nucleic acid chaperone function. *J Mol Biol* **363**, 867–877, 2006.

44. Pant, K., Karpel, R. L., Rouzina, I., and Williams, M. C. Salt dependent binding of T4 gene 32 protein to single and double-stranded DNA: Single molecule force spectroscopy measurements. *J Mol Biol* **349**, 317–330, 2005.

45. Pant, K., Karpel, R. L., Rouzina, L., and Williams, M. C. Mechanical measurement of single-molecule binding rates: Kinetics of DNA helix-destabilization by T4 gene 32 protein. *J Mol Biol* **336**, 851–870, 2004.

46. Davis, W. R., Gabbara, S., Hupe, D., and Peliska, J. A. Actinomycin D inhibition of DNA strand transfer reactions catalyzed by HIV-1 reverse transcriptase and nucleo-capsid protein. *Biochemistry* **37**, 14213–14221, 1998.

47. Guo, J., Wu, T., Bess, J., Henderson, L. E., and Levin, J. G. Actinomycin D inhibits human immunodeficiency virus type 1 minus-strand transfer in in vitro and endogenous reverse transcriptase assays. *J Virol* **72**, 6716–6724, 1998.

48. Jeeninga, R. E., Huthoff, H. T., Gultyaev, A. P., and Berkhout, B. The mechanism of actinomycin D-mediated inhibition of HIV-1 reverse transcription. *Nucleic Acids Res* **26**, 5472–5479, 1998.

49. Chen, F. M., Sha, F., Chin, K. H., and Chou, S. H. Binding of actinomycin D to single-stranded DNA of sequence motifs d(TGTCT(n)G) and d(TGT(n)GTCT). *Biophys J* **84**, 432–439, 2003.

50. Broude, N. E. and Budowsky, E. I. The reaction of glyoxal with nucleic acid compo-nents. 3. Kinetics of the reaction with monomers. *Biochim Biophys Acta* **254**, 380–388, 1971.

51. Broude, N. E. and Budowsky, E. I. The reaction of glyoxal with nucleic acid compo-nents. V. Denaturation of DNA under the action of glyoxal. *Biochim Biophys Acta* **294**, 378–384, 1973.

52. Johnson, D. A new method of DNA denaturation mapping. *Nucleic Acids Res* **2**, 2049–2054, 1975.

53. Lyamichev, V. I., Panyutin, I. G., and Lyubchenko Yu, L. Gel electrophoresis of partially denatured DNA. Retardation effect: Its analysis and application. *Nucleic Acids Res* **10**, 4813–4826, 1982.

54. Lyubchenko Yu, L., Kalambet Yu, A., Lyamichev, V. I., and Borovik, A. S. A comparison of experimental and theoretical melting maps for replicative form of phi X174 DNA. *Nucleic Acids Res* **10**, 1867–1876, 1982.

55. Shokri, L., McCauley, M. J., Rouzina, I., and Williams, M. C. DNA overstretching in the presence of glyoxal: Structural evidence of force-induced DNA melting. *Biophys J* **95**, 1248–1255, 2008.

56. Pant, K., Karpel, R. L., Rouzina, I., and Williams, M. C. Mechanical measurement of single-molecule binding rates: Kinetics of DNA helix-destabilization by T4 gene 32 protein. *J Mol Biol* **336**, 851–870, 2004.

57. Krautbauer, R., Pope, L. H., Schrader, T. E., Allen, S., and Gaub, H. E. Discriminating small molecule DNA binding modes by single molecule force spectroscopy. *FEBS Lett* **510**, 154–158, 2002.

58. Arcamone, F., Penco, S., Orezzi, P., Nicolella, V., and Pirelli, A. Structure and synthesis of distamycin A. *Nature* **203**, 1064–1065, 1964.

59. Bamgbose, S. O. Antagonism of the trypanocide, berenil, and the nematocide, morantel, by tubocurarine. *Naunyn Schmiedebergs Arch Pharmacol* **280**, 103–106, 1973.
60. Shokeir, A. A., al-Hussaini, M. K., and Wasfy, I. A. Methyl green-pyronin stain for the diagnosis of trachoma. *Br J Ophthalmol* **53**, 263–266, 1969.
61. Husale, S., Grange, W., and Hegner, M. DNA mechanics affected by small DNA inter-acting ligands. *Single Mol* **3**, 91–96, 2002.
62. Rocha, M. S., Viana, N. B., and Mesquita, O. N. DNA-psoralen interaction: A single molecule experiment. *J Chem Phys* **121**, 9679–9683, 2004.
63. Lincoln, P. and Norden, B. Binuclear ruthenium(II) phenanthroline compounds with extreme binding affinity for DNA. *Chem Commun* 2145–2146, 1996.
64. Paramanathan, T., Westerlund, F., McCauley, M. J., Rouzina, I., Lincoln, P., and Williams, M. C. Mechanically manipulating the DNA threading intercalation rate. *J Am Chem Soc* **130**, 3752–3753, 2008.
65. Williams, M. C., Rouzina, I., and Bloomfield, V. A. Thermodynamics of DNA interactions from single molecule stretching experiments. *Acc Chem Res* **35**, 159–166, 2002.
66. Rye, H. S., Yue, S., Wemmer, D. E., et al. Stable fluorescent complexes of double-stranded DNA with bis-intercalating asymmetric cyanine dyes: Properties and applications. *Nucleic Acids Res* **20**, 2803–2812, 1992.
67. Bennink, M. L., Scharer, O. D., Kanaar, R., et al. Single-molecule manipulation of double-stranded DNA using optical tweezers: Interaction studies of DNA with RecA and YOYO-1. *Cytometry* **36**, 200–208, 1999.
68. van Mameren, J., Gross, P., Farge, G., et al. Unraveling the structure of DNA during overstretching by using multicolor, single-molecule fluorescence imaging. *Proc Natl Acad Sci USA* **106**, 18231–18236, 2009.
69. Kleimann, C., Sischka, A., Spiering, A., et al. Binding kinetics of bisintercalator Triostin a with optical tweezers force mechanics. *Biophys J* **97**, 2780–2784, 2009.
70. Drew, H. R., Wing, R. M., Takano, T., et al. Structure of a B-DNA dodecamer: Conformation and dynamics. *Proc Natl Acad Sci USA* **78**, 2179–2183, 1981.

7 Fluorescent Nucleoside Analogs for Monitoring RNA–Drug Interactions

Mary S. Noé, Yun Xie, and Yitzhak Tor

CONTENTS

We describe new fluorescent nucleosides that are used to probe the binding of drugs to RNA targets and the interactions between proteins and RNA. Together with sensitive fluorescence-based techniques, such emissive nucleoside analogs can facilitate the study of nucleic acid dynamics, structures, and recognition.

7.1   INTRODUCTION

Data from various analytical techniques have been integrated to advance a fundamental understanding of the structure and properties of nucleic acids. Nuclear magnetic resonance (NMR) spectroscopy and x-ray crystallography, for example, have supplied invaluable information, especially with regard to "static" structures. Nucleic acids, like all other biomolecules, are highly dynamic, with hybridization and melting, as well as strand cleavage, base flipping, and ligand binding, playing important roles in their biological function. Fluorescence-based techniques, due to their sensitivity and short timescale, can facilitate the monitoring of such dynamic events [1–3].

Because of the extremely low emission quantum yields of the native nucleosides [4], nucleic acids must be modified with appropriate chromophores to facilitate fluorescence-based measurements. Although end-labeled oligonucleotides have been extensively used, such approaches rarely provide information at the nucleotide level. In contrast, minimally perturbing fluorescent nucleoside analogs, judiciously replacing selected residues, open a window into otherwise spectroscopically silent molecular events involving nucleic acids [5]. Such "isomorphic" fluorescent nucleobases could ideally replace the native heterocycles without impacting their inherent function, while being proximal to the site of interest. Additionally, tuning the photophysical characteristics of novel nucleobases to match the spectral features of other established fluorophores can facilitate the implementation of Förster resonance energy transfer (FRET)-based assemblies for the study of biological recognition events, with much higher "resolution" compared to assemblies that rely on end-labeled oligonucleotides [6–9]. In this chapter, we focus on advancing RNA as a drug target and understanding the interactions of this important biomolecule with ligands via the implementation of new fluorescent isomorphic nucleosides. We begin with a concise description of relevant fluorescence-based techniques.

7.2   BASICS OF FLUORESCENCE SPECTROSCOPY

7.2.1   A Brief Primer

A detailed overview of photoluminescence is beyond the scope of this chapter and the reader is referred to several excellent texts [1–3,10–12]. A brief introduction to rudimentary principles of fluorescence spectroscopy and commonly used techniques follows. Fluorescence typically refers to the emission of a photon from an excited singlet state of a chromophore. Population of this excited state is normally achieved by the absorption of electromagnetic radiation of a specific energy. In fluorescence spectroscopy, a sample is irradiated with energy in the ultraviolet (UV)–visible range of the electromagnetic spectrum. The extent of energy absorption depends on the molar extinction coefficient (ε), which is directly proportional to the absorption cross section (σ) of the chromophore. After the absorption of a photon, an excited molecule may return to its ground state by a variety of nonradiative and radiative processes. The most common processes of energetic relaxation involve molecular vibrations and rotations as well as intermolecular collisions, especially with solvent molecules. Certain molecules, known as fluorophores, are capable of an alternate process of radiative energy decay in which a photon is emitted.

Several parameters provide insight into the characteristics of a fluorophore and its potential utility. The fluorescence quantum yield (Φ_F) is defined as the fraction of photons emitted versus the number of photons absorbed and ranges from zero to unity. The brightness of a fluorophore is the product of the molar absorptivity (ε) and the fluorescence quantum yield (Φ_F). The fluorescence lifetime of a chromophore (τ_F), which ranges from 0.5 to 20 ns for common organic fluorophores reflects the average amount of time spent in the excited state before photon emission. As apparent from the Jablonski diagram (Figure 7.1), the energy of the emitted photon is lower than the excitation energy. The Franck–Condon state relaxes to a lower vibronic level, leading to emission from a thermally equilibrated excited state, frequently to a higher vibrational level of the ground state (S_0). The resulting energetic difference between the excitation and emission maxima is known as the Stokes shift.

Steady-state fluorescence spectroscopy is the most commonly used technique to measure emission spectra in a laboratory. Although this method is somewhat limited in its scope, it can provide a myriad of useful data with relatively simple benchtop equipment. Steady-state fluorimeters include a light source to provide a constant

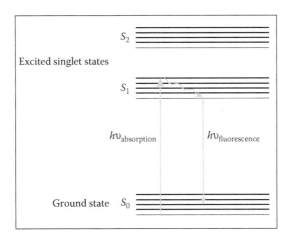

FIGURE 7.1   Simplified Jablonski diagram.

monochromatic photon flux for sample excitation. Photons emitted are typically detected orthogonally to the excitation source. Correlating the number of photons emitted versus wavelength (or energy) yields an emission spectrum that reveals the emission maxima and their relative intensities.

7.2.2 RESONANCE ENERGY TRANSFER

FRET involves the nonradiative dipole–dipole interaction between two chromophores, commonly referred to as a donor and an acceptor. The transfer efficiency of a FRET pair can be predicted through analysis of their spectral properties. Specifically, the absorbance of the acceptor must sufficiently overlap with the emission of the donor. The overlap integral ($J(\lambda)$) is a function of the corrected fluorescence intensity of the donor with the total area under the curve that is normalized to unity ($F_D(\lambda)$) and the extinction coefficient of the acceptor ($\varepsilon_A(\lambda)$) all at wavelength λ.

$$J(\lambda) = \int_0^\infty F_D(\lambda)\varepsilon_A(\lambda)\lambda^4 d\lambda \qquad (7.1)$$

If the optical density of the acceptor is minimal at the wavelength of excitation of the donor, the sensitized emission of the acceptor can be selectively observed.

The efficiency of energy transfer is distance dependent, and the Förster critical radius (R_0) is the interchromophoric distance at which the transfer efficiency is 50% (20–60 Å for common FRET pairs). This critical distance is a function of the overlap integral ($J(\lambda)$), the refractive index of the medium (n), the quantum yield of the donor (Q_D), and the relative orientation factor of the transition dipoles (κ^2).

$$R_0 = 9.78 \times 10^3 \left(\kappa^2 n^{-4} Q_D J(\lambda) \right)^{1/6} \qquad (7.2)$$

When FRET assemblies are studied in aqueous media and the fluorophores are freely rotating, the refractive index (n) is 1.33 and the relative orientation factor (κ^2) is assumed to be 2/3. This value reflects a dynamic random averaging of the donor and acceptor orientation. When the fluorophores are covalently linked in a manner that would impede free rotation, accurate estimation of the orientation factor may present challenges.

The FRET efficiency (E), also known as the energy transfer efficiency, can be determined from the ratio of the fluorescence intensity of the donor alone (F_D) and its intensity in the presence of the acceptor (F'_D).

$$E = 1 - \frac{F'_D}{F_D} = \frac{1}{1 + (r/R_0)^6} \qquad (7.3)$$

This ratio represents the fraction of photons absorbed by the donor that is responsible for energy transfer to the acceptor. As the resonance energy transfer efficiency is proportional to the inverse sixth power of the distance between the donor and the acceptor, sensitized FRET data can provide exceptionally useful information in a variety of biochemical systems and biophysical assays.

7.2.3 FLUORESCENT NUCLEOSIDE ANALOGS

Because the naturally occurring, common nucleosides (A, G, T, U, and C; Figure 7.2) are practically nonemissive [4], oligonucleotides may be selectively modified with fluorescent nucleoside analogs. Ideally, the replacement nucleosides, while providing a spectroscopic handle, must not significantly perturb the native folding, dynamic, and recognition features of the system under study. In addition, the photophysical characteristics of the implemented probes need to be sensitive to changes in their microenvironment to be able to serve as singular responsive probes. Although diverse emissive nucleoside analogs have been made over the years, few can be considered nonperturbing [1]. Perhaps the best-known isomorphic analog is 2-aminopurine (2AP), a UV-emitting isomer of A (Figure 7.3a) [13]. This prototypical analog has been implemented in countless biophysical assays, illustrating the remarkable utility of isomorphic nucleoside analogs [1,14–16]. To generate useful analogs of all letters of the genetic alphabet and across a variety of excitation and emission wavelengths, we have initiated a program aimed at the design and implementation of novel emissive nucleosides, nucleotides, and oligonucleotides.

In designing new fluorescent nucleosides, several principles have been developed and employed. First, analogs should retain the highest possible structural similarity to the native nucleosides, with special consideration given to the maintenance of canonical Watson–Crick hydrogen-bonding faces. Second, the nucleoside analog should possess an absorption spectrum distinct from the native nucleosides (preferably redshifted) to enable a selective excitation of the modified analog in the presence of the overwhelming majority native nucleotides. Third, the quantum efficiency of the modified nucleoside must be adequate for detection by benchtop fluorescence spectrometers. Additionally, the fluorescent nucleoside analogs should exhibit long-wavelength

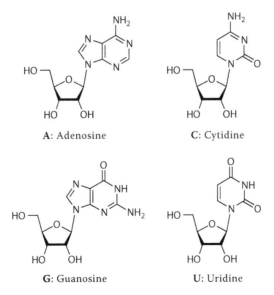

FIGURE 7.2   Naturally occurring common ribonucleosides.

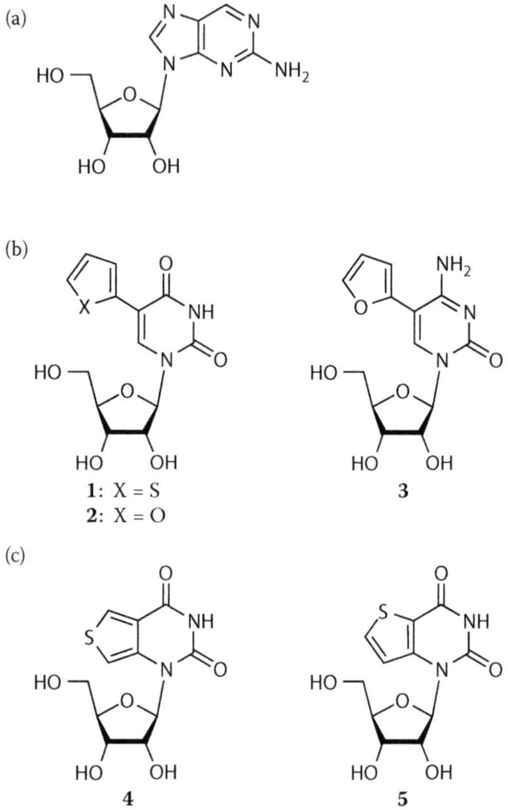

FIGURE 7.3 Isomorphic nucleoside analogs. (a) 2-Aminopurine (2AP); (b) 5-modified nucleosides; (c) fused-ring nucleosides.

emission, ideally in the visible range. Finally, to be implemented as singular probes, the fluorescence emission of the analogs should be sensitive to changes in their microenvironment. These design criteria have led to the development of several useful fluorescent probes and, in turn, biophysical assays.

The first generation of fluorescent nucleoside analogs developed in our program was based on conjugating aromatic five-membered rings to the 5-position of the pyrimidines (Figure 7.3b) [17–25]. These nucleosides were synthesized with ease and incorporated into oligonucleotides enzymatically [20] or through standard solid-phase protocols [22]. Although they possess modest quantum yields, these nucleoside analogs display fluorescence in the visible range, relatively large Stokes shifts, and are quite responsive [23]. Related nucleosides with a fused ring system were also explored (Figure 7.3c) [5,17,21]. The photophysical characteristics were found to depend on the nature of the fused heterocycle and its fusion position. In keeping with previously established design principles, we have also developed a family of nucleosides based on a quinazoline core (Figure 7.4) [6–8,26,27]. This heterocyclic core was decorated with electron-rich groups in positions either *ortho* or *para* to the pyrimidine's C4 carbonyl, to potentially enhance polarization in the excited state. These nucleosides display a

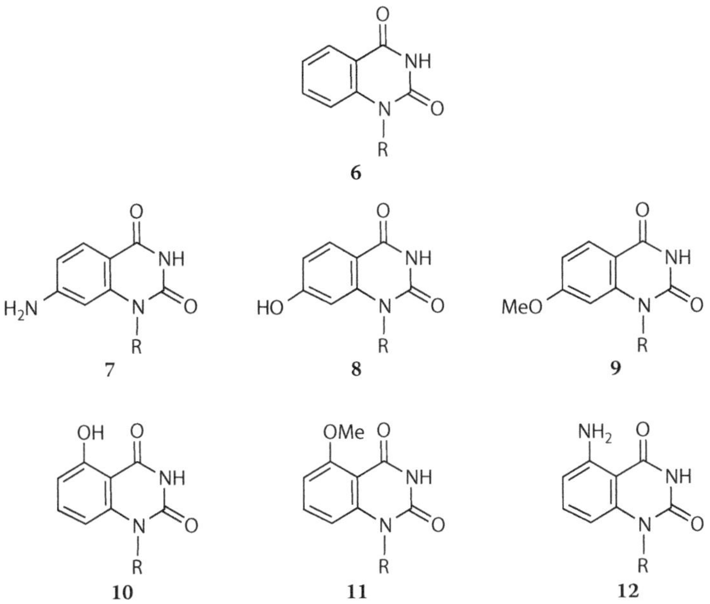

FIGURE 7.4   Quinazoline-based nucleoside family; R = D-ribose.

range of emission maxima and have proven to be useful when applied alone or as members of FRET pairs. Table 7.1 lists selected emissive nucleosides and their photophysical characteristics.

7.2.4   RNA AS A DRUG TARGET

As new and drug-resistant pathogens surface, the need to develop novel antibacterial and antiviral agents becomes more pressing. The central role of RNA in the function of viruses and bacteria has drawn attention to this biomolecule as a promising drug target. Additionally, as the multifaceted cellular roles of RNA become better understood, the potential to use small molecules to modify cellular functions and impact the expression of specific proteins has begun to be realized. Nucleic acid–drug interactions, specifically RNA–drug interactions, involve complex modes of binding that continue to be illuminated. With ever-increasing knowledge of RNA structure, folding, and recognition, potential and confirmed small-molecule binding sites can be explored and exploited, leading ultimately to designer RNA binders.

The bacterial ribosome has served as one of the major and most inspiring natural RNA targets, with numerous naturally occurring and synthetic antibiotics targeting this complex translational ribonucleoprotein machinery. Among them, the aminoglycoside family of highly charged pseudo-oligosaccharides (Figure 7.5) has been the most thoroughly studied. In 1987, Moazed and Noller first established several binding sites of these polycationic, small molecules on the bacterial 16S rRNA through the use of footprinting techniques [28]. In 1994, Purrohit and Stern used a short hairpin loop RNA construct to mimic the decoding site of rRNA [29]. This construct was

TABLE 7.1

Spectroscopic Properties of Isomorphic Fluorescent Nucleosides[a]

Nucleoside	λ_{abs} (nm)	λ_{em} (nm)	Φ_F
2AP	303	370	0.68
1	315	439	0.02
2	316	440	0.04
3	310	443	0.02
4	304	412	0.48
5	292	351	0.06
6	307	371	0.31
7	316	362	0.04
8	308	358	0.04
9	305	357	0.08
10	324	490	≤ 0.01
11	320	395	0.16
12	349	440	0.42

[a] All values are for the free nucleoside in water. Values for λ_{abs} are the local intensity maxima for the most red-shifted absorbance band. Values for λ_{em} are the overall emission intensity maxima when excited at λ_{abs}. All quantum yields are relative [1,6–8,26].

shown to behave as an independent domain with function and recognition capabilities representative of the entire 16S rRNA. Further studies by Puglisi and Yokoyama determined that a 27-mer A-site construct was able to bind neomycin antibiotics in a manner similar to the ribosomal decoding site [30–32].

The use of established shorter constructs enabled high-resolution x-ray and NMR characterization of aminoglycosides bound to the A-site mimic [33–37]. In 2000,

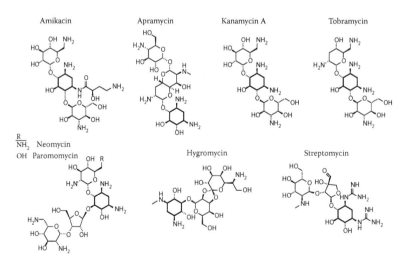

FIGURE 7.5 Representative selection of aminoglycoside antibiotics.

Ramakrishnan was able to crystallize the entire 30S ribosomal subunit containing 16S rRNA in the presence of aminoglycosides [38,39]. The 3 Å resolution structures were then corroborated with molecular interactions observed in complexes with shorter model constructs [40]. These crucial structures confirmed the utility of small RNA constructs outside of the ribosome and inspired the development of discovery assays and new synthetic RNA ligands.

Not only do aminoglycoside antibiotics bind to the bacterial ribosome and interfere with translational fidelity, but also they have been shown to bind to a variety of other RNA targets [39,41–44]. As these natural products have evolved to target the bacterial ribosome, the opportunistic binding of aminoglycosides to other RNA targets reflects their inherent affinity to RNA [44–46]. This underscores the delicate interplay between affinity and selectivity exhibited by these highly cationic molecules, which represents a key challenge for the design of new small-molecule RNA binders. As the overall positive charge of RNA binders is increased, their affinity often rises and is commonly accompanied by an enhanced-binding promiscuity. "Real-time" fluorescent-based biophysical and discovery assays that facilitate the determination of binding affinity and selectivity have therefore become essential for identifying and advancing new RNA-targeting ligands.

To demonstrate the utility of fluorescent nucleoside analogs in advancing biophysical assays, as well as the challenges involved, we briefly discuss singly labeled A-site RNA constructs. One of the early assays involved replacing one of the dynamic A residues (A1492 and 1493) with 2AP, an emissive adenine analog (Figure 7.3a) [47,48]. Although extremely useful, its response was found to be antibiotics dependent, with unreliable response to neomycin, a potent A-site binder [19]. Nucleoside **2** was incorporated into a model bacterial decoding A-site to test the sensitivity of the fluorescence response to antibiotic binding. Replacing U1406, a residue adjacent to the binding pocket, with nucleoside **2** resulted in an emissive A-site construct, which responded to antibiotic binding, including neomycin [19]. The deficient response of 2AP and the utility of **2** in exploring A-site–ligand binding illustrate the scope and limitation of singly labeled oligonucleotides. Assay developers should consider the preparation of multiple constructs and test diverse fluorescent probes to enhance assay reliability and minimize false readings. As described below, FRET-based assays tend to be more robust and less prone to such challenges.

7.3 CASE STUDIES

7.3.1 REAL-TIME DETECTION OF A-SITE–ANTIBIOTICS BINDING USING FRET

7.3.1.1 System Design

The bacterial ribosomal decoding site (A-site) is an appealing target for novel antibacterials and can also function as a model system for RNA–small molecule interactions [49]. As noted above, the use of short A-site constructs has been validated by structural studies, and fluorescent constructs, containing a single fluorescent nucleoside analog, have been successfully utilized to correlate fluorescence responses with antibiotic binding. This response, however, can be dependent on the antibiotic and its mode of binding. To overcome such issues, a novel FRET system has been developed

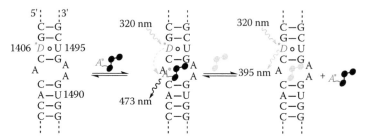

FIGURE 7.6 Schematic of A-site construct modified at position 1406 with a fluorescent nucleoside analog to function as a FRET donor (*D*). When the coumarin-labeled aminoglycoside binds, the sensitized emission of the FRET acceptor (*A*) is observed upon excitation of the donor. If this FRET assembly is disrupted by another antibiotic, the fluorescence of the donor (*D*) is restored.

(Figure 7.6) [6]. Unlike singly labeled RNA constructs, where the fluorophore responds to environmental changes caused by a bound antibiotic, the fluorescence signal of a FRET-based system is sensitive to the distance between a fluorescent RNA construct and a fluorescently labeled RNA binder. Disruption of this assembly, accomplished by competitive A-site binders independent of their specific-binding modes, can be monitored with FRET accuracy.

To retain high structural and functional integrity of this system, an isomorphic fluorescent nucleoside analog functioning as a FRET donor replaced U1406, which is adjacent to the binding site [6]. Additionally, aminoglycosides were covalently linked to a FRET acceptor by modifying hydroxyl groups demonstrated to be inconsequential for binding. Fluorescent nucleobase analog **14** was identified as an excellent donor for the commercially available 7-diethylaminocoumarin-3-carboxylic acid **13** and its derivatives. The emission of **14** centered at 395 nm overlaps well with the absorption band of **14** (Figure 7.7). Additionally, at the absorption maximum of **14**

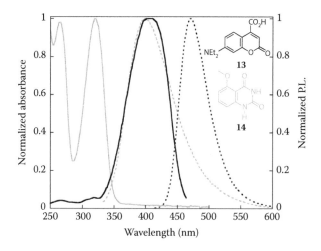

FIGURE 7.7 Absorption and emission spectra of compounds **13** and **14**, demonstrating appropriate spectral characteristics for an effective FRET pair.

(320 nm), the extinction coefficient of **13** is minimal, facilitating the selective excitation of **13**. The critical Förster radius (R_0) was determined to be 27 Å, which is appropriate for this A-site/aminoglycoside assembly (the distance between bound aminoglycosides and **11**, as estimated from crystal structures, is <15 Å) [6].

7.3.1.2 Synthesis and Properties of Fluorescent RNA Construct

Nucleoside **11** was synthesized and converted into its phosphoramidite **15** (Figure 7.8). The phosphoramidite was then employed in the synthesis of a singly modified 27-mer oligonucleotide **17** using standard solid-phase protocols (Figure 7.9). Thermal denaturation studies of the modified (**17**) and control (**16**) constructs demonstrated that replacement of U1406 with the modified nucleobase minimally impacted the overall stability [6]. Incorporation of nucleoside **11** into the RNA construct minimally affected the photophysical characteristics of the nucleoside, maintaining an emission maximum at 395 nm upon excitation at 320 nm, while the fluorescence quantum yield decreased from 16% to 3%, as commonly observed with fluorescent nucleosides. Importantly, the fluorescence intensity variations of the RNA are well above the noise level for typical bench top fluorimeters.

7.3.1.3 Synthesis and Properties of Fluorescently Modified Aminoglycosides

Neomycin B and tobramycin were each labeled with 7-diethylaminocoumarin-3-carboxylic acid **13**, resulting in compounds **20** and **21**, respectively (Figure 7.10). To selectively conjugate neomycin B and tobramycin to fluorophore **13**, the single primary hydroxymethyl groups in the Boc-protected aminoglycosides were activated. Treatment with ammonia yielded the aminomethyl substituted products. Using standard peptide-coupling conditions, the unprotected amine was selectively coupled to the coumarin carboxylic acid. Treatment with trifluoroacetic acid yielded **20** and **21** as their trifluoroacetate (TFA) salts. The photophysical properties of **20** and **21** did not differ significantly from those of the coumarin carboxylic acid **13**,

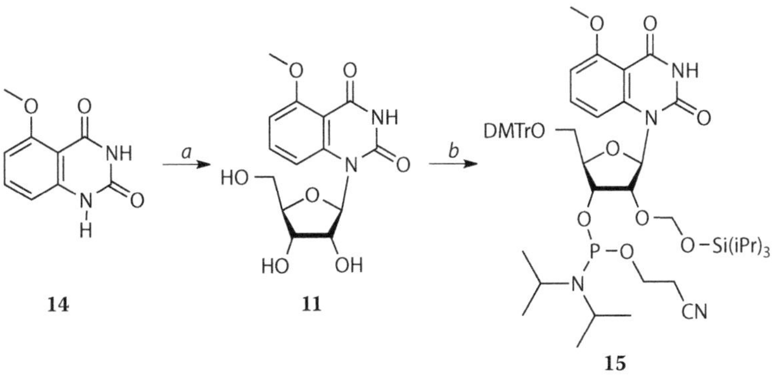

FIGURE 7.8 Synthesis of phosphoramidite **15**. Reagents: (a) (i) *N,O*-bis(trimethylsilyl) acetamide, $CF_3SO_3Si(CH_3)_3$, β-D-ribofuranose 1-acetate-2,3,5-tribenzoate, CH_3CN; (ii) conc. NH_4OH, 81%. (b) (i) DMTrCl, Et_3N, pyridine, 85%; (ii) *i*Pr_2NEt, *n*Bu_2SnCl_2, (*i*Pr_3SiO)CH_2Cl, $ClCH_2CH_2Cl$, 30%; (iii) *i*PrNEt, (*i*Pr_2N)P(Cl)OCH_2CH_2CN, $ClCH_2CH_2Cl$, 60%.

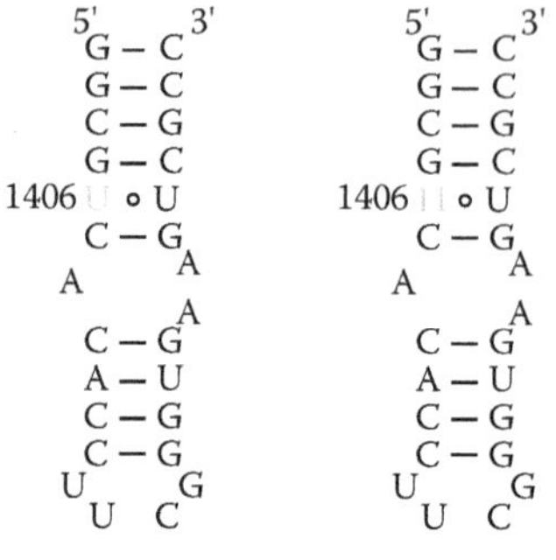

FIGURE 7.9 Unmodified (**16**) and modified (**17**) 27-base RNA A-site model constructs.

with emission maxima of 473 nm upon excitation at 400 nm and a fluorescence quantum yield of 80%. The concentrations of stock solutions containing coumarin-labeled aminoglycosides were determined by measuring the absorbance at 400 nm ($\varepsilon = 20{,}000$ M^{-1} cm^{-1}) [6].

7.3.1.4 Binding and Displacement Studies

To perform binding studies, RNA construct **17** (1.0×10^{-6} M) was annealed in caco-dylate buffer (2.0×10^{-2} M, pH 7.0, 1.0×10^{-1} M NaCl, 5.0×10^{-4} M EDTA) by heating to 75 °C for 5 min and allowing to cool to room temperature over 2 h. After placing the sample on ice for 30 min, coumarin-labeled aminoglycosides 20 or 21

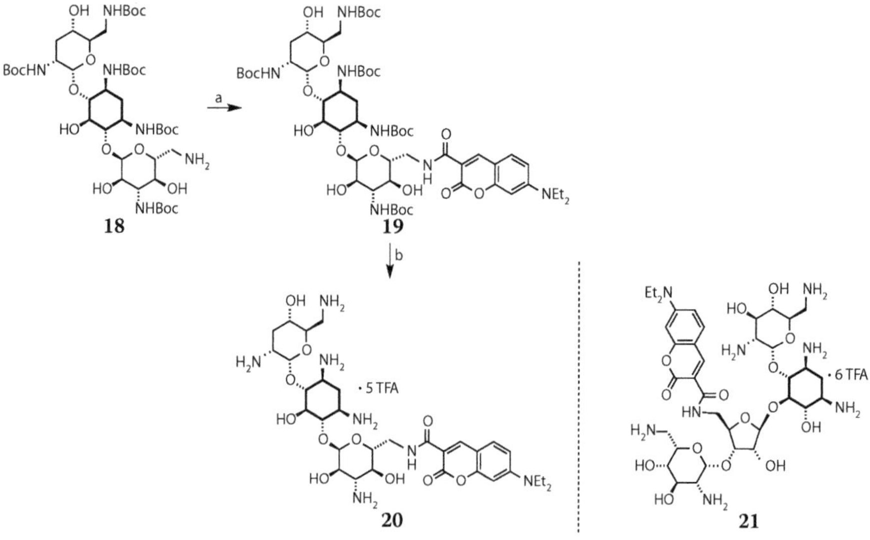

FIGURE 7.10 Synthesis of coumarin-labeled tobramycin (**20**) and structure of coumarin-labeled neomycin (**21**). Reagents: (a) 7-(Et$_2$N)coumarin-3-carboxylic acid, EDC, DMAP, iPr$_2$EtN, Cl$_2$CH$_2$, 84%; (b) trifluoroacetic acid, triisopropylsilane, CH$_2$Cl$_2$ 82%.

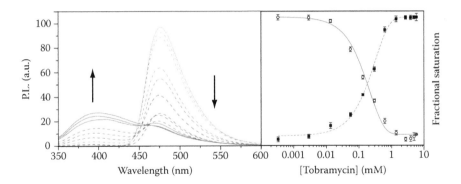

FIGURE 7.11 Representative spectra and titration curves of a displacement assay. Tobramycin titrated into construct **17** saturated with coumarin-labeled tobramycin **20**. Conditions: **17** (1.0×10^{-6} M), cacodylate buffer pH 7.0 (2.0×10^{-2} M), 5.0×10^{-4} M EDTA, 1.0×10^{-1} M NaCl.

were gradually titrated into the RNA solution. Concentrated solutions of 20 or 21 were used to minimize the dilution of the sample (in general, the volume added would not exceed 10% of the starting RNA solution volume). Changes in fluorescence intensity were monitored at 395 nm (emission maximum of the donor) and 473 nm (emission maximum of the acceptor) upon excitation at 320 nm. As the donor's emission decreases, the sensitized emission of the acceptor increases. The fractional fluorescence saturation of the donor and acceptor was plotted against the concentration of the labeled aminoglycoside to determine EC_{50} values of 0.84 ($\pm$0.03) $\times 10^{-6}$ M for 21 and 2.3 ($\pm$0.2) $\times 10^{-6}$ M for 20 (see Figure 7.11 for example spectra and resulting titration curves) [6]. These values correlate with the relative binding affinities of neomycin B and tobramycin as previously reported [50,51].

Relative binding affinities of unlabeled antibiotics were determined through competition experiments where the acceptor-labeled aminoglycoside was displaced and the FRET complex disrupted. Using the same conditions as the binding studies, the saturation of the modified A-site complex with either **20** or **21** resulted in a preformed FRET complex, which could then be disassembled by titrating with unlabeled competitors. As with the previous binding studies, the complex was excited at 320 nm and the emission intensities were monitored at 395 and 473 nm. As displacement of the labeled aminoglycoside occurs, the emission of the donor increases with a concurrent decrease in the emission of the acceptor, corresponding to the separation of the FRET partners. The fractional fluorescence saturation of the donor and acceptor were plotted against the concentration of the unlabeled aminoglycoside to obtain inhibition curves. This curve was used to calculate the IC_{50} values for the displacement of both **20** and **21** by seven different aminoglycosides (Table 7.2) [6].

7.3.1.5 Assay Overview and Summary

The results of the FRET-monitored binding and displacement experiments illustrated above reveal that a FRET-based system, with strategically placed donors and acceptors, unambiguously provides both association and competitive dissociation data. Most importantly, while singly labeled RNA constructs might generate false-positive

TABLE 7.2

IC_{50} Values for Displacing Compounds 20 and 21 by Aminoglycosides[a]

Aminoglycoside	20	21
Neomycin B	0.03 ± 0.01	0.02 ± 0.01
Tobramycin	0.50 ± 0.02	0.16 ± 0.03
Apramycin	3.00 ± 0.09	1.00 ± 0.05
Paromomycin	1.14 ± 0.08	0.06 ± 0.03
Hygromycin	1.46 ± 0.01	1.00 ± 0.05
Amikacin	1.73 ± 0.09	0.56 ± 0.04
Kanamycin A	3.30 ± 0.09	1.61 ± 0.06

[a] Conditions: 17 (1.0×10^{-6} M), cacodylate buffer pH 7.0 (2.0×10^{-2} M), 5.0×10^{-4} M EDTA, 1.0×10^{-1} M NaCl. Aminoglycoside concentration is given in 10^{-3} M.

signals due to remote binding at nonfunctional states (which could alter the environment of a responsive probe), a FRET-based assembly requires specific binding to the A-site pocket. A nonspecific RNA binder would not be able to displace an antibiotic from its cognate recognition site. In the drug discovery context, this increases the chances for the discovery of target-specific and functional antibiotics, while minimizing the risk of false-positive hits.

7.3.2   SINGLE ASSAY FOR DETERMINING BOTH BINDING AND SELECTIVITY OF ANTIBIOTICS

7.3.2.1   System Design

Although the human (18S) and bacterial (16S) A-sites differ at several key residues (Figure 7.12), there is a significant degree of sequence homology between the two targets, and both are capable of binding aminoglycoside antibiotics [45,52–54]. To explore the selectivity and affinity of antibiotics for these two potentially competing sites, a three-component FRET system based on our previously described FRET assay was developed (Figure 7.12), which generated a unique spectral signature for binding at each site [7]. Two orthogonal, yet matched, FRET pairs were selected, building upon the single FRET pair established above (Figure 7.13). As before, the isomorphic fluorescent probe on the bacterial A-site construct at position U1406 (11) serves as a FRET donor to the coumarin-based fluorophore (13) placed on the aminoglycoside antibiotic ($R_0 = 27$ Å). The latter serves as a FRET donor for an acceptor fluorophore (Dy547) placed on the human A-site RNA construct ($R_0 = 45$ Å). Kanamycin A, which has modest affinity to both the 16S and 18S A-sites, was conjugated to coumarin 13 giving the fluorescent derivative 24 to serve as the "place-holder" A-site binder.

When all three FRET components are mixed and equilibrated, the "place-holder" antibiotic bound to the 16S RNA can be visualized by exciting the uracil analog 14

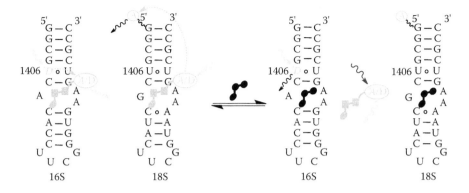

FIGURE 7.12 Fluorescent prokaryotic (16S) and eukaryotic (18S) A-site constructs. The 16S A-site is modified at position 1406 with a fluorescent nucleoside analog (**11**) to function as a FRET donor (***D***). When the "place-holder" is bound to the 16S construct, the fluorescence of the FRET acceptor (***A/D***) is observed. The 18S A-site is 5′-labeled with Dy547 (***A***). When the "place-holder" is bound to the 18S construct, the emission of the acceptor (***A***) is observed. If the bound constructs are disrupted by another antibiotic, the fluorescence of the donor (***D***) is restored concurrently with the loss of fluorescence of (***A/D***), while the fluorescence of the acceptor (***A***) is lost.

and monitoring the emission of coumarin-labeled aminoglycoside 24. The antibiotic bound to the 18S RNA can be visualized by exciting "place-holder" 24 and monitoring the emission of Dy547. When an unlabeled antibiotic is added to the mixture, displacing the "place-holder" from the A-site constructs, the emission of the acceptors is lost along with a concurrent increase in emission intensity of the respective donors.

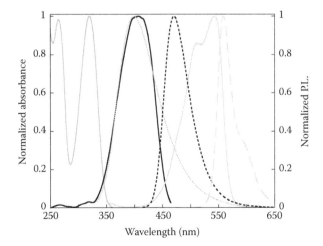

FIGURE 7.13 Absorption (solid line) and emission (dashed line) spectra of compounds **13** (black), **14** (dark gray), and Dy547 (light gray) demonstrating appropriate spectral characteristics for two orthogonal FRET pairs.

7.3.2.2 Component Synthesis

The fluorescent 16S A-site construct was synthesized as described in Section 7.3.1.2, and the 5′-Dy547-labeled 18S A-site was obtained from Dharmacon. Coumarin-labeled kanamycin A was synthesized in a manner similar to the syntheses reported in Section 7.3.1.3 (Figure 7.14) [7]. Boc-protected aminomethyl-substituted kanamy-cin A was coupled to the coumarin carboxylic acid using standard peptide coupling conditions. The resulting compound was deprotected with trifluoroacetic acid, yield-ing "place-holder" 24 as a TFA salt. The photophysical properties of 24 did not differ significantly from those of coumarin-labeled aminoglycosides 20 and 21, and stock solutions were prepared in a similar fashion.

7.3.2.3 Competition Studies

To evaluate the affinity and selectivity of ligands to the two A-sites, equimolar amounts of the 18S and 16S RNA constructs were separately prefolded (as described in Section 7.3.1.4) and mixed, giving a final concentration of 5.0×10^{-7} M of each component. Two-mole equivalent of "place-holder" 24 were added, resulting in a concentration of 2.2×10^{-6} M (>2 eq for each RNA). An unlabeled antibiotic was gradually titrated into this mixture, while fluorescence intensity changes of all fluo-rophores were monitored. The emission intensity of donor 11 increased concurrently with a decrease in emission of coumarin-labeled kanamycin A 24 upon displacement of the "place-holder" from the 16S RNA construct (with excitation at 320 nm). Displacement from the 18S RNA construct by the titrated ligand resulted in a decreas-ing emission intensity of the acceptor Dy547 at 561 nm (upon excitation at 400 nm). Fractional fluorescence intensity was determined at each titration point and plotted against the concentration of the antibiotic to produce titration curves (Figure 7.15). The IC_{50} values were calculated for both the 16S and 18S A-sites and compared with each other to obtain a selectivity ratio (Table 7.3) [7]. This assay was performed for a wide variety of antibiotics, yielding crucial data about affinity and selectivity.

7.3.2.4 Assay Overview and Summary

Not only does this method maintain all of the advantages of the FRET assay described in Section 7.3.1, but it also offers a novel and efficient means of gathering essential information with regard to RNA–small molecule interactions. This powerful assay enables the real-time assessment of the affinity of antibiotics for the human and

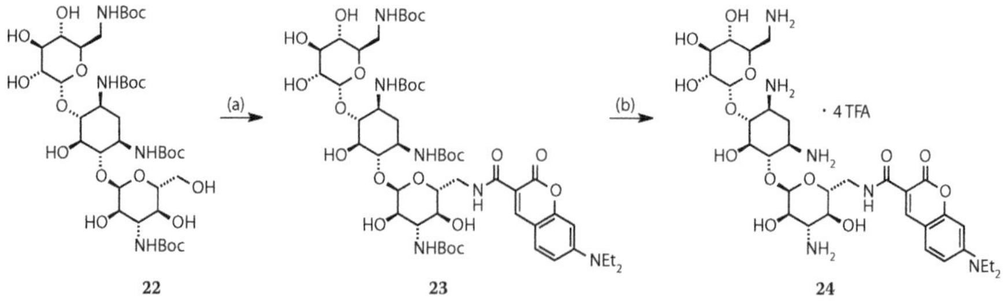

FIGURE 7.14 Synthesis of coumarin-labeled kanamycin A. Reagents: (a) 7-(Et$_2$N)coumarin-3-carboxylic acid, EDC, DMAP, iPr$_2$EtN, CH$_2$Cl$_2$, 87%. (b) TFA, triisopropylsilane, CH$_2$Cl$_2$, 72%.

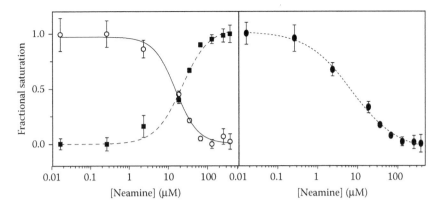

FIGURE 7.15 Example titration curves generated from the fractional fluorescence saturation of donor **11** (solid black squares), acceptor **13** (open circles), and Dy547 (solid black ovals) as neamine is titrated into the preformed 16S and 18S complexes. Conditions: 16S RNA $(5.0 \times 10^{-7} \text{ M})$, 18S RNA $(5.0 \times 10^{-7} \text{ M})$, **24** $(2.2 \times 10^{-6} \text{ M})$, cacodylate buffer pH 7.0 $(2.0 \times 10^{-2} \text{ M})$, 5.0×10^{-4} M EDTA, 1.0×10^{-1} M NaCl.

bacterial A-site. This technique, relying on two orthogonal FRET pairs with unique spectral signatures, is not necessarily limited to A-site constructs and could, in principle, be implemented for combinations of other RNA targets.

7.3.3 FLUORESCENT NUCLEOSIDE ANALOG SERVING AS A FRET DONOR FOR TRYPTOPHAN

7.3.3.1 System Design

Tryptophan is frequently found at the recognition interface of RNA-binding proteins [55]. As such, identifying emissive nucleosides capable of resonance energy transfer

TABLE 7.3
IC$_{50}$ Values for Antibiotics Displacing 25 from 16S and 18S A-Sites[a]

Antibiotic	16S	18S	Selectivity Ratio
Neomycin B	2.8 ± 0.3	4.7 ± 0.2	0.60
Tobramycin	20.2 ± 0.4	19.5 ± 0.3	1.0
Paromomycin	9 ± 1	8.0 ± 0.6	1.1
Kanamycin A	75 ± 3	46 ± 2	1.6
Negamycin	62 ± 5	42 ± 3	1.5
Neamine	18 ± 2	6 ± 1	3
Linezolid	$>9.6 \times 10^3$	$>9.6 \times 10^3$	—

[a] Conditions: 16S RNA $(5.0 \times 10^{-7}$ M), 18S RNA $(5.0 \times 10^{-7}$ M), **24** $(2.2 \times 10^{-6}$ M), cacodylate buffer pH 7.0 $(2.0 \times 10^{-2}$ M), 5.0×10^{-4} M EDTA, 1.0×10^{-1} M NaCl. Concentrations of antibiotics are given in 10^{-6} M.

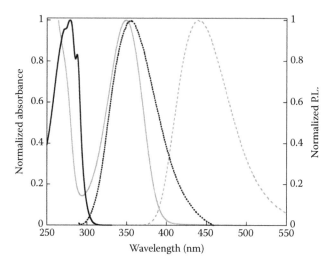

FIGURE 7.16   Absorption (solid line) and emission (dashed line) spectra of tryptophan (black) and **12** (gray) demonstrating appropriate spectral characteristics for an effective FRET pair.

with this naturally occurring, fluorescent amino acid is highly valuable. In water, Trp displays an absorption maximum at 280 nm and an emission maximum at 350 nm with a quantum yield of 12% [8]. We have identified the fluorescent uridine analog **12**, with an absorption maximum at 350 nm and an emission maximum at 440 nm (fluorescence quantum yield of 42%) as an excellent FRET acceptor for Trp (Figure 7.16). The critical Förster radius for this pair was determined to be 22 Å, which is appropriate for monitoring RNA–peptide interactions [8].

The Rev peptide and the Rev Response Element (RRE) serve as an established system for probing RNA–peptide interactions [56–58]. The Rev peptide contains an arginine-rich domain and binds with high affinity and specificity to Stem-loop IIB of the RRE (Figure 7.17). The Rev protein also contains a single tryptophan residue (Trp$_{45}$), which is located within the RNA-binding domain. Based on structural information, residue U66 of the RRE was chosen as the incorporation site for nucleoside

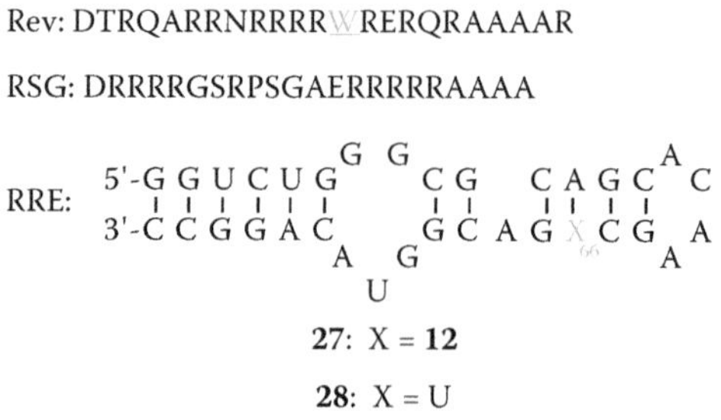

Rev: DTRQARRNRRRRᴡRERQRAAAAR

RSG: DRRRRGSRPSGAERRRRRAAAA

$$
\begin{array}{lllllllllll}
 & & & & & \text{G} & \text{G} & & & & \text{A}\\
 & \text{5'-G G U C U G} & & \text{C G} & & \text{C A G C} & \text{C}\\
\text{RRE:} & \quad|\ |\ |\ |\ |\ | & & |\ | & & |\ |\ |\ | \\
 & \text{3'-C C G G A C} & & \text{G C A G X C G} & \text{A}\\
 & & \text{A} & \text{G} & & & & \text{A}\\
 & & \text{U}
\end{array}
$$

27: X = **12**

28: X = U

FIGURE 7.17   Amino acid sequences of the Rev and RSG peptides along with the secondary structure of the modified (**27**) and unmodified (**28**) RRE RNA model constructs.

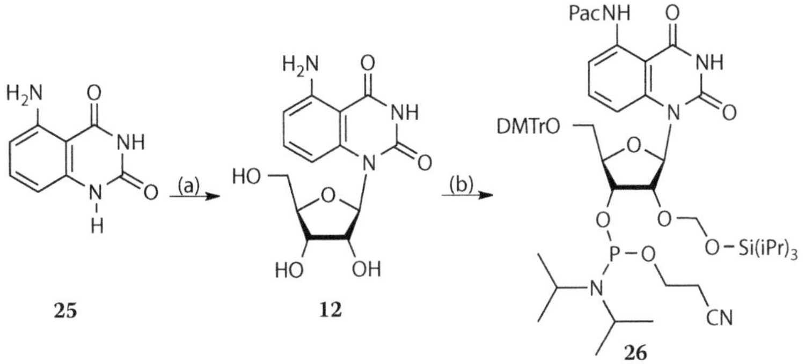

FIGURE 7.18 Synthesis of phosphoramidite **26**. Reagents: (a) (i) *N,O*-bis(trimethylsilyl) acetamide, $CF_3SO_3Si(CH_3)_3$, β-D-ribofuranose 1-acetate 2,3,5-tribenzoate, CH_3CN; (ii) conc. NH_4OH, 81%. (b) (i) $(CH_3)_3SiCl$, phenoxyacetic anhydride, H_2O, conc. NH_4OH, pyridine, 75%; (ii) DMTrCl, Et_3N, pyridine, 82%; (iii) iPr_2NEt, nBu_2SnCl_2, iPr_3SiOCH_2Cl, $ClCH_2CH_2Cl$, 30%; (iv) iPr_2NEt, $(iPr_2N)P(Cl)OCH_2CH_2CN$, $ClCH_2CH_2Cl$, 60%.

12. The estimated distance between Trp_{45} and U66 was 20 Å, which matches well with the calculated critical Förster radius [8].

7.3.3.2 Synthesis and Properties of Fluorescent RNA Construct

Fluorescent uridine analog **12** was synthesized through the glycosylation of nucleobase **25** (Figure 7.18). To incorporate **12** into oligonucleotides, phosphoramidite **26** was synthesized and utilized in standard solid-phase RNA synthesis. Model RRE construct **27** was synthesized and maintained nearly identical stability to an unmodified construct **28** (Figure 7.17), as demonstrated by thermal denaturation studies. Upon excitation at 350 nm, construct **27** maintained an emission maximum of 440 nm, but the fluorescence quantum yield was reduced.

7.3.3.3 Binding and Displacement Studies

Construct **27** was annealed at a concentration of 5×10^{-6} M in 2.0×10^{-2} M cacodylate buffer (pH 7.0, 1.0×10^{-1} M NaCl, 5.0×10^{-4} M EDTA) by heating to 75°C for 5 min and slowly cooling to room temperature over 2 h. This modified RRE solution was titrated into a solution containing 1.0×10^{-5} M Rev peptide in 2.0×10^{-2} M cacodylate buffer, while monitoring fluorescence changes. Samples were excited at 280 nm and emission intensity was monitored at 350 nm (FRET donor emission) and 445 nm (FRET acceptor emission) (Figure 7.19). Plotting the concentration of modified RRE construct **27** against the normalized-sensitized emission at 445 nm generated titration curves and allowed for determination of a K_D value of $(7 \pm 5) \times 10^{-8}$ M [8].

As a potential tool for ligand discovery, this FRET system can be utilized to monitor the displacement of the Rev peptide by competitive binders that do not contain tryptophan residues. To illustrate this capability, the artificially evolved, arginine rich RSG peptide (Figure 7.17), which contains no tryptophan and is known to bind to the RRE with higher affinity than the Rev peptide, was tested. A solution containing 1.0×10^{-5} M RSG peptide in 2.0×10^{-2} M cacodylate buffer was titrated into a solution

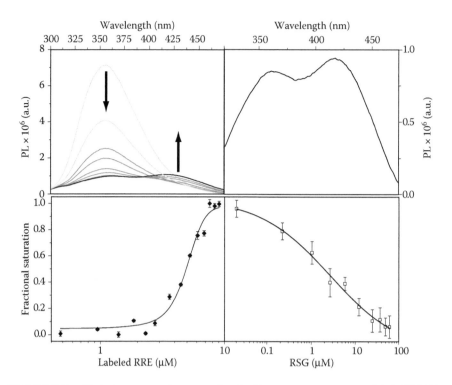

FIGURE 7.19 Emission spectra demonstrating the fluorescence response as labeled RRE is titrated into Rev (top) and resulting titration curve (bottom left). Also shown is the displacement of the RRE-bound Rev by RSG (bottom right).

of Rev peptide saturated with RRE construct 27. Upon excitation at 280 nm, emission was monitored, as before. As the Rev peptide was displaced from the RRE construct, the emission of the acceptor at 445 nm decreased. A plot of the fractional saturation versus the concentration of RSG yielded an IC_{50} value of $(2 \pm 1) \times 10^{-6}$ M [8].

7.3.3.4 Assay Overview and Summary

Cellular functions involve countless RNA–protein interactions. Tryptophan, a fluorescent amino acid, is often found within, or near, the recognition domains of RNA-binding proteins. A fluorescent nucleoside analog, serving as a FRET acceptor for tryptophan, can facilitate a sensitive and versatile means of monitoring important RNA–protein interactions. The FRET system discussed above enables the real-time analysis of these interactions and can also be of use for the development of drug discovery assays.

7.4 CONCLUSIONS

Fluorescent nucleoside analogs with diverse photophysical characteristics offer unique opportunities for developing biophysical and discovery assays. Imparting favorable photophysical features (including distinct absorption maxima, adequate quantum yield, and responsiveness or lack thereof) upon a minimally perturbing heterocyclic core represents, however, a fundamental challenge. To lessen such constraints and facilitate the tunability of the photophysical properties, we have selected the quinazoline core and implemented charge-transfer transitions to generate expanded emissive pyrimidines. This new family of fluorescent nucleoside analogs facilitated the implementation of novel FRET pairs, by identifying fluorophores with matched donor/acceptor features. A single fluorescent uridine analog was used as a FRET donor in two antibiotics discovery assays. Binding, displacement, and selectivity of various RNA ligands were accurately and efficiently determined, relying on the sensitivity of resonance energy transfer. Similarly, a fluorescent uridine analog was paired with Trp, a relatively high energy emitter, serving as a FRET acceptor. This facilitated the monitoring of RNA–protein interactions, without the need to label the protein component.

7.5 PROSPECTS AND OUTLOOK

As the fundamental roles of nucleic acids are further explored and illuminated, RNA is likely to further evolve as a drug target. Efficient and accurate means of determining relative affinity and selectivity of ligands to RNA can facilitate the analysis of large libraries of potential drugs. Judicious incorporation of emissive nucleoside analogs into small RNA constructs can facilitate such high-throughput biophysical and drug discovery assays. Although singly labeled RNA oligonucleotides have been exploited, it is apparent that assays relying on FRET are more robust and likely to minimize the false signals.

The development of new emissive and nonperturbing nucleoside analogs represents in and of itself a fundamental challenge. In particular, predicting the photophysical features of small nucleobase-like structure is, at this stage, impractical. This is especially relevant for emissive nucleoside analogs, since their photophysical features when incorporated into oligonucleotides are "context-specific": rigidification, desolvation, and excited-state processes involving neighboring nucleotides further impact their photophysics. Although this is not always critical for singly labeled oligonucleotides, the identification of useful FRET pairs could be more problematic, as their features need to be matched with respective donors and acceptors. It is likely that approaches related to the one illustrated here, where a polarizable scaffold is further decorated with polar substituents inducing charge transfer transitions, will produce useful cassettes of emissive nucleosides from which candidates with optimal properties can be selected.

ACKNOWLEDGMENTS

We thank the National Institutes of Health for their generous support (grant number GM 069773).

REFERENCES

1. Sinkeldam, R. W., Greco, N. J., and Tor, Y. Fluorescent analogs of biomolecular building blocks: Design, properties, and applications. *Chem. Rev.* **110**, 2579–2619, 2010.

2. Wilson, J. N. and Kool, E. T. Fluorescent DNA base replacements: Reporters and sensors for biological systems. *Org. Biomol. Chem.* **4**, 4265–4274, 2006.

3. Wilhelmsson, L. M. Fluorescent nucleic acid base analogues. *Q. Rev. Biophys.* **43**, 1–25, 2010.

4. Serrano-Andrés, L. and Merchán, M. Are the five natural DNA/RNA base monomers a good choice from natural selection? A photochemical perspective. *J. Photoch. Photobio. C* **10**, 21–32, 2009.

5. Tor, Y., Fluorescent nucleoside analogues: Synthesis, properties and applications. *Tetrahedron* **63**, 3425–3426, 2007.

6. Xie, Y., Dix, A. V., and Tor, Y. FRET enabled real time detection of RNA-small molecule binding. *J. Am. Chem. Soc.* **131**, 17605–17614, 2009.

7. Xie, Y., Dix, A. V., and Tor, Y. Antibiotic selectivity for prokaryotic vs. eukaryotic decoding sites. *Chem. Commun.* **46**, 5542–5544, 2010.

8. Xie, Y., Maxson, T., and Tor, Y. Fluorescent ribonucleoside as a FRET acceptor for tryptophan in native proteins. *J. Am. Chem. Soc.* **132**, 11896–11897, 2010.

9. Borjesson, K., Preus, S., El-Sagheer, A. H., Brown, T., Albinsson, B., and Wilhelmsson, L. M. Nucleic acid base analog FRET-pair facilitating detailed structural measurements in nucleic acid containing systems. *J. Am. Chem. Soc.* **131**, 4288–4293, 2009.

10. Turro, N. J., Ramamurthy, V., and Scaiano, J. C. 2010. *Modern Molecular Photochemistry of Organic Molecules*. New York: University Science Books.

11. Lakowicz, J. R. 2006. *Principles of Fluorescence Spectroscopy*, 3rd edn. New York: Springer.

12. Valeur, B. 2002. *Molecular Fluorescence: Principles and Applications*. Weinheim, Germany: Wiley-VCH.

13. Ward, D. C., Reich, E., and Stryer, L. Fluorescence studies of nucleotides and polynucleotides. *J. Biol. Chem.* **244**, 1228–1237, 1969.

14. Rachofsky, E. L., Osman, R., and Ross, J. B. A. Probing structure and dynamics of DNA with 2-aminopurine: Effects of local environment on fluorescence. *Biochemistry* **40**, 946–956, 2001.

15. Kirk, S. R., Luedtke, N. W., and Tor, Y. 2-aminopurine as a real-time probe of enzymatic cleavage and inhibition of hammerhead ribozymes. *Bioorg. Med. Chem.* **9**, 2295–2301, 2001.

16. Jean, J. M. and Hall, K. B. 2-aminopurine fluorescence quenching and lifetimes: Role of base stacking. *Proc. Natl. Acad. Sci. USA.* **98**, 37–41, 2001.

17. Srivatsan, S. G., Greco, N., J., and Tor, Y. A highly emissive fluorescent nucleoside that signals the activity of toxic ribosome-inactivating proteins. *Angew. Chem. Int. Ed.* **47**, 6661–6665, 2008.

18. Sinkeldam, R. W., Greco, N. J., and Tor, Y. Polarity of major grooves explored by using an isosteric emissive nucleoside. *ChemBioChem* **9**, 706–709, 2008.

19. Srivatsan, S. G. and Tor, Y. Fluorescent pyrimidine ribonucleotide: Synthesis, enzymatic incorporation, and utilization. *J. Am. Chem. Soc.* **129**, 2044–2053, 2007.

20. Srivatsan, S. G. and Tor, Y. Synthesis and enzymatic incorporation of a fluorescent pyrimidine ribonucleotide. *Nat. Protocols* **2**, 1547–1555, 2007.

21. Srivatsan, S. G., Weizman, H., and Tor, Y. A highly fluorescent nucleoside analog based on thieno[3,4-d]pyrimidine senses mismatched pairing. *Org. Biomol. Chem.* **6**, 1334–1338, 2008.

22. Greco, N. J. and Tor, Y. Synthesis and site-specific incorporation of a simple fluorescent pyrimidine. *Nat. Protocols* **2**, 305–316, 2007.

23. Greco, N. J. and Tor, Y. Furan decorated nucleoside analogues as fluorescent probes: Synthesis, photophysical evaluation, and site-specific incorporation. *Tetrahedron* **63**, 3515–3527, 2007.

24. Greco, N. J., Sinkeldam, R. W., and Tor, Y. An emissive C analog distinguishes between G, 8-oxoG, and T. *Org. Lett.* **11**, 1115–1118, 2009.

25. Greco, N. J. and Tor, Y. Simple fluorescent pyrimidine analogues detect the presence of DNA abasic sites. *J. Am. Chem. Soc.* **127**, 10784–10785, 2005.

26. Xie, Y., Maxson, T., and Tor, Y. Fluorescent nucleoside analogue displays enhanced emission upon pairing with guanine. *Org. Biomol. Chem.* **8**, 5053–5055, 2010.

27. Xie, Y. 2010. Synthesis and Applications of Quinazoline-Based Fluorescent Nucleoside Analogues. PhD thesis, University of California, San Diego.

28. Moazed, D. and Noller, H. F. Interaction of antibiotics with functional sites in 16S ribosomal RNA. *Nature* **327**, 389–394, 1987.

29. Purohit, P. and Stern, S. Interactions of a small RNA with antibiotic and RNA ligands of the 30S subunit. *Nature* **370**, 659–662, 1994.

30. Miyaguchi, H., Narita, H., Sakamoto, K., and Yokoyama, S. An antibiotic-binding motif of an RNA fragment derived from the A-site-related region of escherichia coli 16S rRNA. *Nucleic Acids Res.* **24**, 3700–3706, 1996.

31. Recht, M. I., Fourmy, D., Blanchard, S. C., Dahlquist, K. D., and Puglisi, J. D. RNA sequence determinants for aminoglycoside binding to an A-site rRNA model oligonucleotide. *J. Mol. Biol.* **262**, 421–436, 1996.

32. Fourmy, D., Recht, M. I., Blanchard, S. C., and Puglisi, J. D. Structure of the A site of *Escherichia coli* 16S ribosomal RNA complexed with an aminoglycoside antibiotic. *Science* **274**, 1367–1371, 1996.

33. Vicens, Q. and Westhof, E. Crystal structure of paromomycin docked into the eubacterial ribosomal decoding A site. *Structure* **9**, 647–658, 2001.

34. Vicens, Q. and Westhof, E. Crystal structure of a complex between aminoglycoside tobramycin and an oligonucleotide containing the ribosomal decoding A site. *Chem. Biol.* **9**, 747–755, 2002.

35. Vicens, Q. and Westhof, E. Crystal structure of geneticin bound to a bacterial 16S ribosomal RNA A site oligonucleotide. *J. Mol. Biol.* **326**, 1175–1188, 2003.

36. Francois, B., Russell, R. J. M., Murray, J. B., Aboul, N., Masquida, B., Vicens, Q., and Westhof, E. Crystal structures of complexes between aminoglycosides and decoding A site oligonucleotides: Role of the number of rings and positive charges in the specific binding leading to miscoding. *Nucleic Acids Res.* **33**, 5677–5690, 2005.

37. Hermann, T. A-site model RNAs. *Biochimie* **88**, 1021–1026, 2006.

38. Carter, A. P., Clemons, W. M., Brodersen, D. E., Morgan-Warren, R. J., Wimberly, B. T., and Ramakrishnan, V. Functional insights from the structure of the 30S ribosomal subunit and its interactions with antibiotics. *Nature* **407**, 340–348, 2000.

39. Yonath, A. Antibiotics targeting ribosomes: Resistance, selectivity, synergism, and cellular regulation. *Annu. Rev. Biochem.* **74**, 649–679, 2005.

40. Vicens, Q. and Westhof, E. Molecular recognition of aminoglycoside antibiotics by ribosomal RNA and resistance enzymes: An analysis of x-ray crystal structures. *Biopolymers* **70**, 42–57, 2003.

41. Xie, Y., Tam, V. K. and Tor, Y. 2010. The interactions of small molecules with DNA and RNA. In *The Chemical Biology of Nucleic Acids*, ed. G. Mayer, pp. 115–140. Chichester, UK: John Wiley & Sons, Ltd.

42. Puglisi, J. D., Blanchard, S. C., Dahlquist, K. D., Eason, R. G., Fourmy, D., Lynch, S. R., Recht, M. I., and Yoshizawa, S. 2000. Aminoglycoside antibiotics and decoding. In: *The Ribosome: Structure, Function, Antibiotics, and Cellular Interactions* eds. R. A. Garrett, S. A. Douthwaite, A. Liljas, A. T. Matheson, P. B. Moore, and H. F. Noller, pp. 419–429 Washington, DC: ASM Press.

43. Gale, E. F., Cundliffe, E., Renolds, P. E., Richmond, M. H., and Waring, M. J. 1981. *The Molecular Basis of Antibiotic Action*. London, UK: John Wiley & Sons.

44. McCoy, L. S., Xie, Y., and Tor, Y. Antibiotics that target protein synthesis. *Wiley Interdisc. Rev.: RNA* **2**, 209–232, 2010.

45. Kaul, M., Barbieri, C. M., and Pilch, D. S. Defining the basis for the specificity of aminoglycoside-rRNA recognition: A comparative study of drug binding to the A sites of escherichia coli and human rRNA. *J. Mol. Biol.* **346**, 119–134, 2005.

46. Wang, H. and Tor, Y. Electrostatic interactions in RNA aminoglycosides binding. *J. Am. Chem. Soc.* **119**, 8734–8735, 1997.

47. Kaul, M., Barbieri, C. M., and Pilch, D. S. Fluorescence-based approach for detecting and characterizing antibiotic-induced conformational changes in ribosomal RNA: Comparing aminoglycoside binding to prokaryotic and eukaryotic ribosomal RNA sequences. *J. Am. Chem. Soc.* **126**, 3447–3453, 2004.

48. Shandrick, S., Zhao, Q., Han, Q., Ayida, B. K., Takahashi, M., Winters, G. C., Simonsen, K. B., Vourloumis, D., and Hermann, T. Monitoring molecular recognition of the ribosomal decoding site. *Angew. Chem. Int. Ed.* **43**, 3177–3182, 2004.

49. Tor, Y. The ribosomal A-site as an inspiration for the design of RNA binders. *Biochimie* **88**, 1045–1051, 2006.

50. Wang, Y., Hamasaki, K., and Rando, R. R. Specificity of aminoglycoside binding to RNA constructs derived from the 16S rRNA decoding region and the HIV-RRE activator region. *Biochemistry* **36**, 768–779, 1997.

51. Kaul, M., Barbieri, C. M., and Pilch, D. S. Aminoglycoside-induced reduction in nucleotide mobility at the ribosomal RNA A-site as a potentially key determinant of antibacterial activity. *J. Am. Chem. Soc.* **128**, 1261–1271, 2006.

52. Recht, M. I., Douthwaite, S., and Puglisi, J. D. Basis for prokaryotic specificity of action of aminoglycoside antibiotics. *EMBO J.* **18**, 3133–3138, 1999.

53. Lynch, S. R. and Puglisi, J. D. Structure of a eukaryotic decoding region A-site RNA. *J. Mol. Biol.* **306**, 1023–1035, 2001.

54. Hobbie, S. N., Kalapala, S. K., Akshay, S., Bruell, C., Schmidt, S., Dabow, S., Vasella, A., Sander, P., and Bottger, E. C. Engineering the rRNA decoding site of eukaryotic cytosolic ribosomes in bacteria. *Nucl. Acids Res.* **35**, 6086–6093, 2007.

55. Baker, C. M. and Grant, H. G. Role of aromatic amino acids in protein–nucleic acid recognition *Biopolymers* **85**, 456–470, 2007.

56. Battiste, J. L., Mao, H., Rao, N. S., Tan, R., Muhandiram, D. R., Kay, L. E., Frankel, A. D., and Williamson, J. R. Alpha helix-RNA major groove recognition in an HIV-1 Rev peptide-RRE RNA complex. *Science* **273**, 1547–1551, 1996.

57. Gosser, Y., Hermann, T., Majumdar, A. Hu, W., Frederick, R., Jiang, F., Xu, W., and Patel, D. J. Peptide-triggered conformational switch in HIV-1 RRE RNA complexes. *Nat. Struct. Biol.* **8**, 146–150, 2001.

58. Pljevaljčića, G. and Millara, D. P. Single-molecule fluorescence methods for the analysis of RNA folding and ribonucleoprotein assembly *Method Enzymol.* **450**, 233–252, 2008.

8 Atomic Force Microscopy Investigation of DNA–Drug Interactions

Jozef Adamcik and Giovanni Dietler

CONTENTS

This chapter describes the utility of atomic force microscopy (AFM) in studying the structural changes induced by drug binding to single- and double-stranded deoxyribonucleic acid (DNA) at the level of individual molecules. Drugs, like intercalating agents or minor-groove-binding molecules, were incubated with single- or double-stranded DNA and their effect on the DNA conformation was investigated by means of AFM imaging.

8.1 INTRODUCTION

DNA is a molecule of great biological significance. It not only carries genetic information but is also a major target for drugs. Owing to the central role played by DNA in intracellular processes such as replication, transcription, and gene regulation, drug-induced modifications in its mechanical or topological properties can have a profound impact on the metabolism of cells, diminishing, and in some cases terminating, their growth [1]. The investigation of DNA–drug interactions is crucial for an understanding of these biochemical processes and for the rational design of new therapeutic agents [2]. Indispensable to an understanding of DNA–drug interactions is a characterization of the binding modes. In addition to covalent (monofunctional) binding, viz., cross-linkage, there exist several classes of either specific or nonspecific noncovalent binding, including intercalation and groove binding [3,4].

Various physical and chemical techniques have been implemented to study the binding of small ligands to DNA, ranging from thermodynamic and biochemical techniques to structural methods, such as nuclear magnetic resonance spectroscopy and x-ray diffraction [5–9]. Recent advancements in the techniques that are used to manipulate single molecules, such as the introduction of magnetic and optical tweezers and AFM, have permitted a study of DNA–drug interactions at the level of single molecules [10–21].

In this chapter, the applicability of AFM to the study of DNA–drug interactions at the single-molecule level will be discussed. Emphasis will be placed on a characterization of the structural changes thereby induced within single and double strands of the macromolecule.

8.2 BASICS OF AFM FOR BIOMEDICAL RESEARCH

AFM was invented in 1986 [22] as a tool to study the surfaces of nonconducting samples at the atomic level under dry conditions. The principle of AFM is based upon that of the scanning tunneling microscope [23], which is applicable only to conducting samples, likewise under dry conditions. The introduction of light beam-based detection also permitted the study of samples under fluid conditions [24,25]. This facility is important in the study of biological samples, permitting their analysis under quasi-physiological conditions. Figure 8.1 depicts a scheme of the microscope as it is used today. Its key features include the following: (a) a piezo tube, which can move the sample along the three spatial axes within ranges that are appropriate for the study of biological specimens (10 nm–100 μm); (b) a lateral resolution down to ~1 nm suffices to study the shape of, and conformational changes within (but not the three-dimensional atomic organization of) biomolecules on a broad length scale; (c) the spring constant of the lever permits the measurement of intermolecular forces (from 10 to several hundred nano-Newtons); and (d) the tip of the lever (force sensor)

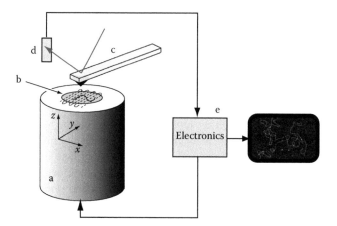

FIGURE 8.1 Scheme of an atomic force microscope: (a) piezo tube; (b) sample; (c) lever with tip; (d) lever-bending detector; (e) control electronics with a screen representing the topography of the sample. The piezo tube is used to scan the sample on a raster beneath the tip, which records height changes in the lever on the screen. This information yields the topography of the sample.

can be used to probe various physical quantities, such as the elasticity of the samples, temperature, pH [26,27], or the interactions between biomolecules [28]. A significant shortcoming of AFM is the necessity to immobilize the biomolecule or any other biological specimen under scrutiny on a very flat surface: usually mica is the substratum of choice for this purpose. The strategy that has to be implemented for the fixation depends upon the sample: cells and other large objects sometimes spontaneously adsorb, while smaller biomolecules, among them proteins and DNA, require more sophisticated methods of attachment (see below for more details for DNA). The choice of the cantilever is also based according to the planned experiments. For imaging, rather stiff cantilevers can be used (from 0.1 N/m up to 40 N/m), while for force measurements, very soft cantilevers are preferred (0.01 N/m) in order to have enough force sensitivity. Presently, a number of software programs are available to further analyze AFM data. DNA images can be analyzed to determine the contour length of the molecules, while force curves are investigated in order to obtain elastic parameters of cells and proteins, or interaction forces between biomolecules.

With a diameter of 2 nm and a length of 10–100 μm, double-stranded DNA (dsDNA) is an "easy" molecule to image. After its deposition upon a substratum of mica, which has been coated with a positively charged layer of APTES (3-aminopropyltriethoxysilane) [29,30] or treated with a divalent cation, such as Mg^{2+} or Ni^{2+}

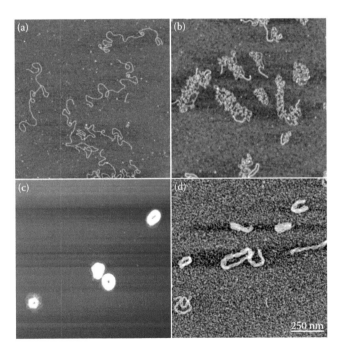

FIGURE 8.2   AFM images of linear molecules of dsDNA and of supercoiled DNA that were exposed to intercalating and minor-groove-binding ligands in the presence of 1 mM TRIS (pH 7.8). (a) Linear pBR322 in the presence of 2 μM daunorubicin; (b) supercoiled DNA in the presence of 2 μM daunorubicin; (c) λ-DNA in the presence of 10 μM netropsin; (d) plasmid pBR322 in the presence of 6 μM netropsin.

[31,32], and subsequently drying, dsDNA has a diameter of ~1.2–2 nm and a contour length that approaches the standard value of 3.4 nm per 10.4 base pairs. Hence, the adsorption process does not significantly distort the double helical conformation [33]. AFM can thus be used to study dsDNA with the confidence of yielding relevant information pertaining to its solution properties. Figure 8.2a depicts an example of one such molecule, namely, the circular pBR322 plasmid (4361 base pairs and a total contour length of 1482 nm), which has been deposited upon Mg^{2+}-pretreated mica (5 mM $MgCl_2$). The imaging of single-stranded DNA (ssDNA) is more challenging. However, we have recently developed a suitable protocol for this that involves depositing the molecules upon a substratum of either modified graphite or mica bearing a thin coating of APTES [34–36]. Furthermore, it is now possible to image bubble formation in dsDNA and its appearance after digestion with exonuclease [37]. And more recently, AFM has been implemented to study the formation of amyloid-like fibrils of β-lactoglobulin [38].

8.3 CASE STUDIES

8.3.1 AFM OF DSDNA–DRUG INTERACTIONS

Drugs interact with DNA via covalent as well as by noncovalent binding, the former being an irreversible process and the latter a reversible one. An example of covalent drug binding is represented by *cis*-diamminedichloroplatinum (cisplatin), which is widely used in cancer therapy. This drug forms intra- and interstrand cross-links via its chloride groups, which interact with nitrogen atoms on the DNA molecule [39,40]. In addition to covalent binding, specific and nonspecific noncovalent interactions, namely, intercalation and groove binding, are also possible. Intercalating drugs often contain planar and hydrophobic closed-ring structures, which inset between two-stacked base pairs in the double helix. Intercalation is often consolidated by hydrogen bonding. Intercalating drugs, which elicit an extension and a partial unwinding of dsDNA, have a profound impact on the structure of the nucleosome. In the case of circularly closed DNA (ccDNA), or of DNA that has been otherwise topologically constrained, one has to consider a conservation of the linking number (Lk), which is defined as the sum of twist (Tw) and writhe (Wr):

$$Lk = Tw + Wr \qquad (8.1)$$

In fact, Lk is strictly conserved, a change in twist causing an opposite change of equal magnitude in writhe [41]. An intercalating drug would be therefore expected to diminish twist and to cause a corresponding increase in writhe. The insertion of an intercalating agent into ccDNA will cause its change from a negatively to a positively supercoiled structure. This change is triggered by a decrease in twist [18,21]. Since the writhe is usually a small number, the change in twist can be readily detected by AFM, even if the strength of the drug intercalation is weak.

The other mode of noncovalent interaction is selective groove binding, which involves the electrostatic association [42] of positively charged drug molecules with the narrow minor groove of AT-rich sequences in dsDNA assisted by hydrogen bonds and van der Waals' interactions [4].

The binding of drugs to dsDNA alters its conformation, which is generally characterized by the contour length, the persistence length, and the linking number. The interaction of cisplatin with dsDNA (by covalent binding) has been recently studied by AFM. At low drug concentrations, local distortions in the molecules of dsDNA were observed. A progressive increase in the concentration of cisplatin successively led to the formation first of microloops, then of aggregates, and finally of globular structures [40]. In the case of intercalating drugs, such as daunorubicin, an increase in the contour length of dsDNA [17] is a hallmark of this type of interaction (Figure 8.2a). On the other hand, drugs that interact with dsDNA via minor-groove binding, such as netropsin, elicit not an increase in the contour length of dsDNA but its condensation (Figure 8.2b).

In addition to causing an increase in the contour length of linear DNA molecules, intercalating agents can also induce conformational changes within ccDNA by decreasing the twist Tw. Using the intercalating drugs daunorubicin and doxorubicin (chemotherapeutic agents), we have demonstrated a dose-dependent transformation of negatively into positively supercoiled DNA (Figure 8.3) [21]. The insertion of an intercalating drug into plasmid DNA elicits a decrease in twist Tw and a corresponding increase in writhe Wr (see Equation 8.1): the ccDNA undergoes a continuous change in phase from a negatively supercoiled configuration (Wr < 0), through a relaxed form (Wr = 0), to a positively supercoiled structure (Wr > 0). In this experiment, the DNA–drug complex was deposited upon a Mg^{2+}-pretreated substratum of

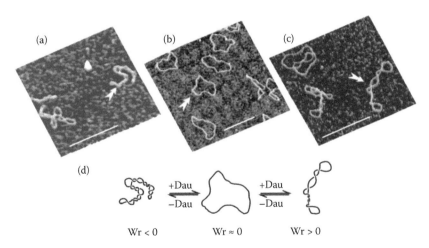

FIGURE 8.3 (a–c) AFM imaging of pUC19 plasmid DNA in air after deposition in the presence of 10 mM TRIS/TRIZMA (pH 7.2) containing 5 mM of $MgCl_2$ at ambient temperature. (a) In the absence of daunorubicin, the plasmids are negatively supercoiled (Wr < 0). (b) In the presence of 3.6 μM daunorubicin and of 5 mM $MgCl_2$, the DNA molecules are in either a fully relaxed state or just a few crossings (Wr ~ 0). (c) Adsorption of DNA in the presence of 37 μM daunorucin and of 5 mM $MgCl_2$ (Wr > 0). Scale bars = 500 nm. (d) Schematic representation of the plasmids denoted with arrows in (a–c), illustrating the transition from a negatively to a positively supercoiled structure. (Reproduced from Viglasky, V. et al. *Electrophoresis* 24, 1703–1711, 2003. With permission.)

mica, and the possibility of competition between the divalent cation and the drug must be borne in mind. The effect of an intercalating agent (ethidium bromide) on supercoiled (plasmid) DNA that has been immobilized on a chemically modified substratum has been studied in more detail by Pope et al. [18]. On raising the concentration of the intercalator, toroidal loops were first formed in the DNA, followed by plectonemic supercoiled structures. At still higher concentrations, a fully plectonemic structure with positive supercoiling was formed (Wr > 0). We have performed similar AFM experiments using negatively supercoiled DNA in conjunction with the intercalating drug daunorubicin and the minor-groove-binding agent netropsin. In accordance with the findings of Pope et al., the intercalator induced the formation of both toroidal and plectonemic structures within positively supercoiled DNA (Figure 8.2b). Netropsin, on the other hand, did not decrease negative supercoiling of the plasmid ccDNA: the molecules formed negatively oversupercoiled structures (Figure 8.2d) [36].

8.3.2 AFM OF ssDNA–DRUG INTERACTIONS

In most published studies dealing with the AFM of DNA–drug interactions, double-stranded DNA molecules have been utilized. However, ssDNA is an important intermediate in processes such as replication and recombination. Unlike dsDNA, ssDNA is a fairly flexible molecule, existing as a random coil or forming helical structures because of the stacking interaction between base pairs. A study of ssDNA–drug interactions is of interest in the context of molecular tagging, identification, and quantification, in the development of ssDNA-based probes, and in the detection of hybridized species. Actinomycin D—a small, widely studied, biologically active molecule—binds not only to dsDNA but also to ssDNA, with the formation of hairpin-like secondary [43–46]. The binding of actinomycin D to ssDNA has been implicated in various biological activities, including the inhibition of viral ligase, helicase, and (–) strand transfer by human immunodeficiency virus (HIV) reverse transcriptase [47–49]. To observe the effects of drugs on ssDNA by AFM, the molecule needs to be in a conformation that lacks secondary structures. The conformation of ssDNA is strongly influenced by the concentration of salts in the bathing medium. At very low salt concentrations, negatively charged phosphate groups on the DNA backbone are electrostatically repulsed, resulting in the formation of an open, random coil. When the salt concentration is raised to physiological levels, the phosphate groups are screened from electrostatic repulsion, thereby permitting the formation of secondary structures that interfere with the ssDNA–drug interaction and frustrate the interpretation of AFM images.

AFM of ssDNA molecules requires their adsorption to a very flat substratum such as mica or highly oriented pyrolytic graphite (HOPG). However, the protocols that are used for the immobilization of dsDNA cannot be implemented without modifications for that of ssDNA, which unavoidably contains secondary structures [34]. Woolley and Kelly [50] have devised a method for producing extended forms of ssDNA, which involves adsorbing the molecules to polylysine-coated mica. We have also devised a method for visualizing ssDNA in an extended form, which involves their adsorption to chemically modified graphite [34,35] or mica [36]. Hence, it is now possible to observe unstructured molecules of ssDNA at low (physiological) ionic strengths.

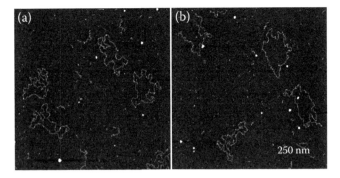

FIGURE 8.4 (a–b) AFM imaging of phiX174 ssDNA in 1 mM TRIS (pH 7.8). The ssDNA molecules are unstructured.

AFM of ssDNA–drug interactions has been performed by us using the intercalating drugs actinomycin D, daunorubicin, and chloroquine, and the minor-groove-binding agents netropsin and berenil. Figure 8.4 depicts AFM images of PhiX174 virion circular ssDNA, which was adsorbed to APTES-coated mica at a low salt concentration (1 mM TRIS). Under these conditions, ssDNA molecules exhibited no secondary

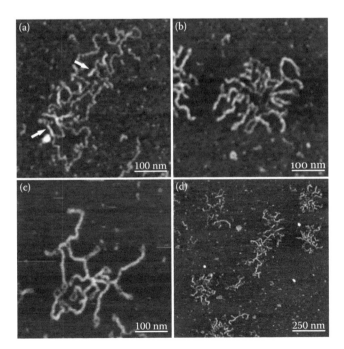

FIGURE 8.5 AFM imaging of phiX174 ssDNA in 1 mM TRIS (pH 7.8) after exposure to actinomycin D at (a) 2.5 μM, (b) 7 μM, and (c) 15 μM. In (d), DNA that had been exposed to 15 μM actinomycin D was heated to 60°C for 1 min prior to its deposition. (Reproduced from Adamcik, J. et al. *Nanotechnology* 19, 384016, 2008. With permission.)

structures [36]. Similarly, when using HOPG-modified graphite as a substratum, ssDNA molecules with an open conformation were observed. In both cases, it was possible to measure the contour length of these molecules, which accorded well with literature values. Owing to the lower resolution limit in DNA length that appertains in AFM, long, heterogeneous circular ssDNA rather than short oligomers with specific sequences have been studied. Figure 8.5 depicts molecules of ssDNA that were incubated with different concentrations of actinomycin D. As the drug concentration was raised, hairpin-like structures were extruded, extended, and stabilized by the agent. These hairpin-like structures have been shown to be stable up to temperatures of 60°C [36].

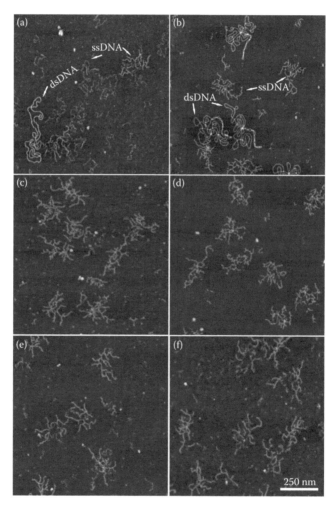

FIGURE 8.6   AFM imaging of phiX174 ssDNA (a–f) and dsDNA (a,b) in 1 mM TRIS (pH 7.8) after exposure to daunorubicin at (a) 0.5 μM, (b) 1.5 μM, and (c) 1 μM, and to (d) 2.5 μM chloroquine, (e) 6 μM netropsin, and (f) 1 μM berenil.

Similar results have been obtained using other intercalating therapeutic drugs, such as daunorubicin and chloroquine (Figure 8.6c and d), even though the chromophore portions of these agents differ from those in actinomycin D: daunorubicin harbors four planar rings, chloroquine only two. We can thus conclude that the planar aromatic structures play an important role in the formation of ssDNA–ligand complexes.

As mentioned in Section 8.3.1, intercalating and minor-groove-binding molecules have different effects on dsDNA. We have also observed the effects of the minor-groove-binding drugs netropsin and berenil on ssDNA (Figure 8.6e and f). As with the intercalators, hairpin-like structures were formed. Hence, intercalating and minor-groove-binding agents have similar effects on ssDNA.

All of these drugs are positively charged and interact electrostatically with the negatively charged ssDNA. The question now arises as to whether the presence of salts in the bathing medium and the interaction of ssDNA with the positively charged substratum influence the ssDNA–drug interaction. This issue was addressed by observing molecules of ssDNA and of supercoiled dsDNA that had been adsorbed to APTES-coated mica at high concentrations of NaCl and $MgCl_2$. At a concentration of 5 mM, $MgCl_2$ has a similar charge-screening effect to 100 mM NaCl. Figure 8.7

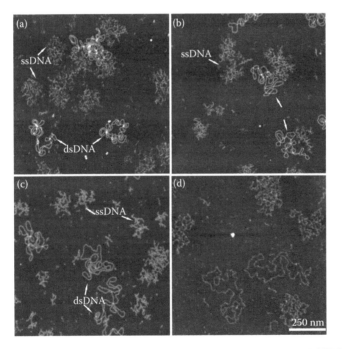

FIGURE 8.7 AFM imaging of phiX174 ssDNA (a–d) and dsDNA (a–c) in 1 mM TRIS (pH 7.8) after exposure to (a) 50 mM NaCl, (b) 100 mM NaCl, (c) 5 mM $MgCl_2$, and (d) simultaneous deposition of ssDNA from 50 mM NaCl solution and of ssDNA from 0 mM NaCl solution; it is clear that the deposition process is fast enough to prevent the rearrangement of the ssDNA conformation. The compact ssDNA molecules are from the 50 mM NaCl while the open conformations are from the 0 mM NaCl solution. (Reproduced from Adamcik, J. et al. *Nanotechnology* 19, 384016, 2008. With permission.)

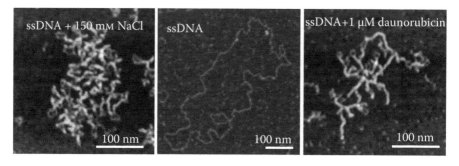

FIGURE 8.8 AFM imaging of ssDNA under different conditions. In the presence of 150 mM NaCl (left-hand image), the hairpin-like structures formed were shorter than those produced after exposure to 1 μM daunorubicin (right-hand image). At a low salt concentration (middle image), no secondary hairpin-like structures were formed. (Reproduced from Adamcik, J. et al. *Nanotechnology* 19, 384016, 2008. With permission.)

depicts AFM images of ssDNA and dsDNA that had been adsorbed to APTES-coated mica in the presence of 50 and 100 mM NaCl or 5 mM MgCl$_2$. These images produced in the presence of NaCl revealed the different conformations of ssDNA to have been elicited not by differences in surface charge but by the NaCl per se. We have also observed differences in the length of hairpin-like secondary structures that permit a distinction between the effects of salts and those of the complexing agents. The findings of these experiments are pictorially summarized in Figure 8.8.

8.4   CONCLUSIONS

In this chapter, we have discussed the imaging of DNA–ligand interactions by AFM. The effects of both intercalators and minor-groove-binding drugs on molecules of ssDNA and dsDNA were described. Our own experiments have permitted a distinction between the effects produced by salts in the bathing medium and those elicited by the drugs themselves. AFM is a powerful tool for studying DNA–drug interactions. In the future, it may prove to be useful in the screening of new therapeutic agents.

8.5   PROSPECTS AND OUTLOOK

Although the imaging of DNA–drug interactions by AFM is as yet in its infancy, the data thus far gleaned [16,20,36,40,51] bode well for a promising future.

ACKNOWLEDGMENTS

We would like to acknowledge F. Valle, G. Witz, and K. Rechendorff for their continuing support and help.

REFERENCES

1. Hurley, L. H. DNA and its associated processes as targets for cancer therapy. *Nat. Rev. Cancer.* **2**, 188–200, 2002.

2. Bischoff, G. and Hoffmann, S. DNA-binding of drugs used in medicinal therapies. *Curr. Med. Chem.* **9**, 312–348, 2002.

3. Graves, D. E. and Velea, L. M. Intercalative binding of small molecules to nucleic acids. *Curr. Org. Chem.* **4**, 915–929, 2000.

4. Reddy, B. S., Sharma, S. K. and Lown, J. W. Recent developments in sequence selective minor groove DNA effectors. *Curr. Med. Chem.* **8**, 475–508, 2001.

5. Chaires, J. B. Energetics of drug-DNA interactions. *Biopolymers* **44**, 201–215, 1997.

6. Haq, I. and Ladbury, J. Drug-DNA recognition: Energetics and implications for design. *J. Mol. Recognit.* **13**, 188–197, 2000.

7. Abu-Daya, A., Brown, P. M. and Fox, K. R. DNA sequence preferences of several AT-selective minor groove binding ligands. *Nucleic Acids Res.* **23**, 3385–3392, 1995.

8. Gelasco, A. and Lippard, S. J. NMR solution structure of a DNA dodecamer duplex containing a *cis*-diamminieplatinum(II) d(GpG) intrastrand cross-link, the major adduct of the anticancer drug cisplatin. *Biochemistry* **37**, 9230–9239, 1998.

9. Coste, F., Malinge, J. M., Serre, L., et al. Crystal structure of a double-stranded DNA containing a cisplatin interstrand cross-link at 1.63 A resolution: Hydration at the platinated site. *Nucleic Acids Res.* **27**, 1837–1846, 1999.

10. Walter, N. G., Huang, C. Y., Manzo, A. J. and Sobhy, M. A. Do-it-yourself guide: How to use the modern single-molecule toolkit. *Nat. Methods* **5**, 475–489, 2008.

11. Neuman, K. C. and Nagy, A. Single-molecule force spectroscopy: Optical tweezers, magnetic tweezers and atomic force microscopy. *Nat. Methods* **5**, 491–505, 2008.

12. Edwardson, J. M. and Henderson, R. M. Atomic force microscopy and drug discovery. *Drug Discov. Today* **9**, 64–71, 2004.

13. Santos, N. C. and Castanho, M. A. An overview of the biophysical applications of atomic force microscopy. *Biophys. Chem.* **107**, 133–149, 2004.

14. Lee, C. K., Wang, Y. M., Huang, L. S. and Lin, S. Atomic force microscopy: Determination of unbinding force, off rate and energy barrier for protein-ligand interaction. *Micron* **38**, 446–461, 2007.

15. Lipfert, J., Klijnhout, S. and Dekker, N. H. Torsional sensing of small molecule binding using magnetic tweezers. *Nucleic Acids Res.* **38**, 7122–7132, 2010.

16. Chaurasiya, K. R., Paramanathan, T., McCauley, M. J. and Williams, M. C. Biophysical characterization of DNA binding from single molecule force measurements. *Phys. Life Rev.* **7**, 299–341, 2010.

17. Coury, J. E., McFail-Isom, L., Williams, L. D. and Bottomley, L. A. A novel assay for drug-DNA binding mode, affinity, and exclusion number: Scanning force microscopy. *Proc. Natl. Acad. Sci. USA* **93**, 12283–12286, 1996.

18. Pope, L. H., Davies, M. C., Laughton, C. A., Roberts, C. J., Tendler, S. J. and Williams, P. M. Atomic force microscopy studies of intercalation-induced changes in plasmid DNA tertiary structure. *J. Microsc.* **199**, 68–78, 2000.

19. Krautbauer, R., Pope, L. H., Schrader, T. E., Allen, S. and Gaub, H. E. Discriminating small molecule DNA binding modes by single molecule force spectroscopy. *FEBS Lett.* **510**, 154–158, 2002.

20. Sischka, A., Toensing, K., Eckel, R., et al. Molecular mechanisms and kinetics between DNA and DNA binding ligands. *Biophys. J.* **88**, 404–411, 2005.

21. Viglasky, V., Valle, F., Adamcik, J., Joab, I., Podhradsky, D. and Dietler, G. Anthracycline-dependent heat-induced transition from positive to negative supercoiled DNA. *Electrophoresis* **24**, 1703–1711, 2003.

22. Binnig, G., Quate, C. F. and Gerber, C. Atomic force microscope. *Phys. Rev. Lett.* **56**, 930–933, 1986.

23. Binning, G., Rohrer, H., Gerber, C. and Weibel, E. Surface studies by scanning tunneling microscopy. *Phys. Rev. Lett.* **49**, 57–61, 1982.

24. Marti, O., Drake, B. and Hansma, P. K. Atomic force microscopy of liquid-covered surfaces—Atomic resolution images. *Appl. Phys. Lett.* **51**, 484–486, 1987.
25. Putman, C. A., van der Werf, K. O., de Grooth B. G., van Hulst, N. F. and Greve, J. Tapping mode atomic force microscopy in liquid. *Appl. Phys. Lett.* **64**, 2454–2456, 1994.
26. Roduit, C., Sekatski, S., Dietler, G., Catsicas, S., Lafont, F. and Kasas, S. Stiffness tomography by atomic force microscopy. *Biophys. J.* **97**, 674–677, 2009.
27. Hinterdorfer, P. and Dufrêne, Y. F. Detection and localization of single molecular recognition events using atomic force microscopy. *Nat. Methods* **3**, 347–355, 2006.
28. Cappella, B. and Dietler, G. Force-distance curves by atomic force microscopy. *Surf. Sci. Rep.* **34**, 1–104, 1999.
29. Lyubchenko, Y., Gall, A. A., Shlyakhtenko, L. S., et al. Atomic force microscopy imaging of double stranded DNA and RNA. *J. Biomol. Struct. Dyn.* **9**, 589–606, 1992.
30. Lyubchenko, Y. and Shlyakhtenko, L. S. Visualization of supercoiled DNA with atomic force microscopy in situ. *Proc. Natl. Acad. Sci. USA* **94**, 496–501, 1997.
31. Hansma, H. G. and Laney, D. E. DNA binding to mica correlates with cationic radius: Assay by atomic force microscopy. *Biophys. J.* **70**, 1933–1939, 1996.
32. Hansma, H. G., Kasuya, K. and Oroudjev, E. Atomic force microscopy imaging and pulling of nucleic acids. *Curr. Opin. Struct. Biol.* **14**, 380–385, 2004.
33. Witz, G., Rechendorff, K., Adamcik, J. and Dietler, G. Conformation of circular DNA in two dimensions. *Phys. Rev. Lett.* **101**, 148103, 2008.
34. Adamcik, J., Klinov, D. V., Witz, G., Sekatskii, S. K. and Dietler, G. Observation of single-stranded DNA on mica and highly oriented pyrolytic graphite by atomic force microscopy. *FEBS Lett.* **580**, 5671–5675, 2006.
35. Rechendorff, K., Witz, G., Adamcik, J. and Dietler, G. Persistence length and scaling properties of single-stranded DNA adsorbed on modified graphite. *J. Chem. Phys.* **131**, 095103, 2009.
36. Adamcik, J., Valle, F., Witz, G., Rechendorff, K. and Dietler, G. The promotion of secondary structures in single-stranded DNA by drugs that bind to duplex DNA: An atomic force microscopy study. *Nanotechnology* **19**, 384016, 2008.
37. Jeon, J. H., Adamcik, J., Dietler, G. and Metzler, R. Supercoiling induces denaturation bubbles in circular DNA. *Phys. Rev. Lett.* **105**, 208101, 2010.
38. Adamcik, J., Jung, J. M., Flakowski, J., De Los Rios, P., Dietler, G. and Mezzenga, R. Understanding amyloid aggregation by statistical analysis of atomic force microscopy images. *Nat. Nanotechnol.* **5**, 423–428, 2010.
39. Takahara, P. M., Rosenzweig, A. C., Frederick, C. A. and Lippard, S. J. Crystal structure of double-stranded DNA containing the major adduct of the anticancer drug cisplatin. *Nature* **377**, 649–652, 1995.
40. Hou X. M., Zhang, X. H., Wei, K. J., et al. Cisplatin induces loop structures and condensation of single DNA molecules. *Nucleic Acids Res.* **37**, 1400–1410, 2009.
41. Bates, A. D and Maxwell, A. *DNA Topology.* Oxford University Press, New York, 2005.
42. Eckel, R., Ros, R., Ros, A., Wilking, S. D., Sewald, N. and Anselmetti, D. Identification of binding mechanisms in single molecule-DNA complexes. *Biophys. J.* **85**, 1968–1973, 2003.
43. Wadkins, R. M., Jares-Erijman, E. A., Klement, R., Rudiger, A. and Jovin, T. M. Actinomycin D binding to single-stranded DNA: Sequence specificity and hemi-intercalation model from fluorescence and 1H NMR spectroscopy. *J. Mol. Biol.* **262**, 53–68, 1996.
44. Chou, S. H., Chin, K. H. and Chen, F. M. Looped-out and perpendicular: Deformation of Watson-Crick base pair associated with actinomycin D binding. *Proc. Natl. Acad. Sci. USA* **99**, 6625–6630, 2002.
45. Chen, F. M., Sha, F., Chin, K. H. and Chou, S. H. The nature of actinomycin D binding to d(AACCAXYG) sequence motifs. *Nucleic Acids Res.* **32**, 271–277, 2004.

46. Chen, F. M., Sha, F., Chin, K. H. and Chou, S. H. Unique actinomycin D binding to self-complementary d(CXYGGCCY'X'G) sequences: Duplex disruption and binding to a nominally base-paired hairpin. *Nucleic Acids Res.* **31**, 4238–4246, 2003.
47. Shuman, S. Vaccinia virus DNA ligase: Specificity, fidelity and inhibition. *Biochemistry* **34**, 16138–16147, 1995.
48. Jeeninga, R. E., Huthoff, H. T., Gultyaev, A. P. and Berkhout, B. The mechanism of actinomycin D-mediated inhibition of HIV-1 reverse transcription. *Nucleic Acids Res.* **26**, 5472–5479, 1998.
49. Gabbara, S., Davis, W. R., Hupe, L. and Peliska, J. A. Inhibitors of DNA strand transfer reactions catalyzed by HIV-1 reverse transcriptase. *Biochemistry* **38**, 13070–13076, 1999.
50. Woolley, A. T. and Kelly, R. T. Deposition and characterization of extended single-stranded DNA molecules on surfaces. *Nano Lett.* **1**, 345–348, 2001.
51. Cassina, V., Seruggia, D., Beretta, G. L., et al. Atomic force microscopy study of DNA conformation in the presence of drugs. *Eur. Biophys. J.* **40**, 59–68, 2011.

9 Characterizing RNA–Ligand Interactions Using Two-Dimensional Combinatorial Screening

Sai Pradeep Velagapudi and Matthew D. Disney

CONTENTS

This chapter describes a microarray-based method that is used to identify RNA motifs that bind to small molecules. This method, termed two-dimensional combinatorial screening (2DCS), screens chemical and RNA secondary structure spaces simultaneously. A key feature of the approach is the statistical analysis of the selected RNA sequences. By analyzing all statistically significant trends displayed by a singular RNA sequence/structure, we can qualitatively predict both the selectivity and the affinity of RNA motif–ligand interactions (structure–activity relationships through sequencing (StARTS)). The information from these studies can then be utilized to rationally design the ligands that target RNA. Two examples are provided in which modularly assembled ligands that target the RNAs that cause myotonic muscular dystrophy types 1 and 2 are described. Collectively, these studies show that the affinity and the specificity of modularly assembled small molecules targeting RNA can be controlled by both the nature of an RNA-binding molecule and the spacing between ligand modules on an assembly scaffold. This approach has the potential to enable the rational design of small molecules targeting genomic RNAs.

9.1   INTRODUCTION

Ribonucleic acid (RNA) has important and essential biological functions beyond the classical roles in protein biosynthesis including catalysis (Stark et al. 1978; Zaug and Cech 1986), spliceosomal splicing (Kiss-Laszlo et al. 1996), editing and modification (Blum et al. 1990; Kiss-Laszlo et al. 1996), and cellular localization (Bishop et al. 1970; Shan and Walter 2005). In addition, the $5'$ and $3'$ untranslated regions (UTRs) of numerous mRNAs regulate gene expression through interactions with proteins (Kurokawa et al. 2009; Mazumder et al. 2001; Sevo et al. 2004; Takagi et al. 2005) and metabolites (Doudna 2000). The vast array of RNA's biological functions can be attributed to its ability to adopt various three-dimensional (3D) structures.

There are numerous examples of RNA-mediated diseases that are caused by misfolding, mutations, additions, and deletions (Altman and Guerrier-Takada 1986; Duncan and Weeks 2008; Emerick and Woodson 1993; Hogan et al. 1984a,b; Hogan and Noller 1978; Nikolcheva and Woodson 1999; Pan and Woodson 1998; Pan and Sosnick 1997; Woese et al. 1980; Woodson and Cech 1991; Zarrinkar et al. 1996). Mutations can lead to changes in splicing patterns and cause diseases such as cystic fibrosis and β-thalassemia (Faustino and Cooper 2003). RNA-gain-of-function due to microsatellite expansions leads to disorders such as myotonic dystrophy types 1 (DM1) (Brook et al. 1992; Fu et al. 1992) and 2 (DM2) (Liquori et al. 2001), fragile X-associated tremor ataxia (FXTAS) (Caskey et al. 1992), and Huntington's disease-like 2 (Holmes et al. 2001). Misexpression of RNAs can cause disease as well. For example, over- and underexpression of microRNAs are implicated in cancer (Calin and Croce 2006; Chen 2005; Esquela-Kerscher and Slack 2006; Hammond 2006).

Because RNA performs essential cellular functions and mediates disease, it is an important target for chemical genetics probes or small-molecule therapeutics. The most well-studied RNA drug target is the bacterial ribosome. Antibacterials bind different regions of the ribosome and inhibit protein synthesis, including the mRNA tunnel (Brodersen et al. 2000; Pioletti et al. 2001), the nascent protein exit tunnel (Berisio et al. 2003a,b; Hansen et al. 2003; Schlunzen et al. 2003, 2001), and the rRNA aminoacyl-site (A-site) (Carter et al. 2000; Ogle et al. 2001; Schlunzen et al. 2001; Vicens and Westhof 2002; Walter et al. 1999; Yoshizawa et al. 1999). Aminoglycoside antibiotics bind the bacterial A-site and interfere with the decoding process by causing conformational changes within the RNA (Kaul et al. 2006). Importantly, high-resolution crystal structures of aminoglycosides in complex with an oligonucleotide mimic of the A-site and the entire ribosome have shown that aminoglycosides bind the A-site similarly in both contexts. Therefore, understanding how ligands interact with secondary structural elements, such as the 4×3 nucleotide internal loop present in the A-site, could be useful for designing compounds that target the same secondary structural elements in a large, biologically active RNA.

Many methods are available to explore the chemical space (or the small molecules) that binds RNA or to explore the RNA space that binds small molecules. Systematic evolution of ligands by exponential enrichment (SELEX) (Tuerk and Gold 1990) finds an RNA that binds a ligand with high affinity using multiple rounds of selection. SELEX has been useful in the development of biosensors (Tombelli et al. 2005), analytical techniques (Tombelli et al. 2005), and therapeutics (Ellington and Conrad 1995). It is rare, however, to find the output of SELEX, an RNA aptamer, in a biologically relevant RNA due to the large size of the randomized region. RNA aptamer–ligand pairs, however, have been used to enable the rational design of ligands targeting RNA. For example, Ellington used the output of a SELEX experiment to identify RNAs from pathogenic organisms that are potential targets for small molecules (Lato and Ellington 1996), and the Beal group has completed extensive studies to target rRNAs and pre- and pri-microRNAs with threading intercalator compounds (Carlson et al. 2003).

Structure–activity relationships (SAR) by mass spectrometry (Griffey et al. 1999; Swayze et al. 2002) and by nuclear magnetic resonance (NMR) (Johnson et al. 2003) and high-throughput screening of small molecules are complementary methods to SELEX. That is, they are used to screen libraries of chemicals to find a ligand that binds an RNA of interest. In high-throughput screening, a library of chemical ligands is probed for binding to a validated RNA drug target such as the hepatitis C internal ribosomal entry site (Parsons et al. 2009; Seth et al. 2005) or the A-site (Swayze et al. 2002). Although high-throughput screening has been successful in various instances (Jefferson et al. 2002; Parsons et al. 2009; Seth et al. 2005; Swayze et al. 2002), the hit rates for identifying ligands that bind to RNA is lower than the hit rates for proteins. This is likely due to our lack of understanding of the chemical space that is privileged to bind RNA and the RNA structures that are privileged to bind small molecules.

Efforts have been made by our group to combine selection methods and high-throughput screening into one single method, potentially combining the advantages of both. The method, 2DCS (Aminova et al. 2008; Childs-Disney et al. 2007; Disney and Childs-Disney 2007; Disney et al. 2008), screens both RNA and chemical spaces simultaneously and allows us to understand features in both spaces that are important for molecular recognition. The overall goal is to use the output of 2DCS to improve our understanding of RNA–ligand interactions such that rational design of small molecules targeting RNA is as simple as the modular assembly of polyamides used to target DNA (Dervan and Burli 1999).

The 2DCS method uses a microarray platform to display a small-molecule library that is hybridized with a radioactively labeled RNA library displaying a discrete secondary structural element (Figure 9.1). Small secondary structural elements such as 3×3 nucleotide internal loops or 6-nucleotide hairpin loops are chosen for these studies because library members are likely to be found in larger, biologically relevant RNAs. By utilizing this approach, all possible interactions between a secondary structural element and each ligand are probed. For example, a 6-nucleotide hairpin

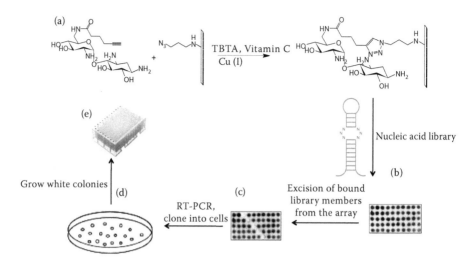

FIGURE 9.1   **(See color insert.)** Schematic of the microarray-based selection process. (a) Immobilization of small molecules on an activated agarose microarray via a Huisgen dipolar cycloaddition reaction. (b) Image of the slide after hybridization with radioactively labeled RNA. (c) Image of the same slide after mechanical removal of bound RNA. (d) Schematic of an agar plate after transformation with a vector containing selected members of the library. (e) Colonies that are white from blue/white screening are grown in medium and sequenced to deconvolute the selected members of the library.

loop library has 4096 unique sequences that form 4- and 6-nucleotide hairpins. If that RNA library is hybridized with a microarray of 50 compounds, then 204,800 interactions are probed simultaneously. This is in contrast to high-throughput screens where a single or only a few biologically important RNAs are probed for binding.

The sequences of the RNAs that bind to each ligand are then determined using standard-sequencing methods. Trends within the selected RNA sequences that are important for binding are determined via statistical analysis. In order to facilitate the fast and accurate analysis of 2DCS output, a computer program, RNA-Privileged Space Predictor (RNA-PSP), was developed to automate this process. We recently reported that statistical analysis can qualitatively predict the selectivity (Paul et al. 2009) and affinity (Velagapudi et al. 2010) of RNA–ligand complexes.

The output of 2DCS, RNA–ligand partners, is deposited into a database. The database can then be compared to the secondary structure motifs present in genomic RNAs. Once an RNA is identified that contains two or more secondary structural motifs that overlap with our database, the two motifs can be targeted simultaneously with a single multivalent ligand. This is accomplished by linking, or modularly assembling, the two ligands together (Figure 9.2). Modular assembly has been shown to increase both affinity and specificity (Gestwicki et al. 2002; Goringer and Wagner 1986; Kitov et al. 2000; Mammen et al. 1998; Shuker et al. 1996), and this approach has been used to rationally design cell-permeable ligands that target the RNAs that cause DM1 (Disney et al. 2010; Pushechnikov et al. 2009), DM2 (Lee et al. 2009b), and spinocerebellar ataxia type 3 (SCA3)

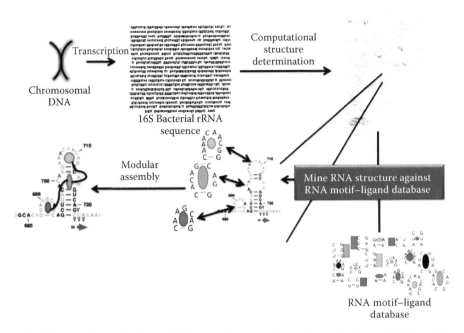

FIGURE 9.2 **(See color insert.)** The overall approach to using a database of RNA motif–ligand partners to rationally design small molecules targeting RNA. First, a target RNA sequence is subjected to structure determination (free energy minimization, phylogenic comparison, or selective 2′-hydroxyl acylation analyzed by primer extension (SHAPE)). Next, the secondary structures within the target RNA and their relative orientation are extracted and mined against the RNA motif–ligand database. Once overlaps between the database and the secondary structures are found, modular assembly routes are used to target multiple motifs in the RNA simultaneously.

(Pushechnikov et al. 2009). Herein, we describe the 2DCS method and present two case studies of its use.

9.2 BASICS OF THE TECHNIQUE

9.2.1 MICROARRAY CONSTRUCTION

In order to successfully merge nucleic acid selections and high-throughput screening using a microarray platform (i.e., 2DCS), a robust array surface that allows for ligand screening and for harvesting of bound RNAs is required. An agarose-coated microarray surface is ideal for this purpose (Afanassiev et al. 2000; Disney and Childs-Disney 2007; Dufva et al. 2004). Agarose provides a 3D surface that affords high ligand loading, resists nonspecific binding, and can be easily functionalized to display chemical handles for site-specific small-molecule immobilization (Afanassiev et al. 2000; Childs-Disney et al. 2007).

A variety of factors should be taken into consideration when choosing the small molecules that are screened for binding RNA secondary structures. In order to extract SAR data from the 2DCS platform, it is imperative that the small molecules are immobilized in a site-specific manner. Thus, each small molecule should have the same singular functional group that is used for immobilization.

Since agarose is a versatile surface, many different immobilization chemistries can be used to construct the chemical microarray. A library of amine-containing small molecules can be immobilized by first oxidizing the agarose surface to afford aldehydes (Afanassiev et al. 2000). Azide-containing compounds can also be immobilized on the slide surface by first oxidizing the agarose and then reacting the surface with an amine linker containing an alkyne. The alkyne and azide then form a triazole ring via a Cu-catalyzed Huisgen dipolar cycloaddition reaction (Kolb et al. 2001). Alkyne-containing compounds can be immobilized analogously on azide-displaying surfaces.

The diversity of the library should also be considered. Often, a compromise must be made between how focused the library is (target-oriented synthesis, TOS) and how diverse it is (diversity-oriented synthesis, DOS) (Burke and Schreiber 2004; Fergus et al. 2005; Schreiber 2000). If a focused library is used, perhaps of small molecules containing privileged scaffolds or moieties thereof, then the best ligand within this small window may be identified. Novel binders, however, could be overlooked. In contrast, if a diverse library is used, then the best compound may not be identified.

Each ligand probed by 2DCS is spotted on the array surface at different concentrations, affording a dose response upon hybridization with a radioactively labeled RNA library (Aminova et al. 2008; Childs-Disney et al. 2007; Disney et al. 2008; Paul et al. 2009; Tran and Disney 2010). RNAs harvested from the lowest ligand loading that yields signal over background are the highest-affinity interactions (Childs-Disney et al. 2007). Spotting solution without ligand is also delivered to the slide surface to serve as a negative control.

9.2.2 Design of the RNA Library

The RNA libraries used in the 2DCS selection experiments are designed to display discrete secondary structural element such as symmetric (Childs-Disney et al. 2007; Disney et al. 2008) or asymmetric (Tran and Disney 2010) internal loops or hairpins (Aminova et al. 2008; Paul et al. 2009). The randomized region in the pattern of interest is embedded in a cassette (Bevilacqua and Bevilacqua 1998; Shu and Bevilacqua 1999). The cassette in which internal loop motifs are embedded contains an ultrastable GNRA (where "N" is any nucleotide and "R" is a purine) tetraloop to ensure proper folding (9, Figure 9.3). The cassettes for internal loops and hairpins contain primer-binding sites for reverse transcription-polymerase chain reaction (RT-PCR) amplification and encode for restriction endonuclease sites. It is important to note that the randomized regions and thus the secondary structural elements are intentionally small, <10 nucleotides, such that it is likely they will also be present in biological targets. The RNA library is synthesized by run-off transcription using a randomized DNA template and radioactively labeled (either internally or on the 5′ end). If a 2DCS selection is successful, the selected RNAs will bind with higher affinity than the cassette in which they were embedded.

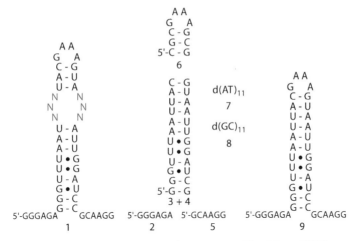

FIGURE 9.3 The secondary structures and sequences utilized in a 2DCS experiment to identify the members of an RNA 3×3 nucleotide internal loop library that bind a small molecule. The oligonucleotides used are (1) the 3×3 nucleotide internal loop library and (2–8) competitor oligonucleotides. Oligonucleotide 9 is the cassette used to display the 3×3 nucleotide internal loop library and is not used in the selection experiments.

9.2.3 Hybridization of the RNA Library with the Chemical Microarray

Since we are only interested in the interactions of the ligands with the randomized region, hybridization of the microarray is completed under conditions of high oligonucleotide stringency using mimics of regions constant to all library members (Figure 9.3). Oligonucleotides 2 and 5 are used to compete off interactions with single-stranded regions; oligos $3 + 4$ mimic the base-paired region; oligo 6 competes off interactions with hairpin loop. DNA oligonucleotides 7 and 8 ensure that interactions are RNA specific. A 1000-fold molar excess of competitor oligonucleotides over the moles of ligand delivered (not necessarily immobilized) to the microarray surface is used. Detailed procedures have been previously published by Disney and coworkers (Aminova et al. 2008; Childs-Disney et al. 2007; Paul et al. 2009; Tran and Disney 2010).

9.2.4 Identification of the RNAs that Bind Each Ligand

After washing and imaging the microarrays, bound RNAs are harvested from the surface using manual excision and RT-PCR amplified. (A background or negative control is also harvested from a region where only spotting solution was delivered.) RT-PCR reactions are analyzed on a denaturing 15% polyacrylamide gel stained with SYBR gold (Invitrogen) or ethidium bromide. Only the experiments where negative controls yield no product are carried over to subsequent cloning steps.

After RT-PCR, the cDNA obtained for each mixture is cloned and sequenced to identify the individual members of the selected RNA mixture using one of two methods. In the first standard cloning procedure, the RT-PCR product can be cloned directly into pGEM-T (Promega) or TOPO-TA (Invitrogen) vectors. Alternatively, the RT-PCR product can be digested with restriction enzymes and ligated into a complementary vector.

In the second procedure, the number of cDNAs ligated into a single vector is increased using a ligation-based concatenation (Childs-Disney and Disney 2008; Velagapudi et al. 2010). Briefly, the selected RNAs are RT-PCR amplified using biotinylated primers. Both the 5′ and 3′ ends of the resulting RT-PCR product contain a BamHI restriction endonuclease recognition site. After digestion, the released, biotinylated sticky ends are captured with streptavidin resin. The doubly digested products are then ligated together to form oligomers. After adding adenosine residues using Taq DNA polymerase, the concatenated products are ligated into a pGEM-T or TOPO-TA vector. On average, each plasmid contains 2.5 sequences, increasing the information content in each sequencing run (Figure 9.4) (Childs-Disney and Disney 2008; Velagapudi et al. 2010). After cDNAs are sequenced, the secondary structures of all selected RNAs are predicted by free energy minimization using the RNA structure program (Eddy 2004; Mathews et al. 2004, 1999; Turner et al. 1988).

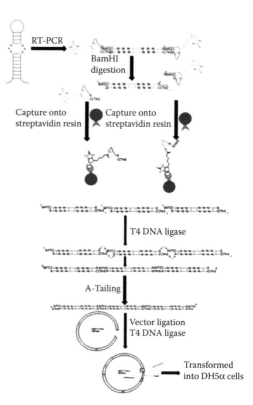

FIGURE 9.4 **(See color insert.)** The schematic of a concatenation approach that is used to ligate multiple copies of the selected members of the RNA motif library together. Key features of the method are the installation of two BamHI restriction sites into the RT-PCR product and the use of biotinylated primers. After RT-PCR amplification, the cDNAs are digested with BamHI and purified using streptavidin resin. The products are ligated together using T4 DNA ligase.

9.2.5 IDENTIFICATION OF TRENDS IN THE SELECTED RNAS USING STATISTICAL ANALYSIS

Statistical analysis of selected sequences allows for the identification of statistically significant trends within the variable region of the selected RNAs. Some of the trends observed in 2DCS experiments include the presence of an adenine across from a cytosine (Childs-Disney et al. 2007), 5′GC steps (Aminova et al. 2008), and guanosine across from an adenine (Disney et al. 2008). If indeed a ligand prefers to bind a specific domain of RNA space, then the statistical significance of this preference can be quantified by computing Z-scores. A Z-score compares the percentage of the selected RNAs that display a trend of interest to the percentage of the RNAs in the entire library that contain the trend. It is computed using the following equations:

$$\phi = \frac{n_1 p_1 + n_2 p_2}{n_2 + p_2} \tag{9.1}$$

$$Z_{obs} = \frac{(p_1 - p_2)}{\sqrt{\phi(1 - \phi)\left(\left(\dfrac{1}{n_1}\right) + \left(\dfrac{1}{n_2}\right)\right)}} \tag{9.2}$$

where ϕ is the pooled sample proportion, n_1 is the size of the population of the selected mixture (population 1), n_2 is the size of the population of the entire RNA library subjected to 2DCS (population 2), p_1 is the observed proportion of population 1 (selected mixture) displaying the trend, and p_2 is the observed proportion for population 2 (entire library) displaying the trend.

Two-tailed p-values are calculated manually from the Z-scores from standard statistical tables. The two-tailed p-value describes the probability that the trend observed is a true preference and not a random event. For example, a two-tailed p-value equal to 0.01 would indicate that there is a 1% chance that the trend identified occurred randomly or 99% confidence. Only trends that have ≥95% confidence are considered statistically significant.

Since searching for statistically significant trends manually is time consuming, a computer program, RNA-PSP, was developed to automate statistical analysis of the selections (Paul et al. 2009). The program allows for input of sequence files from any selection. The sequence files are then processed to extract the nucleotides from the randomized region. The output of the program is a list of statistically significant trends and the corresponding Z-scores. Interestingly, both positive and negative Z-scores are reported corresponding to features that predispose the RNA for binding to the ligand of interest and those that contain features that deleteriously affect affinity, respectively. RNA-PSP is freely available on the web at http://www.scripps.edu/disney/.

The statistical analysis performed by RNA-PSP on a 2DCS selection can be used to predict qualitatively the specificity of RNA–ligand interactions identified by 2DCS (Paul et al. 2009). When a selection experiment for multiple ligands is completed, cross-analysis identifies overlapping trends. If overlap is present, it

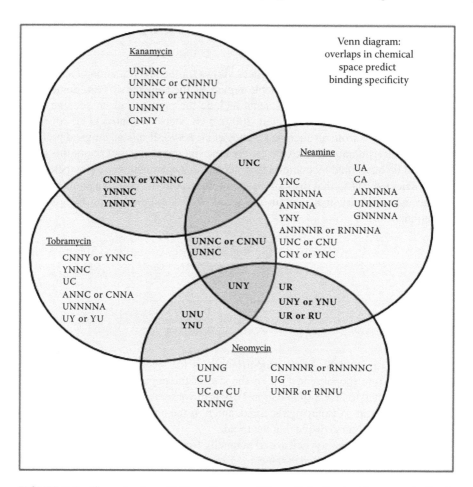

FIGURE 9.5 **(See color insert.)** Venn diagram of the statistically significant trends identified in the RNA sequence space from a 6-nucleotide hairpin library selected for four aminoglycoside derivatives. Overlapping trends are shown in bold. The most statistically significant trend for the kanamycin A derivative is 5′UNNNC3′; for the tobramycin derivative, 5′UNNC3′; for the neamine derivative, 5′UNC3′; and for the neomcyin B derivative, 5′UNNG3′. (Reproduced from Paul, D. J., Seedhouse, S. J. and Disney, M. D. *Nucleic Acids Res.* 37, 5894–5907, 2009. With permission.)

qualitatively indicates that the RNAs would not be specific to one ligand. In contrast, if no overlap is observed, the RNAs displaying the trend bind selectively. Such analysis is best represented using a Venn diagram, an example of which is shown in Figure 9.5.

Because a subset of RNAs has been selected with common features that predispose them for binding a ligand, the RNAs that best represent the overall statistical analysis of the selected pool should have the highest affinity. That is, the RNAs that contain the largest number of statistically significant trends should bind the ligand most tightly. The relative "fit" of a particular RNA within the identified privileged

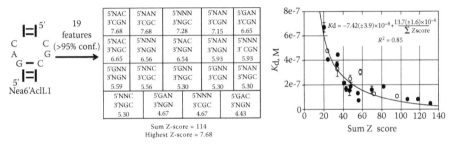

FIGURE 9.6 Illustration of the StARTS method to develop scoring functions for RNA motif–ligand partners. (Right) The number of sequence trends in an internal loop selected to bind 6′-N-5-hexynoate neamine, and the associated Z-scores for each motif. (Left) A plot of the Sum Z-scores for a series of RNA internal loops derived from a 3×3 nucleotide internal loop library that are selected to bind 6′-N-5-hexynoate neamine. Internal loops selected to bind the ligand are shown as filled circles and loops that are members of the original library but were not identified in the sequencing data are shown as open circles. All data can be fit to a rather simple inverse first-order equation to accurately predict the affinity of RNA motif–ligand interactions. (Reproduced from Velagapudi, S. P., Seedhouse, S. J., and Disney, M. D. *Angew. Chem. Int. Ed. Engl.* 49, 3816–3818, 2010. With permission.)

RNA space can be quantified by calculating the sum of the individual Z-scores (Sum Z-scores) of every statistically significant trend found within that RNA. This is illustrated in Figure 9.6. If the privileged space for a ligand is a combination of the RNA space trends found in the selected pool, then the sum of the Z-scores (Sum Z-scores) of these trends should scale with affinity of the ligand–RNA complex. Indeed this is the case (Velagapudi et al. 2010), and RNA-PSP has been modified to compute Sum Z-scores. This analysis, which we term structure–activity relationships through sequencing (StARTS), provides a way to assess not only the selected RNAs for binding a ligand but also all of the RNAs that are members of the library (Figure 9.6). See Case Study I for further details.

9.2.6 AFFINITY OF RNA–LIGAND INTERACTIONS

The affinity of RNA–ligand interactions can be determined using a variety of methods. Typically, fluorescently labeled small molecules are used to monitor the decrease in fluorescence intensity as a function of oligonucleotide concentration. For these experiments, the dye that is used to report on RNA binding is conjugated to the small-molecule ligand using the same immobilization chemistry that was used in the 2DCS selection. Competition-binding experiments are also completed with an unlabeled ligand to determine how the dye affects affinity. Typically, the dyes contribute little (positively or negatively; ~2-fold), if any, to the affinity of these complexes (Aminova et al. 2008; Disney et al. 2008; Paul et al. 2009).

In many cases, the RNAs identified are selective for the ligand for which they were selected over the other arrayed ligands, even when the ligands are structurally similar (Aminova et al. 2008; Disney et al. 2008; Tran and Disney 2010). The observed selectivity is likely due to nature of the experiment which creates a competition between all arrayed ligands on the surface. During hybridization, the ligands are present in excess while the RNA is limiting (Disney et al. 2008). Moreover, the selected RNA motifs are portable. That is, the RNA in which they are embedded does not significantly affect the affinity of the RNA–ligand interaction (Disney et al. 2008).

9.2.7   A DATABASE OF RNA MOTIF–LIGAND INTERACTIONS

The RNA motif–ligand partners that are identified by 2DCS are deposited into a database that currently contains ~1000 entries. The database is constantly updated as new RNA motif–ligand partners are discovered via 2DCS or reported in the literature. At present, the database is mined manually for similarities with genomic RNAs in order to identify potential targets. However, we are developing computational methods to automate this process. If two or more motifs are identified, then the corresponding ligands can be linked together using modular assembly. Modular assembly has been previously reported as a strategy to increase the affinity of ligands for binding to a variety of targets (Gestwicki et al. 2002; Goringer and Wagner 1986; Kitov et al. 2000; Mammen et al. 1998; Shuker et al. 1996). In our approach, we display ligand modules on a peptoid scaffold, although other scaffolds are also amendable to this strategy (Dervan 2001; Kitov et al. 2000).

A peptoid scaffold was chosen because the synthesis is modular, straightforward, and high yielding. Moreover, peptoids are stable to cellular proteases and some have been proven to be bioactive (Bremner et al. 2010; Gocke et al. 2009; Goodson et al. 1999; Lee et al. 2010; Lynn et al. 2010; Mas-Moruno et al. 2007; Masip et al. 2005; Tran et al. 1998; Zuckermann and Kodadek 2009). In all of our investigations, ligand modules are anchored on the peptoid backbone using the same immobilization chemistry that was used in the 2DCS studies that identified the RNA motif–ligand interaction. For example, alkyne-containing small molecules are conjugated to peptoids that display 3-azidopropylamine. Analogously, azide-containing small molecules are conjugated to peptoids that display propargylamine. Therefore, the number of 3-azidopropylamine or propargylamine submonomers incorporated in the peptoid backbone control valency. The distance between ligand modules is controlled by the number of spacing submonomers, in our case propylamine, coupled between ligand-displaying modules. Interestingly, the identity of the spacing submonomer can affect binding affinity, cellular uptake, and cellular localization (Goun et al. 2006a,b; Yu et al. 2005, 2007).

The nomenclature for our modularly assembled compounds has the general format $mL - n$ where m is the valency, L indicates the ligand module, and n is the number of spacing submonomers coupled between ligand-displaying modules (Figure 9.7). For example, 3K-1 describes a trimeric peptoid (valency = 3) that displays kanamycin ligand modules each separated by one propylamine-spacing submonomer. We used modular assembly to target the RNAs that causes myotonic (DM1 and DM2). Both RNAs contain regularly repeating internal loops that are similar to a motif identified from 2DCS. Thus, it was an ideal model system to test our strategy (Case Study II).

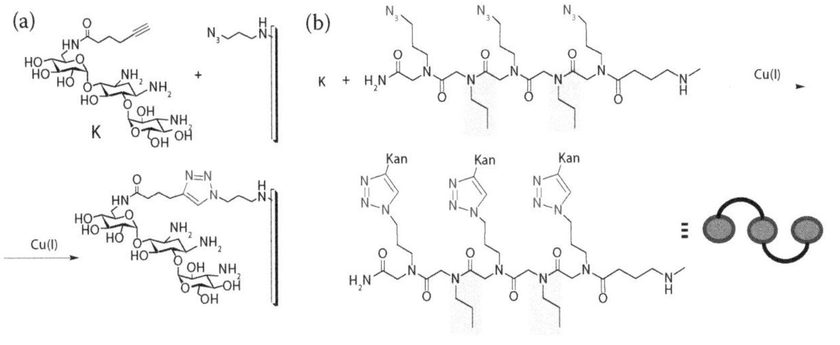

3K-1

Nomenclature: 3K-1 is a peptoid that displays three kanamycin ligand modules (3K) each separated by one propylamine-spacing module.

FIGURE 9.7 The 2DCS immobilization chemistry and anchoring of ligand modules onto a peptoid scaffold. (a) Immobilization of 6′-N-5-hexynoate kanamycin A onto azide-functionalized microarray surfaces via a Huisgen dipolar cycloaddition reaction. (b) Conjugation of 6′-N-5-hexynoate kanamycin A onto a peptoid. The number of RNA-binding modules conjugated is controlled by the number of azido submonomers conjugated while the spacing between modules is controlled by the number of propylamine submonomers installed between each azido submonomer.

9.3 CASE STUDIES

9.3.1 IDENTIFYING THE INTERNAL LOOPS THAT BIND 6′-N-5-HEXYNOATE KANAMYCIN A AND THE USE OF StARTS TO DEVELOP SCORING FUNCTIONS FOR RNA–LIGAND INTERACTIONS

In our first 2DCS selection, a 3 × 3 nucleotide internal loop library (1, Figure 9.3) was probed for binding to 6′-N-5-hexynoate kanamycin (K, Figure 9.7) (Childs-Disney et al. 2007). The 6′ position of kanamycin A was acylated to mimic the product of modification by the aminoglycoside 6′-N-acetyltransferase AAC(6′) (Magnet and Blanchard 2005) and to provide a chemical handle for immobilization. Although it is known that kanamycin A binds the rRNA A-site, the 6′ acylated product of AAC(6′) has about 1000-fold reduced affinity (Magnet and Blanchard 2005). It was therefore likely that the internal loops that prefer 6′-N-5-hexynoate kanamycin A would have a different sequence space than the bacterial A-site.

The RNA library chosen for this selection experiment has six randomized positions displayed in a 3 × 3 nucleotide internal loop pattern (Figure 9.8) (Bevilacqua and Bevilacqua 1998). The library consists of 4096 members: 1600 unique 3 × 3 nucleotide internal loops (including RNAs with potential two 1-nucleotide bulges), 1200 unique 2 × 2 nucleotide internal loops, 1080 unique 1 × 1 nucleotide internal loops, and 216 unique base-paired regions (Figure 9.8) (Childs-Disney et al. 2007).

Agarose arrays functionalized with 3-azidopropylamine were conjugated to serially diluted 6′-N-5-hexynoate kanamycin A via a Huisgen dipolar cycloaddition reaction. Upon hybridization with the radioactively labeled RNA internal loop library (1) and a 1000-fold excess of oligonucleotide competitors over 1 (Figure 9.3), a dose

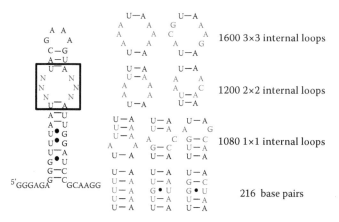

FIGURE 9.8 The number of different RNA structural motifs that are present in an RNA 3 × 3 nucleotide internal loop library.

response was observed (Childs-Disney et al. 2007). The bound RNAs from different ligand loadings were harvested from the surface and RT-PCR amplified. The RT-PCR products were used as templates for *in vitro* transcription to afford a mixture of RNAs. The affinities of the mixtures for 6'-*N*-5-hexynoate kanamycin A were then determined using a fluorescence-based assay. The RNAs harvested from the spot where 5 nmol of compound were delivered have a dissociation constant of 65 nM. As the amount of ligand delivered decreases, the affinity of the RNAs harvested from that position increases. For example, a 10-fold decrease in the moles of ligand delivered to the surface results in the selection of mixtures of RNA that bind with ca. sixfold higher affinity (K_d = 11 nM) (Childs-Disney et al. 2007).

The RT-PCR product amplified from the RNAs harvested where 0.5 nmol of ligand was delivered was cloned and sequenced. Analysis of the sequences identified various trends that predispose RNAs for binding the kanamycin derivative. There was a clear preference for adenine across from cytosine independent of loop size (Childs-Disney et al. 2007). Of the 16 loops identified from the selection, 10 (62.5%) contain at least one adenine across cytosine (compared to 33% of all library members), corresponding to a two-tailed *p*-value of 0.0124 (Childs-Disney et al. 2007). The other important trend identified from our selection was the preference of pyrimidines neighbored by adenine across from cytosine in 2 × 2 nucleotide loops (i.e., pyrimidine-rich loops) (Childs-Disney et al. 2007). This trend had a two-tailed *p*-value of 0.0093 and is similar to the preference observed from a resin-based selection (Disney and Childs-Disney 2007). Interestingly, many 1 × 1 and 2 × 2 nucleotide loops have G•U closing base pairs (Childs-Disney et al. 2007).

The affinities of the individual loops identified from the selection were also determined. The dissociation constants ranged from 5 to 22 nM, >12-fold higher affinity than the entire 3 × 3 nucleotide internal loop library (1, Figure 9.3), >60-fold higher affinity than the cassette in which the randomized region was embedded (9, Figure 9.3), and >230-fold higher affinity than the base-paired region and the DNA competitor oligonucleotides 7 and 8 (Figure 9.3) (Childs-Disney et al. 2007). Thus, the interactions that were probed on the surface were to the randomized region and not to the regions common to all library members. Interestingly, the loops are portable; that is, the affinities of the loops are similar when they are displayed within different RNAs (Childs-Disney et al. 2007).

Since determining the affinities of all library members for a ligand of interest is intractable, we developed a method, StARTS. StARTS predicts the affinities of all library members based on statistical analysis and experimentally determined affinities for a subset of selected RNAs with a range of Sum Z-scores. To do so, we identified the RNA internal loops that bind 6′-N-5-hexynoate neamine using 2DCS. The neamine derivative was chosen for this selection for several reasons, primarily because 1 was screened previously for binding to 5-O-(2-azidoethyl) neamine and no statistically significant trends were observed (Disney et al. 2008).

The 6′ position of neamine was acylated to mimic the product of modification by AAC(6′) (Magnet and Blanchard 2005). Using 6′-N-5-hexynoate neamine could therefore be advantageous if the privileged RNA space that binds the neamine derivative was present in pathogenic RNA targets; the use of this ligand would not be limited by the presence of AAC(6′) resistance-causing enzymes. It is likely that neamine can accommodate various RNA folds due to its small size and large cationic charge density. If, however, the privileged RNA space for neamine could be determined, then it could have interesting implications.

We used a concatenation approach to increase the number of selected RNAs cloned into each vector in order to provide better sequence coverage for statistical analysis (Velagapudi et al. 2010). A modified procedure was followed during RT-PCR amplification of the selected RNAs, which involves the use of biotinylated primers (Figure 9.4). After amplification, the RT-PCR product is digested with BamHI, releasing the biotinylated sticky ends that are captured with streptavidin resin. The RT-PCR products are then concatenated via ligation (Childs-Disney and Disney 2008), increasing the information density in the sequencing reactions.

Statistical analysis was performed on the selected pool of RNAs using RNA-PSP (Paul et al. 2009). Two of the most statistically significant features in the internal loops were 5′NAC/3′CGN and 5′NAN/3′CGC (Z score= 7.68; two-tailed p-value is <0.0001). In an attempt to correlate the results of the statistical analysis with binding affinity, the affinities of 15 selected loops for the ligand were determined using a fluorescein-labeled derivative of 6′-N-5-hexynoate neamine (Childs-Disney et al. 2007; Disney et al. 2008). The average affinity of the mixture of RNAs selected to bind the ligand was 315 nM, while the affinities of the 3 × 3 nucleotide loop library (1, Figure 9.3) and the empty cassette in which the loop was embedded (9, Figure 9.3) were >> 3000 nM. The sum of the Z-scores (Sum Z-scores) for all trends displayed by a loop was calculated using RNA-PSP (Velagapudi et al. 2010) and then plotted against binding affinity (Figure 9.6). These data could then be fit to an inverse first-order relationship (Equation 9.3, $R^2 = 0.85$):

$$K_d = -7.42(\pm 3.9) \times 10^{-8} + \frac{13.7(\pm 1.6) \times 10^{-6}}{\sum Z - score} \tag{9.3}$$

To test the functionality of the equation for predicting the affinity of all library members for 6′-N-5-hexynoate neamine, the binding affinities of six random RNAs derived from the original library but not observed in the sequence data were determined (Childs-Disney et al. 2007). These affinities were plotted as a function of their Sum Z-scores, and the data fit well to Equation 9.3 (Figure 9.6). This analysis allowed a predictive model to be derived to estimate the affinities of all members of an RNA motif library, thus allowing quick annotation of the RNA motif–ligand database.

9.3.2 RATIONAL AND MODULAR DESIGN OF SMALL MOLECULES TARGETING THE RNAS THAT CAUSE MYOTONIC MUSCULAR DYSTROPHY

We sought to use the RNA motif–ligand interactions identified via 2DCS to guide the rational design of a modularly assembled ligand to target an RNA associated with disease. The results of 2DCS studies showed that 2×2 nucleotide, pyrimidine-rich internal loops (5′UU/3′CU or 5′CU/3′UU) bind 6′-N-5-hexynoate kanamycin with the highest affinity (Childs-Disney et al. 2007; Disney and Childs-Disney 2007). Fortuitously, the RNA that causes myotonic DM2 contains periodically repeating 2×2 nucleotide, pyrimidine-rich internal loops that are similar to those identified from our selection (Figure 9.9). Thus, it provided an ideal model system to apply our modular assembly strategy.

DM2 is caused by an expansion of an rCCUG tetranucleotide repeat in intron 1 of the zinc-finger 9 protein (ZNF9) pre-mRNA. The expanded repeats fold into a hairpin structure with pyrimidine-rich internal loops that sequester muscleblind-like 1 protein (MBNL1), an RNA-splicing regulator (Figure 9.9). Sequestration of MBNL1 causes myotonic dystrophy through a variety of mechanisms including the aberrant splicing of the main muscle chloride channel (Mankodi et al. 2002) and insulin receptor (Dansithong et al. 2005; Savkur et al. 2001) pre-mRNAs. Aberrant

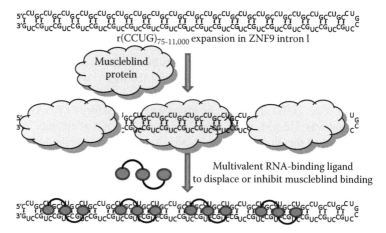

FIGURE 9.9 Schematic of the interaction of MBNL1 with the DM2 rCCUG repeat hairpin and the use of modularly assembled ligands to inhibit the complex. (From Lee, M. M., Pushechnikov, A. and Disney, M. D. *ACS Chem. Biol.* 4, 345–355, 2009b. With permission.)

splicing of these RNAs explains the muscle atrophy and insulin insensitivity observed in DM patients.

The disease model described above points to a potential therapeutic strategy in which a small molecule binds the toxic repeats and inhibits formation of the DM RNA–MBNL1 complex. If this were to occur *in vivo*, then there would be an increase in the amount of free MBNL1 and could result in the restoration of normal splicing patterns. Two pieces of evidence suggest that this strategy is viable: (1) correction of splicing defects in DM-affected cells by overexpression of MBNL1 (Kanadia et al. 2006); and (2) correction of splicing defects and altered muscle excitability in a DM mouse model by a complementary oligonucleotide (Wheeler et al. 2009).

A series of peptoids was synthesized that display two 6′-*N*-5-hexynoate kanamycin A ligand modules separated by different distances (Figure 9.7). Peptoids were chosen as a scaffold because their synthesis is straightforward, and the valency and the distance between ligand modules can be precisely controlled. The 6′-*N*-5-hexynoate kanamycin A ligand modules were anchored to the peptoid backbone using the same chemistry that was used to anchor it to the slide surface in 2DCS—a Huisgen dipolar cycloaddition reaction. By using the same chemistry, potential decreases in affinity due to conjugation effects can be avoided. Therefore, 3-azido-propylamine was reacted with the growing peptoid chain to control valency. The distances between ligand modules were controlled by coupling different numbers of propylamine-spacing submonomers between ligand-displaying modules.

Using a modified ELISA assay in which the DM2 RNA is immobilized in the well of a streptavidin-coated plate (Gareiss et al. 2008; Lee et al. 2009b), the potency of each compound for displacing an MBNL1-β-galactosidase fusion protein from the RNA was determined. The optimal compound, 2K-4 (where 2 indicates the valency, K indicates the kanamycin module, and 4 indicates the number of propyl-amine-spacing submonomers), has four propylamine-spacing submonomers separating ligand modules and has an IC_{50} of ~90 nM (Figures 9.7 and 9.10) (Lee et al. 2009b). This compound binds to the DM2 RNA with a dissociation constant of 50 nM (Lee et al. 2009b). In order to increase affinity and specificity of our modularly assembled compounds, the valency was increased. The corresponding compound, 3K-4, has an IC_{50} of ~2 nM for displacing MBNL1 from the DM2 RNA, an affinity of 8 nM, and is at least 30-fold specific over related RNAs (Lee et al. 2009b). Importantly, the K module is required for potency as a peptoid displaying for neamine derivatives is a poor inhibitor of DM2 RNA–MBNL1 complex formation (Lee et al. 2009a,b).

One of the related RNAs that were studied to determine the binding selectivity of the modularly assembled compounds for the DM2 RNA was the RNA that causes myotonic dystrophy type 1 (DM1). Like DM2, DM1 is caused by a repeat expansion in a noncoding region. The DM1 triplet repeat, rCUG, is found in the 3′ UTR of the dystrophia myotonica protein kinase (DMPK) mRNA (Figure 9.10). When the expanded repeats fold into a hairpin structure, it forms regularly repeating 1×1 nucleotide internal loops with UU mismatches. The kanamycin module binds the DM1 repeats, but with a lower affinity than the DM2 repeats (Lee et al. 2009b). Since the regularly repeating internal loops that cause DM1 and DM2 are different

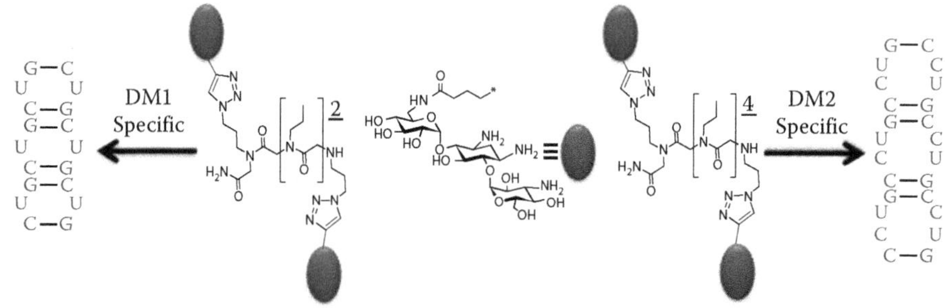

FIGURE 9.10 The spacing between ligand modules can be used to control the specificity of designed small molecules targeting the RNAs that cause myotonic dystrophy types 1 and 2. (From Lee, M. M. et al. *J. Am. Chem. Soc.* 131, 17464–17472, 2009a. With permission.)

sizes (1×1 vs. 2×2 nucleotide loops, respectively; Figure 9.10), we hypothesized that the distance between ligand modules could be altered to afford a high affinity, DM1-specific ligand.

The potencies of the same series of dimeric peptoids synthesized for the DM2 studies were determined for the DM1 RNA–MBNL1 interaction. The most potent ligand had two propylamine spacers separating K modules, consistent with the smaller size of the internal loops (Figure 9.10). As also observed in the DM2 studies, the potency of the modularly assembled compound increases as valency increases. 2K-2 has an IC_{50} of 300 nM while 3K-2 and 4K-2 have IC_{50}s of 40 and 7 nM, respectively. These values correspond to 350-, 1750-, and 7500-fold increases in potency compared to the ligand module K.

The affinities and selectivities of the ligand were also studied. The 2K-2, 3K-2, and 4K-2 modularly assembled ligands bind to the DM1 RNA with dissociation constants of 120, 65, 25 nM, respectively. Moreover, 4K-2 binds the DM1 repeats 63-fold more tightly than MBNL1 and on average with more selectivity. The trimer, 3K-2, is threefold selective for the DM1 RNA over the DM2 RNA even though 6′-N-5-hexynoate kanamycin A binds 2.5-fold more tightly to the DM2 motif (Lee et al. 2009b). Since 3K-4 is 20-fold selective for the DM2 RNA over the DM1 RNA (Lee et al. 2009b), proper spacing can affect selectivity by up to 60-fold.

The sequestration of MBNL1 by the expanded DM1 and DM2 repeats leads to formation of nuclear foci (Fardaei et al. 2002; Jiang et al. 2004; Mankodi et al. 2001). The mRNAs containing the repeats are not exported efficiently to the cytoplasm for translation. Therefore, potential therapeutics should be cell permeable and localize to the nucleus to displace MBNL1 from the repeats. The uptake and cellular localization properties of the modularly assembled ligands into a mouse myoblast cell line (C2C12) were determined using flow cytometry and fluorescence microscopy (Lee et al. 2009a; Lee et al. 2009b). Fortuitously, the modularly assembled compounds are cell permeable in the absence of transfection vehicles. After treatment with fluorescently labeled modular ligands, ~75% of the cells are fluorescent compared to 20% of cells treated with the monomer (Lee et al. 2009a,b). Approximately, threefold more modularly assembled compound enters the cells than fluorescently labeled K (Lee et al. 2009a,b).

Evidently, the peptoid scaffold improves uptake. Moreover, 2K-2, 3K-2, and 4K-2 are nontoxic as assayed by propidium iodide staining. The cellular localization of the compounds was studied by fluorescence microscopy. The modularly assembled compounds localize in the cytoplasm and perinuclear region (Lee et al. 2009a,b).

Various ligand modules that inhibit DM1 RNA–MBNL1 complex formation have been reported including a Hoechst derivative (Pushechnikov et al. 2009), short peptides displaying natural and unnatural amino acids linked by disulfide bonds (Gareiss et al. 2008), pentamidine (Warf et al. 2009), and a triaminotriazine–acridine conjugate (Arambula et al. 2009). Although many of the ligands bind the RNA with dissociation constants in the nM to low μM range, they are mid-micromolar inhibitors of the DM1 RNA–MBNL1 interaction (Arambula et al. 2009; Gareiss et al. 2008). This is likely due to the inability of the ligand module to sequester enough surface area to preclude MBNL1 binding. It has been shown that MBNL1 binds to six base pairs (Teplova and Patel 2008; Warf and Berglund 2007), or two rCUG repeats (Warf and Berglund 2007). Thus, a modular assembly approach affords high-affinity small-molecule binders that are also potent inhibitors of RNA–protein complexes (Disney et al. 2010; Lee et al. 2009a,b; Pushechnikov et al. 2009).

9.4 CONCLUSIONS

The rational design of ligands targeting RNA is difficult, mostly due to an incomplete understanding of the chemical space that is privileged to bind RNA and the RNA motifs that are privileged to bind small molecules. Therefore, we developed the 2DCS method in order to study RNA and chemical spaces simultaneously. The output of 2DCS is a set of RNA motif–ligand interactions. In the two case studies presented herein, we described how the affinities of these RNA–ligand interactions can be predicted and how the interactions can be used to rationally design modularly assembled ligands that bind tightly to the RNAs that cause myotonic muscular dystrophy. As more diverse chemical space is probed, the database expands, and computational tools are developed to utilize the database, modularly assembled compounds can be designed to bind other RNA targets.

9.5 PROSPECTS AND OUTLOOK

In order to most effectively target genomic RNAs, the diversity of the chemical space probed by 2DCS should be expanded. Small molecules with and without known RNA-binding properties should be explored to identify privileged scaffolds. It will also be important to experimentally determine the optimal distances between ligand modules displayed on the peptoid scaffold to bind different secondary structures and different distances between those structures. Moreover, computational methods that search the RNA motif–small molecule partners against genomic RNAs need to be developed to exploit the wealth of information in the database. Since RNA structure prediction is generally accurate (Mathews et al. 2004), rational design could be completed without the need to determine experimentally the secondary structure of an RNA.

REFERENCES

Afanassiev, V., Hanemann, V., and Wolfl, S. Preparation of DNA and protein micro arrays on glass slides coated with an agarose film. *Nucleic Acids Res.* 28, E66, 2000.

Altman, S. and Guerrier-Takada, C. M1 RNA, the RNA subunit of *Escherichia coli* ribonuclease P, can undergo a pH-sensitive conformational change. *Biochemistry* 25, 1205–1208, 1986.

Aminova, O., Paul, D. J., Childs-Disney, J. L., and Disney, M. D. Two-dimensional combinatorial screening identifies specific 6′-acylated kanamycin A- and 6′-acylated neamine–RNA hairpin interactions. *Biochemistry* 47, 12670–12679, 2008.

Arambula, J. F., Ramisetty, S. R., Baranger, A. M., and Zimmerman, S. C. A simple ligand that selectively targets CUG trinucleotide repeats and inhibits MBNL protein binding. *Proc. Natl. Acad. Sci. USA.* 106, 16068–16073, 2009.

Berisio, R., Schluenzen, F., Harms, J., Bashan, A., Auerbach, T., Baram, D., and Yonath, A. Structural insight into the antibiotic action of telithromycin against resistant mutants. *J. Bacteriol.* 185, 4276–4279, 2003a.

Berisio, R., Schluenzen, F., Harms, J., Bashan, A., Auerbach, T., Baram, D., and Yonath, A. Structural insight into the role of the ribosomal tunnel in cellular regulation. *Nat. Struct. Biol.* 10, 366–370, 2003b.

Bevilacqua, J. M. and Bevilacqua, P. C. Thermodynamic analysis of an RNA combinatorial library contained in a short hairpin. *Biochemistry* 37, 15877–15884, 1998.

Bishop, J. M., Levinson, W. E., Sullivan, D., Fanshier, L., Quintrell, N., and Jackson, J. The low molecular weight RNAs of Rous sarcoma virus. II. The 7 S RNA. *Virology* 42, 927–937, 1970.

Blum, B., Bakalara, N., and Simpson, L. A model for RNA editing in kinetoplastid mitochondria: "Guide" RNA molecules transcribed from maxicircle DNA provide the edited information. *Cell* 60, 189–198, 1990.

Bremner, J. B., Keller, P. A., Pyne, S. G., Boyle, T. P., Brkic, Z., David, D. M., Robertson, M. et al. Synthesis and antibacterial studies of binaphthyl-based tripeptoids. Part 1. *Bioorg Med. Chem.* 18, 2611–2620, 2010.

Brodersen, D. E., Clemons, W. M., Jr., Carter, A. P., Morgan-Warren, R. J., Wimberly, B. T., and Ramakrishnan, V. The structural basis for the action of the antibiotics tetracycline, pactamycin, and hygromycin B on the 30S ribosomal subunit. *Cell* 103, 1143–1154, 2000.

Brook, J. D., McCurrach, M. E., Harley, H. G., Buckler, A. J., Church, D., Aburatani, H., Hunter, K. et al. Molecular basis of myotonic dystrophy: Expansion of a trinucleotide (CTG) repeat at the 3′ end of a transcript encoding a protein kinase family member. *Cell* 68, 799–808, 1992.

Burke, M. D. and Schreiber, S. L. A planning strategy for diversity-oriented synthesis. *Angew. Chem. Int. Ed. Engl.* 43, 46–58, 2004.

Calin, G. A. and Croce, C. M. MicroRNA–cancer connection: The beginning of a new tale. *Cancer Res.* 66, 7390–7394, 2006.

Carlson, C. B., Vuyisich, M., Gooch, B. D., and Beal, P. A. Preferred RNA binding sites for a threading intercalator revealed by *in vitro* evolution. *Chem. Biol.* 10, 663–672, 2003.

Carter, A. P., Clemons, W. M., Brodersen, D. E., Morgan-Warren, R. J., Wimberly, B. T., and Ramakrishnan, V. Functional insights from the structure of the 30S ribosomal subunit and its interactions with antibiotics. *Nature* 407, 340–348, 2000.

Caskey, C. T., Pizzuti, A., Fu, Y. H., Fenwick, R. G., Jr., and Nelson, D. L. Triplet repeat mutations in human disease. *Science* 256, 784–789, 1992.

Chen, C. Z. MicroRNAs as oncogenes and tumor suppressors. *N. Engl. J. Med.* 353, 1768–1771, 2005.

Childs-Disney, J. L. and Disney, M. D. A simple ligation-based method to increase the information density in sequencing reactions used to deconvolute nucleic acid selections. *RNA* 14, 390–394, 2008.

Childs-Disney, J. L., Wu, M., Pushechnikov, A., Aminova, O., and Disney, M. D. A small molecule microarray platform to select RNA internal loop–ligand interactions. *ACS Chem. Biol.* 2, 745–754, 2007.

Dansithong, W., Paul, S., Comai, L., and Reddy, S. MBNL1 is the primary determinant of focus formation and aberrant insulin receptor splicing in DM1. *J. Biol. Chem.* 280, 5773–3780, 2005.

Dervan, P. B. Molecular recognition of DNA by small molecules. *Bioorg. Med. Chem.* 9, 2215–2235, 2001.

Dervan, P. B. and Burli, R. W. Sequence-specific DNA recognition by polyamides. *Curr. Opin. Chem. Biol.* 3, 688–693, 1999.

Disney, M. D. and Childs-Disney, J. L. Using selection to identify and chemical microarray to study the RNA internal loops recognized by 6'-*N*-acylated kanamycin A. *Chembiochem* 8, 649–656, 2007.

Disney, M. D., Labuda, L. P., Paul, D. J., Poplawski, S. G., Pushechnikov, A., Tran, T., Velagapudi, S. P., Wu, M., and Childs-Disney, J. L. Two-dimensional combinatorial screening identifies specific aminoglycoside–RNA internal loop partners. *J. Am. Chem. Soc.* 130, 11185–11194, 2008.

Disney, M. D., Lee, M. M., Pushechnikov, A., and Childs-Disney, J. L. The role of flexibility in the rational design of modularly assembled ligands targeting the RNAs that cause the myotonic dystrophies. *Chembiochem* 11, 375–382, 2010.

Doudna, J. A. Structural genomics of RNA. *Nat. Struct. Biol.* 7(Suppl), 954–956, 2000.

Dufva, M., Petronis, S., Jensen, L. B., Krag, C., and Christensen, C. B. Characterization of an inexpensive, nontoxic, and highly sensitive microarray substrate. *Biotechniques* 37, 286–296, 2004.

Duncan, C. D. and Weeks, K. M. SHAPE analysis of long-range interactions reveals extensive and thermodynamically preferred misfolding in a fragile group I intron RNA. *Biochemistry* 47, 8504–8513, 2008.

Eddy, S. R. How do RNA folding algorithms work? *Nat. Biotechnol.* 22, 1457–1458 (2004).

Ellington, A. D. and Conrad, R. Aptamers as potential nucleic acid pharmaceuticals. *Biotechnol. Annu. Rev.* 1, 185–214, 1995.

Emerick, V. L. and Woodson, S. A. Self-splicing of the Tetrahymena pre-rRNA is decreased by misfolding during transcription. *Biochemistry* 32, 14062–14067, 1993.

Esquela-Kerscher, A. and Slack, F. J. Oncomirs—MicroRNAs with a role in cancer. *Nat. Rev. Cancer* 6, 259–269, 2006.

Fardaei, M., Rogers, M. T., Thorpe, H. M., et al. Three proteins, MBNL, MBLL and MBXL, co-localize *in vivo* with nuclear foci of expanded-repeat transcripts in DM1 and DM2 cells. *Hum. Mol. Genet.* 11, 805–814, 2002.

Faustino, N. A. and Cooper, T. A. Pre-mRNA splicing and human disease. *Genes Dev.* 17, 419–437, 2003.

Fergus, S., Bender, A. and Spring, D. R. Assessment of structural diversity in combinatorial synthesis. *Curr. Opin. Chem. Biol.* 9, 304–309, 2005.

Fu, Y. H., Pizzuti, A., Fenwick, R. G., King, J., Rajnarayan, S., Dunne, P. W., Dubel, J. et al. An unstable triplet repeat in a gene related to myotonic muscular dystrophy. *Science* 255, 1256–1258, 1992.

Gareiss, P. C., Sobczak, K., McNaughton, B. R., Palde, P. B., Thornton, C. A., and Miller, B. L. Dynamic combinatorial selection of molecules capable of inhibiting the (CUG) repeat RNA–MBNL1 interaction *in vitro*: Discovery of lead compounds targeting myotonic dystrophy (DM1). *J. Am. Chem. Soc.* 130, 16254–16261, 2008.

Gestwicki, J. E., Cairo, C. W., Strong, L. E., Oetjen, K. A., and Kiessling, L. L. Influencing receptor–ligand binding mechanisms with multivalent ligand architecture. *J. Am. Chem. Soc.* 124, 14922–14933, 2002.

Gocke, A. R., Udugamasooriya, D. G., Archer, C. T., Lee, J., and Kodadek, T. Isolation of antagonists of antigen-specific autoimmune T cell proliferation. *Chem. Biol.* 16, 1133–1139, 2009.

Goodson, B., Ehrhardt, A., Ng, S., Nuss, J., Johnson, K., Giedlin, M., Yamamoto, R. et al. Characterization of novel antimicrobial peptoids. *Antimicrob. Agents Chemother.* 43, 1429–1434, 1999.

Goringer, H. U. and Wagner, R. Does 5S RNA from *Escherichia coli* have a pseudoknotted structure? *Nucleic Acids Res.* 14, 7473–7485, 1986.

Goun, E. A., Pillow, T. H., Jones, L. R., Rothbard, J. B., and Wender, P. A. Molecular transporters: Synthesis of oligoguanidinium transporters and their application to drug delivery and real-time imaging. *Chembiochem* 7, 1497–1515, 2006a.

Goun, E. A., Shinde, R., Dehnert, K. W., Adams-Bond, A., Wender, P. A., Contag, C. H., and Franc, B. L. Intracellular cargo delivery by an octaarginine transporter adapted to target prostate cancer cells through cell surface protease activation. *Bioconjug. Chem.* 17, 787–796, 2006b.

Griffey, R. H., Hofstadler, S. A., Sannes-Lowery, K. A., Ecker, D. J., and Crooke, S. T. Determinants of aminoglycoside-binding specificity for rRNA by using mass spectrometry. *Proc. Natl. Acad. Sci. USA.* 96, 10129–10133, 1999.

Hammond, S. M. MicroRNAs as oncogenes. *Curr. Opin. Genet. Dev.* 16, 4–9, 2006.

Hansen, J. L., Moore, P. B., and Steitz, T. A. Structures of five antibiotics bound at the peptidyl transferase center of the large ribosomal subunit. *J. Mol. Biol.* 330, 1061–1075, 2003.

Hogan, J. J., Gutell, R. R., and Noller, H. F. Probing the conformation of 18S rRNA in yeast 40S ribosomal subunits with kethoxal. *Biochemistry* 23, 3322–3330, 1984a.

Hogan, J. J., Gutell, R. R., and Noller, H. F. Probing the conformation of 26S rRNA in yeast 60S ribosomal subunits with kethoxal. *Biochemistry* 23, 3330–3335, 1984b.

Hogan, J. J. and Noller, H. F. Altered topography of 16S RNA in the inactive form of *Escherichia coli* 30S ribosomal subunits. *Biochemistry* 17, 587–593, 1978.

Holmes, S. E., O'Hearn, E., Rosenblatt, A., Callahan, C., Hwang, H. S., Ingersoll-Ashworth, R. G., Fleisher, A. A repeat expansion in the gene encoding junctophilin-3 is associated with Huntington disease-like 2. *Nat. Genet.* 29, 377–378, 2001.

Jefferson, E. A., Arakawa, S., Blyn, L. B., Miyaji, A., Osgood, S. A., Ranken, R., Risen, L. M., and Swayze, E. E. New inhibitors of bacterial protein synthesis from a combinatorial library of macrocycles. *J. Med. Chem.* 45, 3430–3439, 2002.

Jiang, H., Mankodi, A., Swanson, M. S., Moxley, R. T., and Thornton, C. A. Myotonic dystrophy type 1 is associated with nuclear foci of mutant RNA, sequestration of muscleblind proteins and deregulated alternative splicing in neurons. *Hum. Mol. Genet.* 13, 3079–3088, 2004.

Johnson, E. C., Feher, V. A., Peng, J. W., Moore, J. M., and Williamson, J. R. Application of NMR SHAPES screening to an RNA target. *J. Am. Chem. Soc.* 125, 15724–15725, 2003.

Kanadia, R. N., Shin, J., Yuan, Y., Beattie, S. G., Wheeler, T. M., Thornton, C. A., and Swanson, M. S. Reversal of RNA missplicing and myotonia after muscleblind overexpression in a mouse poly(CUG) model for myotonic dystrophy. *Proc. Natl. Acad. Sci. USA.* 103, 11748–11753, 2006.

Kaul, M., Barbieri, C. M., and Pilch, D. S. Aminoglycoside-induced reduction in nucleotide mobility at the ribosomal RNA A-site as a potentially key determinant of antibacterial activity. *J. Am. Chem. Soc.* 128, 1261–1271, 2006.

Kiss-Laszlo, Z., Henry, Y., Bachellerie, J. P., Caizergues-Ferrer, M., and Kiss, T. Site-specific ribose methylation of preribosomal RNA: A novel function for small nucleolar RNAs. *Cell* 85, 1077–1088, 1996.

Kitov, P. I., Sadowska, J. M., Mulvey, G., et al. Shiga-like toxins are neutralized by tailored multivalent carbohydrate ligands. *Nature* 403, 669–672, 2000.

Kolb, H. C., Finn, M. G., and Sharpless, K. B. Click chemistry: Diverse chemical function from a few good reactions. *Angew. Chem. Int. Ed. Engl.* 40, 2004–2021, 2001.

Kurokawa, R., Rosenfeld, M. G., and Glass, C. K. Transcriptional regulation through noncoding RNAs and epigenetic modifications. *RNA Biol.* 6, 233–236, 2009.

Lato, S. M. and Ellington, A. D. Screening chemical libraries for nucleic-acid-binding drugs by *in vitro* selection: A test case with lividomycin. *Mol. Divers.* 2, 103–110, 1996.

Lee, J., Udugamasooriya, D. G., Lim, H. S., and Kodadek, T. Potent and selective photo-inactivation of proteins with peptoid–ruthenium conjugates. *Nat. Chem. Biol.* 6, 258–260, 2010.

Lee, M. M., Childs-Disney, J. L., Pushechnikov, A., French, J. M., Sobczak, K., Thornton, C. A., and Disney, M. D. Controlling the specificity of modularly assembled small molecules for RNA via ligand module spacing: Targeting the RNAs that cause myotonic muscular dystrophy. *J. Am. Chem. Soc.* 131, 17464–17472, 2009a.

Lee, M. M., Pushechnikov, A., and Disney, M. D. Rational and modular design of potent ligands targeting the RNA that causes myotonic dystrophy 2. *ACS Chem. Biol.* 4, 345–355, 2009b.

Liquori, C. L., Ricker, K., Moseley, M. L., Jacobsen, J. F., Kress, W., Naylor, S. L., Day, J. W., and Ranum, L. P. Myotonic dystrophy type 2 caused by a CCTG expansion in intron 1 of ZNF9. *Science* 293, 864–867, 2001.

Lynn, K. D., Udugamasooriya, D. G., Roland, C. L., Castrillon, D. H., Kodadek, T. J., and Brekken, R. A. GU81, a VEGFR2 antagonist peptoid, enhances the anti-tumor activity of doxorubicin in the murine MMTV-PyMT transgenic model of breast cancer. *BMC Cancer* 10, 397, 2010.

Magnet, S. and Blanchard, J. S. Molecular insights into aminoglycoside action and resistance. *Chem. Rev.* 105, 477–498, 2005.

Mammen, M., Choi, S. K., and Whitesides, G. M. Polyvalent interactions in biological systems: Implications for design and use of multivalent ligands and inhibitors. *Angew. Chem. Int. Ed. Engl.* 37, 2755–2794, 1998.

Mankodi, A., Takahashi, M. P., Jiang, H., Beck, C. L., Bowers, W. J., Moxley, R. T., Cannon, S. C., and Thornton, C. A. Expanded CUG repeats trigger aberrant splicing of ClC-1 chloride channel pre-mRNA and hyperexcitability of skeletal muscle in myotonic dystrophy. *Mol. Cell* 10, 35–44, 2002.

Mankodi, A., Urbinati, C. R., Yuan, Q. P., Moxley, R. T., Sansone, V., Krym, M., Henderson, D., Schalling, M., Swanson, M. S., and Thornton, C. A. Muscleblind localizes to nuclear foci of aberrant RNA in myotonic dystrophy types 1 and 2. *Hum. Mol. Genet.* 10, 2165–2170, 2001.

Mas-Moruno, C., Cruz, L. J., Mora, P., Francesch, A., Messeguer, A., Perez-Paya, E., and Albericio, F. Smallest peptoids with antiproliferative activity on human neoplastic cells. *J. Med. Chem.* 50, 2443–2449, 2007.

Masip, I., Perez-Paya, E., and Messeguer, A. Peptoids as source of compounds eliciting antibacterial activity. *Comb. Chem. High Throughput Screen.* 8, 235–239, 2005.

Mathews, D. H., Disney, M. D., Childs, J. L., Schroeder, S. J., Zuker, M., and Turner, D. H. Incorporating chemical modification constraints into a dynamic programming algorithm for prediction of RNA secondary structure. *Proc. Natl. Acad. Sci. USA.* 101, 7287–7292, 2004.

Mathews, D. H., Sabina, J., Zuker, M., and Turner, D. H. Expanded sequence dependence of thermodynamic parameters improves prediction of RNA secondary structure. *J. Mol. Biol.* 288, 911–940, 1999.

Mazumder, B., Seshadri, V., Imataka, H., Sonenberg, N., and Fox, P. L. Translational silencing of ceruloplasmin requires the essential elements of mRNA circularization: Poly(A) tail, poly(A)-binding protein, and eukaryotic translation initiation factor 4G. *Mol. Cell. Biol.* 21, 6440–6449, 2001.

Nikolcheva, T. and Woodson, S. A. Facilitation of group I splicing *in vivo*: Misfolding of the Tetrahymena IVS and the role of ribosomal RNA exons. *J. Mol. Biol.* 292, 557–567, 1999.

Ogle, J. M., Brodersen, D. E., Clemons, W. M., Jr., Tarry, M. J., Carter, A. P., and Ramakrishnan, V. Recognition of cognate transfer RNA by the 30S ribosomal subunit. *Science* 292, 897–902, 2001.

Pan, J. and Woodson, S. A. Folding intermediates of a self-splicing RNA: Mispairing of the catalytic core. *J. Mol. Biol.* 280, 597–609, 1998.

Pan, T. and Sosnick, T. R. Intermediates and kinetic traps in the folding of a large ribozyme revealed by circular dichroism and UV absorbance spectroscopies and catalytic activity. *Nat. Struct. Biol.* 4, 931–938, 1997.

Parsons, J., Castaldi, M. P., Dutta, S., Dibrov, S. M., Wyles, D. L., and Hermann, T. Conformational inhibition of the hepatitis C virus internal ribosome entry site RNA. *Nat. Chem. Biol.* 5, 823–825, 2009.

Paul, D. J., Seedhouse, S. J., and Disney, M. D. Two-dimensional combinatorial screening and the RNA Privileged Space Predictor program efficiently identify aminoglycoside–RNA hairpin loop interactions. *Nucleic Acids Res.* 37, 5894–5907, 2009.

Paul, N. and Joyce, G. F. A self-replicating ligase ribozyme. *Proc. Natl. Acad. Sci. USA.* 99, 12733–12740, 2002.

Pioletti, M., Schlunzen, F., Harms, J., Zarivach, R., Gluhmann, M., Avila, H., and Bashan, A. Crystal structures of complexes of the small ribosomal subunit with tetracycline, edeine and IF3. *EMBO J.* 20, 1829–1839, 2001.

Pushechnikov, A., Lee, M. M., Childs-Disney, J. L., Sobczak, K., French, J. M., Thornton, C. A., and Disney, M. D. Rational design of ligands targeting triplet repeating transcripts that cause RNA dominant disease: Application to myotonic muscular dystrophy type 1 and spinocerebellar ataxia type 3. *J. Am. Chem. Soc.* 131, 9767–9779, 2009.

SantaLucia, J., Jr. and Turner, D. H. Structure of (rGGCGAGCC)$_2$ in solution from NMR and restrained molecular dynamics. *Biochemistry* 32, 12612–12623, 1993.

Savkur, R. S., Philips, A. V., and Cooper, T. A. Aberrant regulation of insulin receptor alternative splicing is associated with insulin resistance in myotonic dystrophy. *Nat. Genet.* 29, 40–47, 2001.

Schlunzen, F., Harms, J. M., Franceschi, F., Hansen, H. A., Bartels, H., Zarivach, R., and Yonath, A. Structural basis for the antibiotic activity of ketolides and azalides. *Structure (Cambridge)* 11, 329–338, 2003.

Schlunzen, F., Zarivach, R., Harms, J., Bashan, A., Tocilj, A., Albrecht, R., Yonath, A., and Franceschi, F. Structural basis for the interaction of antibiotics with the peptidyl transferase centre in eubacteria. *Nature* 413, 814–821, 2001.

Schreiber, S. L. Target-oriented and diversity-oriented organic synthesis in drug discovery. *Science* 287, 1964–1969, 2000.

Seth, P. P., Miyaji, A., Jefferson, E. A., Sannes-Lowery, K. A., Osgood, S. A., Propp, S. S., Ranken, R. SAR by MS: Discovery of a new class of RNA-binding small molecules for the hepatitis C virus: Internal ribosome entry site IIA subdomain. *J. Med. Chem.* 48, 7099–7102, 2005.

Sevo, M., Buratti, E., and Venturi, V. Ribosomal protein S1 specifically binds to the 5' untranslated region of the *Pseudomonas aeruginosa* stationary-phase sigma factor rpoS mRNA in the logarithmic phase of growth. *J. Bacteriol.* 186, 4903–4909, 2004.

Shan, S. O. and Walter, P. Co-translational protein targeting by the signal recognition particle. *FEBS Lett.* 579, 921–926, 2005.

Shu, Z. and Bevilacqua, P. C. Isolation and characterization of thermodynamically stable and unstable RNA hairpins from a triloop combinatorial library. *Biochemistry* 38, 15369–15379, 1999.

Shuker, S. B., Hajduk, P. J., Meadows, R. P., and Fesik, S. W. Discovering high-affinity ligands for proteins: SAR by NMR. *Science* 274, 1531–1534, 1996.

Stark, B. C., Kole, R., Bowman, E. J., Altman, S., and Altman, S. Ribonuclease P: An enzyme with an essential RNA component. *Proc. Natl. Acad. Sci. USA.* 75, 3717–3721, 1978.

Swayze, E. E., Jefferson, E. A., Sannes-Lowery, K. A., et al. SAR by MS: A ligand based technique for drug lead discovery against structured RNA targets. *J. Med. Chem.* 45, 3816–3819, 2002.

Takagi, M., Absalon, M. J., McLure, K. G., and Kastan, M. B. Regulation of p53 translation and induction after DNA damage by ribosomal protein L26 and nucleolin. *Cell* 123, 49–63, 2005.

Teplova, M. and Patel, D. J. Structural insights into RNA recognition by the alternative-splicing regulator muscleblind-like MBNL1. *Nat. Struct. Mol. Biol.* 15, 1343–1351, 2008.

Thomas, J. R. and Hergenrother, P. J. Targeting RNA with small molecules. *Chem. Rev.* 108, 1171–1224, 2008.

Tombelli, S., Minunni, M., and Mascini, M. Analytical applications of aptamers. *Biosens. Bioelectron.* 20, 2424–2434, 2005.

Tran, T. and Disney, M. D. Two-dimensional combinatorial screening of a bacterial rRNA A-site-like motif library: Defining privileged asymmetric internal loops that bind aminoglycosides. *Biochemistry* 49, 1833–1842, 2010.

Tran, T. A., Mattern, R. H., Afargan, M., et al. Design, synthesis, and biological activities of potent and selective somatostatin analogues incorporating novel peptoid residues. *J. Med. Chem.* 41, 2679–2685, 1998.

Tuerk, C. and Gold, L. Systematic evolution of ligands by exponential enrichment: RNA ligands to bacteriophage T4 DNA polymerase. *Science* 249, 505–510, 1990.

Turner, D. H., Sugimoto, N., and Freier, S. M. RNA structure prediction. *Annu. Rev. Biophys. Biophys. Chem.* 17, 167–192, 1988.

Velagapudi, S. P., Seedhouse, S. J., and Disney, M. D. Structure–activity relationships through sequencing (StARTS) defines optimal and suboptimal RNA motif targets for small molecules. *Angew. Chem. Int. Ed. Engl.* 49, 3816–3818, 2010.

Vicens, Q. and Westhof, E. Crystal structure of a complex between the aminoglycoside tobramycin and an oligonucleotide containing the ribosomal decoding a site. *Chem. Biol.* 9, 747–755, 2002.

Walter, F., Vicens, Q., and Westhof, E. Aminoglycoside–RNA interactions. *Curr. Opin. Chem. Biol.* 3, 694–704, 1999.

Warf, M. B. and Berglund, J. A. MBNL binds similar RNA structures in the CUG repeats of myotonic dystrophy and its pre-mRNA substrate cardiac troponin T. *RNA* 13, 2238–2251, 2007.

Warf, M. B., Nakamori, M., Matthys, C. M., Thornton, C. A., and Berglund, J. A. Pentamidine reverses the splicing defects associated with myotonic dystrophy. *Proc. Natl. Acad. Sci. USA.* 106, 18551–18556, 2009.

Wheeler, T. M., Sobczak, K., Lueck, J. D., et al. Reversal of RNA dominance by displacement of protein sequestered on triplet repeat RNA. *Science* 325, 336–339, 2009.

Woese, C. R., Magrum, L. J., Gupta, R., et al. Secondary structure model for bacterial 16S ribosomal RNA: Phylogenetic, enzymatic and chemical evidence. *Nucleic Acids Res.* 8, 2275–2293, 1980.

Woodson, S. A. and Cech, T. R. Alternative secondary structures in the 5' exon affect both forward and reverse self-splicing of the Tetrahymena intervening sequence RNA. *Biochemistry* 30, 2042–2050, 1991.

Wu, M. and Turner, D. H. Solution structure of $(rGCGGACGC)_2$ by two-dimensional NMR and the iterative relaxation matrix approach. *Biochemistry* 35, 9677–9689, 1996.

Yoshizawa, S., Fourmy, D., and Puglisi, J. D. Recognition of the codon-anticodon helix by ribosomal RNA. *Science* 285, 1722–1725, 1999.

Yu, P., Liu, B., and Kodadek, T. A high-throughput assay for assessing the cell permeability of combinatorial libraries. *Nat. Biotechnol.* 23, 746–751, 2005.

Yu, P., Liu, B., and Kodadek, T. A convenient, high-throughput assay for measuring the relative cell permeability of synthetic compounds. *Nat. Protoc.* 2, 23–30, 2007.

Zarrinkar, P. P., Wang, J., and Williamson, J. R. Slow folding kinetics of RNase P RNA. *RNA* 2, 564–573, 1996.

Zaug, A. J. and Cech, T. R. The intervening sequence RNA of Tetrahymena is an enzyme. *Science* 231, 470–475, 1986.

Zuckermann, R. N. and Kodadek, T. Peptoids as potential therapeutics. *Curr. Opin. Mol. Ther.* 11, 299–307, 2009.

10 EPR Spectroscopy for the Study of RNA–Ligand Interactions

Snorri Th. Sigurdsson

CONTENTS

Electron paramagnetic resonance (EPR) spectroscopy is a valuable technique for the study of biopolymer dynamics and structure. Specific examples illustrate how continuous wave (CW) EPR can be used to study RNA structural dynamics, that is, how information about motion can give insight into RNA–small molecule and RNA–protein interactions.

10.1 INTRODUCTION

Given its importance in various biological processes, RNA has become an important target for drug discovery. The design of compounds that interfere with specific functions of RNA relies on fundamental understanding of the structure and dynamics of RNA and how these relate to interactions with other molecules. High-resolution structures, obtained by x-ray crystallography [1] and nuclear magnetic resonance (NMR) spectroscopy [2], give location of atoms in three-dimensional space and are valuable for gaining mechanistic insights. However, high-resolution data can be difficult to obtain for RNA and its complexes [3]. Furthermore, the conditions under which the data is collected or the sample is prepared, for example, during crystallization, are often very different from native conditions [3] and may affect the RNA conformation.

Small-molecule binding to RNA has been studied by chemical and enzymatic footprinting, where a binding site is inferred from changes in cleavage/modification of the RNA upon binding [4]. However, a conformational change in the RNA upon binding may affect the rates of RNA modification at a location removed from the binding site. Circular dichroism (CD) is useful for distinguishing RNA groove binding from intercalation but cannot be used to locate the binding site [5]. Fluorescence [6], mass spectrometry [7], and surface plasmon resonance [8] are useful for the determination of binding constants, but give limited structural information.

This chapter describes the use of EPR spectroscopy to study RNA–ligand interactions. First, the basics of EPR spectroscopy will be described, including how EPR can be used to study both structure and dynamics. Second, case studies will show how EPR spectroscopy can be used to study structural dynamics, that is, extraction of structural information from changes in internal dynamics of the RNA. Specifically, the use of EPR to study the binding of the trans-activation responsive (TAR) RNA to small molecules and derivatives of the Tat protein, its biological partner, will be described in detail.

10.2 THE BASICS

10.2.1 EPR SPECTROSCOPY

EPR spectroscopy, also called electron spin resonance (ESR) spectroscopy, is a magnetic resonance technique similar to NMR spectroscopy. EPR enables the study of electron spins, while NMR is used to interrogate nuclear spins. Unpaired electrons are found in free radicals, often associated with biological processes, and in many transition metal ions. EPR spectroscopy has been used extensively to study paramagnetic metal ions in various inorganic complexes, but has increasingly been used to study biopolymers [9–15]. Some biopolymers contain paramagnetic metal ions that can be directly studied by EPR methods [11,14,16], but often stable free radicals (spin labels) need to be incorporated into the oligomer. Therefore, ready access to spin-labeled material is a prerequisite for facile EPR studies [10,13]. Development of spin-labeling techniques, in addition to advances in EPR theory and instrumentation, has significantly advanced the field of EPR spectroscopy in the last two decades [9,12,13,16–21].

EPR spectroscopy can probe dynamics on a variety of timescales (ps–ms) [20] and extract information about structure. EPR also has practical advantages when it comes to the study of nucleic acids. For example, a small amount of material is required (nmols), there is no molecular size limit, and the samples studied by EPR contain the biomolecule in biological-like conditions. A potential drawback, like for some other spectroscopic techniques that require reporter groups (probes), is the requirement for incorporation of a spin label that may affect the function of the nucleic acid. However, potential pitfalls can be avoided by a careful choice of a spin label and by performing appropriate control experiments.

Section 10.2.1 describes spin-labeling of nucleic acids, where the choice of label will be discussed with regard to its structure and the chemistry required for incorporation. Section 10.2.2 introduces the basic theory of EPR spectroscopy. Finally, specific examples of EPR applications for the study of RNA–ligand interactions will be described in Section 10.3.

10.2.2 SITE-DIRECTED SPIN-LABELING OF RNA

As mentioned above, most nucleic acids are intrinsically diamagnetic and, therefore, require site-specific incorporation of spin labels for EPR studies. This is referred to as site-directed spin-labeling (SDSL), or site-specific spin-labeling. The functional group of choice for labeling nucleic acids is a nitroxide, which is a stable radical that can be readily prepared and/or modified using the tools of organic synthesis (Figure 10.1). Many nitroxides are also commercially available.

There are three main questions/factors that should be considered when selecting a nitroxide spin label. First, are there some structural features that are desirable for a particular experiment? For example, a label for probing dynamics of individual sites by EPR spectroscopy should have limited motion, independent of the nucleotide itself, to enable the spin label to report the dynamics of the nucleotide to which it is attached. Second, should the label be attached to the base, sugar, or the phosphate backbone of the RNA? Here, the answer may depend on what kind of interaction is to be probed and any structure–function data available about the RNA. Third, how should the label be incorporated into the nucleic acid? Several different methods have been utilized for spin-labeling nucleic acids, but they fall into two main categories, incorporation of spin labels during synthesis of the nucleic acid (the phosphoramidite approach) and postsynthetic spin-labeling (Figure 10.2). These two approaches will be described below and examples of spin-labeled nucleotides in RNA given.

10.2.2.1 The Phosphoramidite Method

Nucleic acids are readily prepared by automated chemical synthesis using commercially available reagents. The synthesis of nucleic acids is based on 3′-phosphoramidite chemistry (Figure 10.2a). The nucleic acid is synthesized on a solid support, with the 3′-end of the growing chain attached to the resin. The reaction of a free 5′-hydroxyl group on the growing end of the solid support with a 3′-phosphoramidite is the coupling step that is used to stitch the nucleotides together. For incorporation of spin labels, a phosphoramidite that contains a spin label is prepared and used to incorporate

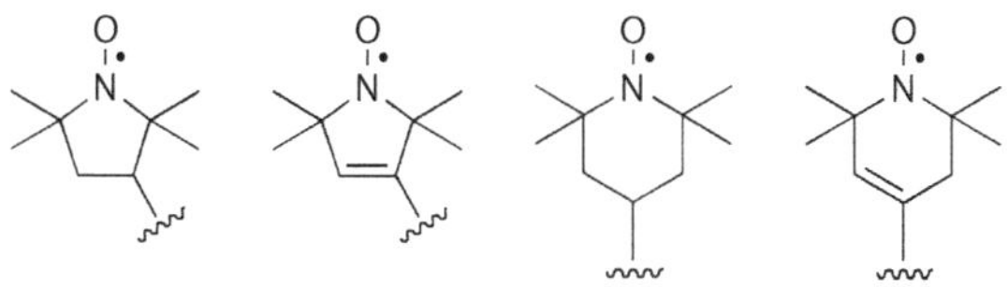

FIGURE 10.1 Examples of nitroxides that have been used for RNA spin-labeling.

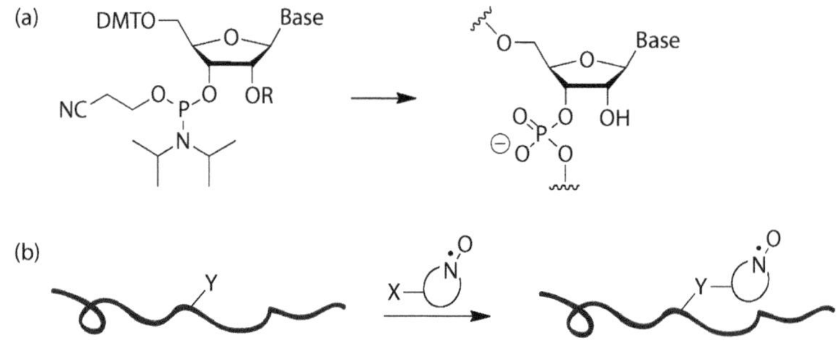

FIGURE 10.2 Strategies for incorporation of spin labels into nucleic acids. (a) The phosphoramidite method. (b) Postsynthetic spin-labeling.

the spin-labeled nucleotide at a specific location in the sequence. The main advantage of the phosphoramidite approach for spin-labeling is that labels with sophisticated structures can be prepared and incorporated into the chosen location(s). A potential drawback is that the synthesis of the labeled phosphoramidite can be laborious and requires a skill set in organic synthesis. Furthermore, nitroxides are partially reduced to amines, due to exposure to the chemicals used in the oligomer synthesis [22–24], and the purification of the labeled material can be nontrivial.

10.2.2.2 Postsynthetic Labeling

The reaction of a uniquely reactive group in the nucleic acid with a spin-labeling reagent is called postsynthetic labeling because the label is added after the preparation of the biopolymer. It should be noted that the nucleotide that contains the reactive functional group to be modified needs to be incorporated into the oligomer using the phosphoramidite approach. However, such phosphoramidites are often commercially available. Advantages of the postsynthetic labeling include the simplicity (usually a single reaction) and the use of oligomers and labeling reagents that are frequently commercially available.

Despite advances in spin-labeling of oligonucleotides, most of the labeling methods described thus far have been used to spin-label DNA, primarily due to the greater ease of synthesizing DNA with the phosphoramidite approach. However, a number of spin-labeling methods exist for RNA, through attachment to the base, the sugar, and the phosphate backbone (Figure 10.3). Spin labels have been incorporated into nucleoside bases using acetylene chemistry (**1–3**) [23,25], by linking to the exocyclic amino groups (**4–7**) [26] or by attachment to a thio-modified base (**8, 9**) [27–29]. The 2′-position of the sugar has been used for spin label attachment, through either a urea linkage (**10**) [30] or an amide (**11**) [31]. The phosphate backbone has been modified internally (**12, 13**) [32,33] or at the 5′-end (**12–14**) [34–36] utilizing phosphorothioates. RNA has also been labeled at the 3′-end by oxidation and reductive amination (**15**) [37].

10.2.3 THEORY OF EPR SPECTROSCOPY

This section describes EPR studies of nitroxides, the EPR reporter groups of choice to study nucleic acids. The theory of EPR spectroscopy will be touched upon, starting

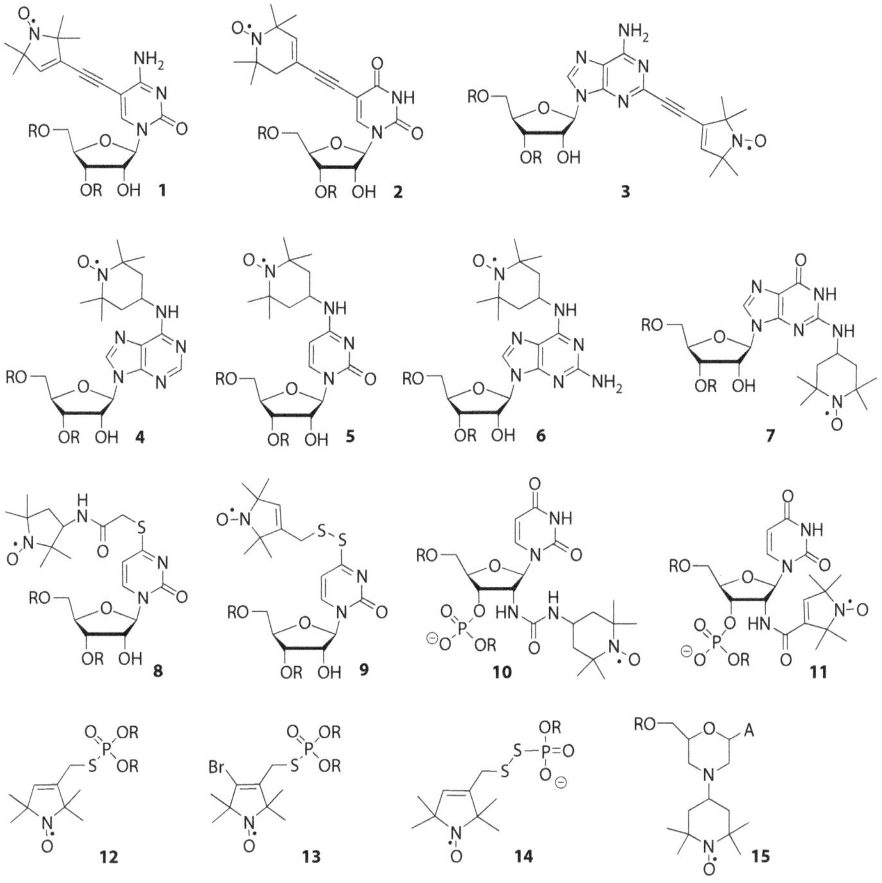

FIGURE 10.3   Examples of spin-labeled nucleotides in RNA.

with the basics of CW-EPR spectroscopy, which is useful for obtaining information about motion (dynamics) and for the determination of shorter distances (<20 Å). The focus will be on CW-EPR spectra collected at X-band (ca. 9.5 GHz, or 0.338 T) because this is the most common frequency used. Pulsed EPR spectroscopy will also be briefly explained, in particular pulsed electron–electron double resonance (PELDOR), also called double electron–electron resonance (DEER), which is a technique that is increasingly being used for measuring long-range distances in nucleic acids and other biopolymers [12,14,38].

10.2.3.1   Continuous Wave EPR Spectroscopy

10.2.3.1.1   The Zeeman Effect

A single unpaired electron, for example, in a nitroxide, has a spin quantum number of $S = 1/2$ and two spin states corresponding to the magnetic components $M_s = -1/2$ and $M_s = +1/2$. In the absence of an external magnetic field, there is no difference in the energy of the two spin states; they are degenerate. However, due to its spin angular momentum, the electron has a spin magnetic moment that interacts with an external magnetic field (B_0) generated by a magnet in the laboratory (the Zeeman effect), and results in two allowed states (energy levels) for the electron. In the lower energy state, the magnetic dipole is aligned with the magnetic field ($M_s = -1/2$) and in the higher energy state the magnetic dipole is aligned against the magnetic field ($M_s = +1/2$). The difference between the energy states (ΔE), caused by the interaction of the electron with the magnetic field, is shown in the following equation:

$$\Delta E = g\mu_B B_0 \Delta m_S = g\mu_B B_0 \tag{10.1}$$

where g is the g-factor, a proportionality constant, which is ca. 2 for an electron, μ_B is the Bohr magneton, and Δm_s is the change in spin state. The difference in the energy of the two levels is proportional to the magnetic field strength and is the energy required to produce a transition from one state to the other (Planck's law, Equation 10.2).

$$\Delta E = h\nu \tag{10.2}$$

where h is the Planck's constant and ν is the frequency of the microwave radiation (the resonance frequency) at which the sample can absorb the radiation to produce a signal. For technical reasons, the frequency is held constant while the magnetic field strength is varied during acquisition of a continuous-wave EPR spectrum, unlike an NMR experiment where the magnetic field is held constant and the frequency is changed.

10.2.3.1.2　Hyperfine Interactions

Nuclei that have magnetic moments and are proximal to an electron can affect the local magnetic field of the electron. Such an effect is called a hyperfine interaction, or hyperfine splitting, which can give valuable information about the local surroundings of an electron. The hyperfine interaction in EPR spectroscopy is analogous to J-coupling in NMR spectroscopy. For nitroxides, the nitrogen has a strong hyperfine interaction with the electron. The most abundant isotope of nitrogen is ^{14}N, which has three spin states ($I = 1$). Therefore, the nitrogen atom splits the EPR signal of the nitroxide's unpaired electron into three lines, each of which corresponds to an ^{14}N spin state (Figure 10.4a).

10.2.3.1.3　Anisotropy and Dynamics

Both the g-factor and the hyperfine interaction have an anisotropic component, that is, they are dependent on the orientation of the nitroxide with respect to the external magnetic field. For a nitroxide that is moving rapidly, these anisotropic components average to zero and only the isotropic (independent on orientation of a radical in the magnetic field) g-factor and the isotropic hyperfine constant are visible (Figure 10.4a). However, when the nitroxide motion slows down, the anisotropic components start to manifest and the anisotropy becomes visible in the EPR spectrum; the signal broadens and becomes asymmetric, with extra features becoming visible in the high- and low-field region of the spectrum (Figure 10.4b). This correlation between inherent motion of the nitroxide and spectral shape can be used to extract information about dynamics from the spectra. This gives indirect information about the local environment of probes that are attached to biopolymers.

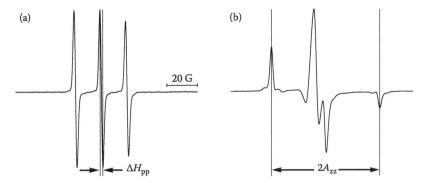

FIGURE 10.4 EPR spectra of nitroxides in the fast (a) and slow (b) motion regime. The width of the EPR spectrum yields information about motion and is determined from the spectral width, either of the central line (ΔH_{pp}) or the whole spectrum ($2A_{zz}$).

10.2.3.1.4 Dynamics from EPR Spectral Shape

Several approaches have been applied for extracting information about dynamics from EPR spectra. A simple, semiquantitative approach is to measure the spectral width, either of the whole spectrum, from the crest of the low-field peak to the trough of the high-field peak ($2A_{zz}$) [39], or the central line (ΔH_{pp}) [10] (Figure 10.4). Alternatively, the dynamics can be expressed by a rotational correlation time (τ_R), determined by the ΔS method [39] or by calculating the scaled relative mobility factor (M_S), from the width of the central line (ΔH_{pp}) [10]. The parameters defining the dynamics can also be determined by detailed spectral simulation [40].

The motion of the spin label that is reflected in the spectrum consists of three main components. First, the biomolecule as a whole tumbles with a rotational correlation time, determined by the size and shape of the molecule. Second, there is internal motion in the oligomer where the spin label is connected. Third, the label may move independent of the biopolymer because most spin labels are attached to the biopolymer with a tether that has some degree of flexibility. Therefore, it can be difficult to determine the dynamics of the residue to which the spin label is conjugated. Spin labels that are connected with a large, flexible tether have an EPR spectrum similar to a small molecule in solution. Thus, it is important that the label have limited or no flexibility independent of the nucleic acid to which it is attached when determining internal dynamics. Changes in dynamics by CW-EPR have, for example, been used to study metal-ion-dependent folding of the hammerhead ribozyme [41,42] and folding of the cocaine aptamer [43].

10.2.3.1.5 Distance Measurements by CW-EPR

EPR spectroscopy is useful for determining the distances between two radicals; the degree of dipolar coupling between the radicals is inversely related to r^3, where r is the distance between two radicals [38]. For CW-EPR, the dipolar coupling causes a line-broadening in the spectrum, which can be used to determine the distance between the radicals [31,44]. At distances longer than 25 Å, the line-broadening caused by the dipolar coupling is small relative to the width of the EPR line. Therefore, CW-EPR is usually used for the determination of distances between 8 and

20 Å. Such analysis can be nontrivial due to the effects of relative orientation of the two spins on the lineshape [45], but the relative orientation can provide additional valuable structural information. Distance measurements between two radicals by CW-EPR have been used to determine RNA conformational changes upon binding to metal ions [31] and should also be applicable for studying small-molecule binding to nucleic acids.

10.2.4 PULSED EPR SPECTROSCOPY: PELDOR/DEER

Recent advances in high-field instrumentation and the development of pulse sequences have advanced the field of EPR spectroscopy and facilitated the investigation of multispin systems [9,11,12,16,18,19]. In these experiments, the magnetic field is held constant while short pulses are applied to the sample. Pulsed EPR methods that are used to interrogate electron coupling with nuclei include electron spin echo envelope modulation (ESEEM), along with the two-dimensional variant 2D hyperfine sublevel correlation (HYSCORE), and electron nuclear double resonance (ENDOR). These experiments provide detailed information about the surroundings of radical centers and have, for example, been used extensively to study metal-ion binding sites in both proteins and RNA [11,16,46–50].

Pulsed EPR methods have also been developed to study interactions between radical centers in nucleic acids, specifically to measure distances [9,12,38,51,52]. PELDOR can be used to distinguish a dipolar coupling from other line-broadening effects. As a consequence, PELDOR is useful for the determination of distances up to 80 Å. Moreover, distance measurements by PELDOR do not have some of the limitations of distance measurements by CW-EPR, such as having to fully account for incomplete incorporation of spin labels.

In a four-pulse PELDOR experiment [53], microwave pulses (typically 5–30 ns long) at two different frequencies are applied to the sample. A refocused spin echo, generated with a pulse sequence in the observer frequency, is affected by a π-pulse at the second frequency if the radicals are dipolar coupled. When the intensity of the refocused spin echo is recorded as a function of time-position of the second frequency pulse, a time-domain signal is obtained that oscillates with the frequency of the dipolar coupling. From the Fourier-transformed PELDOR time-domain signal or from signal inversion by Tikhonov regularization, a distance between the two radicals can be directly calculated [54,55]. PELDOR can also be used to extract structural information beyond the determination of distances. For example, relative orientation of rigid spin labels embedded within nucleic acid duplexes has been determined [56].

10.3 CASE STUDIES: STRUCTURAL DYNAMICS OF THE TAR RNA

EPR spectroscopy has infrequently been used to probe RNA–ligand interactions. This section will describe the use of EPR to study changes in internal dynamics of the human immunodeficiency virus (HIV) TAR RNA (Figure 10.5b) upon binding to ligands and show how the dynamics give insight into the interaction. Binding of the TAR RNA to the Tat protein is vital for efficient transcription during viral replication [57] and has, therefore, been extensively studied as a drug target.

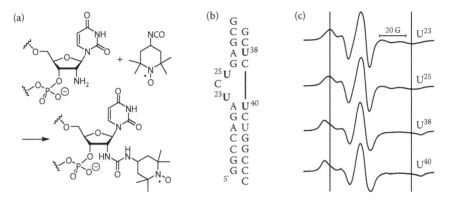

FIGURE 10.5 TAR RNA spin-labeling and EPR analysis. (a) Postsynthetic spin-labeling at the 2′-position of selected nucleotides in the TAR RNA. (b) The TAR RNA. The nucleotides that were spin-labeled (one label per TAR RNA) are numbered and bold. (c) EPR analysis of the four spin-labeled TAR RNAs. (From Edwards, T. E., *J. Am. Chem. Soc.* **123**, 1527–1528, 2001. With permission.)

After developing a general spin labeling approach for RNA, in which 2′-amino groups were postsynthetically modified with an aliphatic and nitroxide-containing isocyanate (Figure 10.5a), four spin-labeled TAR oligonucleotides were prepared (Figure 10.5b), where each TAR RNA contained a single spin label [30] (Figure 10.5b). The EPR spectra showed a variation in spectral shape (Figure 10.5c). Specifically, the width of the U23 and U25 spectra were narrower than observed for the U38 and U40 spectra. The narrower spectral width of U23 and U25 indicated that those nucleotides had higher mobility than U38 and U40, which correlated well with the structural context: U23 and U24 were located in a flexible loop, whereas U38 and U40 were placed in a duplex region.

The fact that the EPR spectra identified the more mobile nucleotides showed that the nitroxide was reporting the nucleotide mobility although it was connected to the nucleoside through a semiflexible linker. This result indicated that it could be possible to use EPR to identify structural motifs based on the internal mobility of specific nucleotides in the TAR RNA. In fact, several EPR studies have yielded information about specific structural motifs of ligands bound to the TAR RNA, such as small organic molecules [58], peptides [59,60], and metal ions [61]. For example, the internal dynamics of the TAR RNA were different for Mg^{2+} than for Na^+ [61], later shown by NMR spectroscopy to induce different conformations of the RNA [62]. Based on a crystallographic analysis of the TAR RNA, calcium ions had been suggested to induce a calcium-specific structural change in the TAR bulge [63], but EPR data showed that the internal dynamics of the TAR were the same for calcium and sodium ions [59].

The study of small-molecule interactions with the TAR RNA is detailed in Section 10.3.1 as an example of using the internal dynamics of RNA to obtain information about binding sites of drug-like ligands. It will also be shown how changes in the internal dynamics of the TAR RNA upon binding to different peptides identify amino acids in the Tat protein that are involved in specific complex formation with the TAR RNA (Section 10.3.2).

10.3.1   TAR RNA-BINDING MODE OF SMALL MOLECULES IDENTIFIED BY EPR

Given the biological importance of the TAR RNA, an active area of research has been to discover small-molecule inhibitors of the interaction of TAR with the Tat protein [64]. A few examples of such compounds are shown in Figure 10.6a. To determine their effect on the internal dynamics of the TAR RNA, the compounds were individually incubated with four TAR RNAs, each one containing a single spin label, at position U23, U25, U38, or U40 (Figure 10.5b) [58]. The EPR spectra of the resulting RNA–ligand complexes were subsequently collected. A convenient representation of the EPR data is to plot the change in spectral width as a function of spin label position for each complex, which gives a dynamic signature, or a fingerprint, for each ligand (Figure 10.6b).

For the multicyclic dyes, Hoechst 33258, DAPI, and berenil, the pattern is the same: the mobility of U23, U25, and U38 is higher (decrease in spectral width),

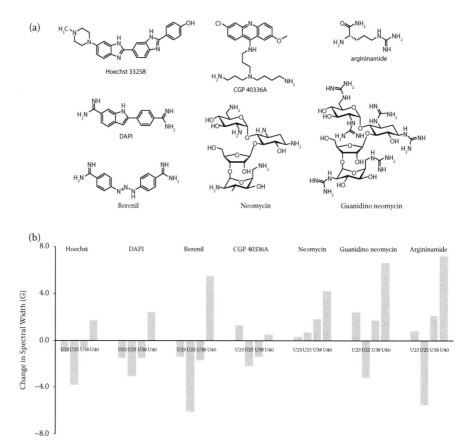

FIGURE 10.6   (a) Structures and names of compounds that bind to the TAR RNA. (b) Dynamic signatures of TAR RNA, bound to the ligands. The change in spectral width is plotted as a function of nucleotide position (U23, U25, U38, and U40) for each compound. (From Edwards, T. E. and Sigurdsson, S. T., *Biochemistry* **41**, 14843–14847, 2002. With permission.)

whereas the mobility of U40 increases. The same dynamic signature for these compounds suggests that they bind to the TAR RNA in a similar way. In fact, electric linear dichroism, UV, and CD spectroscopy, in addition to nuclease footprinting experiments, indicate that all of these compounds bind in the major groove pocket created by the trinucleotide bulge [65,66].

The acridine–spermidine complex CGP 40336A, a tight-binding ligand of the TAR RNA, has a dynamic signature different from all the other ligands. This compound, which has displayed strong antiviral activity *in vitro* [67], has been shown to bind to the G26•C39 base pair without distorting the RNA [67,68]. Therefore, it is not surprising that small changes are observed in the dynamics of the TAR RNA, in spite of tight binding.

Both neomycin and argininamide have specific binding motifs and unique dynamic signatures that are different from the others. Neomycin has been shown to bind to the minor groove of the lower stem [69], while argininamide, a simple analog of the Tat protein, has been shown to induce a conformational change in which the trinucleotide bulge is inverted and U23 forms a base triple with U38•A27 [70].

The aforementioned data is evidence for specific RNA internal dynamics for a particular RNA–ligand structure and suggests EPR spectroscopy as a useful and straightforward technique for identification of binding sites for RNA ligands. One of the compounds in Figure 10.6a, guanidino neomycin [71,72], binds to the TAR RNA, but its binding site is not known. This compound is a hybrid between neomycin and argininamide, in which the amino groups of neomycin have been replaced with guanidino groups. A guanidino group is the recognition moiety in argininamide that induces a large conformational change in the RNA [70]. It was of interest to determine if the same binding mode was observed for guanidino neomycin as for either neomycin or argininamide, or if an entirely new dynamic signature was obtained. The EPR data clearly shows the same signature for guanidino neomycin as for argininamide (Figure 10.6b), which strongly indicates that guanidino neomycin binds like argininamide. Thus, the guanidino groups govern the binding location of guanidino neomycin, rather than the overall shape and charge that would direct it to the same binding site as neomycin.

10.3.2 Internal Dynamics of the TAR RNA Identify Amino Acids in the Tat Protein that Are Important for Specific Complex Formation

The EPR studies of small-molecule binding to the TAR RNA demonstrated the ability of EPR spectroscopy to detect changes in structure-dependent RNA dynamics upon ligand binding. Therefore, this approach seemed amenable to study the interaction of the TAR RNA with its biological partner, the Tat protein.

It has been shown that the basic domain of the Tat protein, containing residues 47–57, is responsible for binding to the RNA. In particular, arginine 52 (R52) has been identified as a critical amino acid for the TAR–Tat complex formation, while mutation of amino acids flanking R52 to lysine had limited effects on either the binding affinity of short peptides containing the basic sequence or transactivation activity of the Tat protein [64,73,74]. Although a high-resolution structure for the TAR–Tat complex has not been solved yet, NMR spectroscopy has given insight into the interaction of the TAR RNA with the Tat protein. A structure of the TAR RNA bound to argininamide, the simplest analog of the Tat protein, revealed a major conformational change in the RNA [70]. NMR studies with Tat-derived peptides revealed two levels of molecular recognition, a specific recognition by the wild-type (wt) peptide RKKRRQRRR and nonspecific complex formation by the sequence RKKRKQKKK, where a number of arginines had been mutated to lysines [75].

The question remained whether the loss of specific complex formation for the mutant peptide RKKRKQKKK was due to mutation of a specific amino acid or multiple mutations. Determining which amino acids are responsible for the specific complex formation by NMR spectroscopy requires a substantial effort, whereas EPR spectroscopy enables rapid screening of peptide mutants through the collection of EPR dynamic signatures. To determine the feasibility of using EPR for distinguishing the specific from the nonspecific binding, the dynamic signatures of argininamide, the wt peptide YGRKKRRQRRR, and the mutant YKKKKRKKKKA were obtained [59] (Figure 10.7a). All three compounds have the same basic signature, with a decrease in the mobility of U23, U38, and U40, whereas U25 was more mobile. However, it was striking that the change in spectral width of U23 and particularly U38 showed a much larger decrease in mobility for the wt peptide than the mutant. This data clearly showed that EPR spectroscopy could be used to detect the specific complex formation. Therefore, a series of mutant peptides were prepared and their dynamic signatures determined by EPR [60].

There was only a minor change in the spectral width of U25 and U40 for all the mutant peptides. In contrast, a large variation in spin-label mobility was observed for U38 (Figure 10.7b). Thus, the U38 data enabled identification of amino acids that are important for specific complex formation. A few mutations did not have any effect, in particular G48K and R53K (peptides I and IV). Only a minor change was observed for R55K and R57K (peptides V and VII). Three peptides containing C-terminal

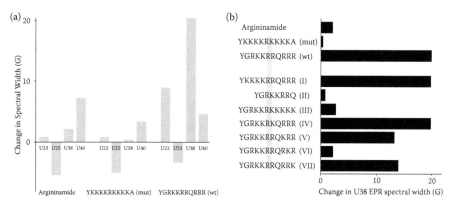

FIGURE 10.7 EPR analysis of TAR RNA binding to Tat analogs. (a) Dynamic signatures for argininamide, the mutant (mut), and the wild-type (wt) Tat peptides. (From Edwards, T. E., Okonogi, T. M., and Sigurdsson, S. T., *Chem. Biol.* **9**, 699–706, 2002. With permission.) (b) Change in spectral width of U38 in the TAR RNA upon binding to a series of Tat mutant peptides. Amino acids that were mutated are bold. A gray bar overlaying the peptide sequences identifies the essential amino acid R52. (From Edwards, T. E., Robinson, B. H., and Sigurdsson, S. T., *Chem. Biol.* **12**, 329–337, 2005. With permission.)

mutations showed dramatically reduced mobility for U38, similar to what was obtained for argininamide. Peptide II contained a C-terminal deletion and in peptide III, amino acids 53–57 were all converted to lysines. The most striking result was that of peptide VII, which contained a single amino acid mutation and showed non-specific binding, thereby identifying R56 as an important amino acid for the specific complex formation.

In an NMR structure of the Tat protein, the basic region forms one face of the protein, specifically amino acids 52–57 [76]. More importantly, R52, the essential amino acid in the TAR–Tat interaction, and R56, identified by EPR spectroscopy, protrude into the solution from the binding surface. Thus, the structure of the Tat protein provides indirect support for R56, identified by EPR spectroscopy, being an important amino acid that contributes to specific complex formation.

10.4   CONCLUSIONS

The use of dynamic signatures to study RNA–ligand interactions reduces the analysis of each EPR spectrum to a simple measure of spectral width. This straightforward approach relies on the selection of a sufficient number of spin-labeled nucleotides at or near the binding sites to give a unique signature for each compound. Dynamic signatures, obtained in this manner, provided insights into structural determinants of TAR RNA recognition by small molecules and peptides. Specifically, the small-molecule studies showed that a unique dynamic signature is obtained for compounds that bind in a particular fashion to the RNA. For the protein study, information about internal dynamics of the RNA, while bound to a series of mutant peptides, was used to identify amino acids in the Tat protein that are important for specific complex formation.

10.5   PROSPECTS AND OUTLOOK

RNA–ligand interactions are increasingly being studied by EPR spectroscopy and continued advances in instrumentation and spin-labeling techniques bode well for further applications. CW-EPR spectroscopy, using readily available spectrometers, can yield information about internal dynamics of RNAs upon binding to small molecules and peptides, as shown in the examples above, or to other RNAs, such as the GNRA binding to its receptor [29,32,77]. Other possible strategies for investigating conformational changes that are associated with ligand binding by EPR spectroscopy includes distance measurements between two spin labels. Distances of up to 20 Å can be determined using CW-EPR spectroscopy and measurements of distances in the range of 20–80 Å are possible using pulsed EPR spectroscopy, the latter ideally suited to study long-range, ligand-induced conformational changes [78,79]. PELDOR/DEER has, for example, been used to show that the neomycin riboswitch has a preorganized structure prior to ligand binding [80] and to reveal conformational changes associated with metal-ion binding to the hammerhead ribozyme [42]. Spin-labeled ligands can also be used to study ligand binding, especially of small molecules, since the mobility of the ligand can change dramatically upon binding [81]. A combination of spin-labeled ligands and spin-labeled RNA can also be envisioned for mapping binding sites and ligand orientation when bound to the

RNA through distance measurements. EPR is also uniquely suited to study specific metal-ion binding sites in RNA [47,48]. Thus, EPR spectroscopy holds great promise to make a further impact in the study of RNA structure and dynamics.

ACKNOWLEDGMENTS

I thank Dr. E. Hustedt, Dr. O. Schiemann, and Dr. T. F. Prisner for helpful discussions and S. A. Shelke for assistance with the preparation of figures.

REFERENCES

1. Holbrook, S. R. Structural principles from large RNAs, *Annu. Rev. Biophys.* **37**, 445–464, 2008.
2. Scott, L. G. and Hennig, M. RNA structure determination by NMR, *Methods Mol. Biol.* **452**, 29–61, 2008.
3. Reyes, F. E., Garst, A. D., and Batey, R. T. 2009. Strategies in RNA crystallography, In: *Methods in Enzymology, Vol 469: Biophysical, Chemical, and Functional Probes of RNA Structure, Interactions and Folding, Pt B*, ed. D. Herschlag, pp. 119–139. San Diego: Elsevier Academic Press Inc.
4. McPike, M. P., Goodisman, J., and Dabrowiak, J. C. 2001. Drug-RNA footprinting, In: *Methods in Enzymology, Vol 340: Drug-Nucleic Acid Interactions*, eds. J. B. Chaires and M. J. Waring, pp. 431–449. San Diego: Elsevier Academic Press Inc.
5. Eriksson, M. and Norden, B. 2001. Linear and circular dichroism of drug-nucleic acid complexes, In: *Methods in Enzymology, Vol 340: Drug-Nucleic Acid Interactions*, eds. J. B. Chaires and M. J. Waring, pp. 68–98. San Diego: Elsevier Academic Press Inc.
6. Luedtke, N. W. and Tor, Y. A novel solid-phase assembly for identifying potent and selective RNA ligands, *Angew. Chem. Int. Ed.* **39**, 1788–1790, 2000.
7. Hofstadler, S. A. and Griffey, R. H. Analysis of noncovalent complexes of DNA and RNA by mass spectrometry, *Chem. Rev.* **101**, 377–390, 2001.
8. Davis, T. M. and Wilson, W. D. 2001. Surface plasmon resonance biosensor analysis of RNA-small molecule interactions, In: *Methods in Enzymology, Vol 340: Drug-Nucleic Acid Interactions*, eds. J. B. Chaires and M. J. Waring, pp. 22–51. San Diego: Elsevier Academic Press Inc.
9. Prisner, T., Rohrer, M. and MacMillan, F. Pulsed EPR spectroscopy: Biological applications, *Annu. Rev. Phys. Chem.* **52**, 279–313, 2001.
10. Columbus, L. and Hubbell, W. L. A new spin on protein dynamics, *Trends Biochem. Sci.* **27**, 288–295, 2002.
11. Ubbink, M., Worrall, J. A. R., Canters, G. W., Groenen, E. J. J., and Huber, M. Paramagnetic resonance of biological metal centers, *Annu. Rev. Biophys. Biomol. Struct.* **31**, 393–422, 2002.
12. Schiemann, O. and Prisner, T. F. Long-range distance determinations in biomacro-molecules by EPR spectroscopy, *Q. Rev. Biophys.* **40**, 1–53, 2007.
13. Sowa, G. Z. and Qin, P. Z. Site-directed spin labeling studies on nucleic acid structure and dynamics, *Prog. Nucleic Acid Res. Mol. Biol.* **82**, 147–197, 2008.
14. Hunsicker-Wang, L., Vogt, M., and DeRose, V. J. 2009. EPR methods to study specific metal-ion binding sites in RNA, In: *Methods in Enzymology, Vol 468: Biophysical, Chemical, and Functional Probes of RNA Structure, Interactions and Folding, Pt A*, ed. D. Herschlag, pp. 335–367. San Diego: Elsevier Academic Press Inc.
15. Plonka, P. M. and Elas, M. Application of the electron paramagnetic resonance spectros-copy to modern biotechnology, *Curr. Top. Biophys.* **26**, 175–189, 2002.

16. Calle, C., Sreekanth, A., Fedin, M. V., Forrer, J., Garcia-Rubio, I., Gromov, I. A., Hinderberger, D. et al. Pulse EPR methods for studying chemical and biological samples containing transition metals, *Helv. Chim. Act.* **89**, 2495–2521, 2006.

17. Qin, P. Z. and Dieckmann, T. Application of NMR and EPR methods to the study of RNA, *Curr. Opin. Struct. Biol.* **14**, 350–359, 2004.

18. Bennati, M. and Prisner, T. F. New developments in high-field electron paramagnetic resonance with applications in structural biology, *Rep. Prog. Phys.* **68**, 411–448, 2005.

19. Mobius, K., Savitsky, A., Schnegg, A., Plato, M., and Fuchs, M. High-field EPR spectroscopy applied to biological systems: Characterization of molecular switches for electron and ion transfer, *Phys. Chem. Chem. Phys.* **7**, 19–42, 2005.

20. Zhang, X. J., Cekan, P., Sigurdsson, S. T., and Qin, P. Z. 2009. Studying RNA using site-directed spin-labeling and continuous-wave electron paramagnetic resonance spectroscopy, In: *Methods in Enzymology, Vol 469: Biophysical, Chemical, and Functional Probes of RNA Structure, Interactions and Folding, Pt B*, ed. D. Herschlag, pp. 303–328. San Diego: Elsevier Academic Press Inc.

21. Cruickshank, P. A. S., Bolton, D. R., Robertson, D. A., Hunter, R. I., Wylde, R. J., and Smith, G. M. A kilowatt pulsed 94 GHz electron paramagnetic resonance spectrometer with high concentration sensitivity, high instantaneous bandwidth, and low dead time, *Rev. Sci. Instrum.* **80**, 103102, 2009.

22. Gannett, P. M., Darian, E., Powell, J., Johnson, E. M., 2nd, Mundoma, C., Greenbaum, N. L., Ramsey, C. M., Dalal, N. S., and Budil, D. E. Probing triplex formation by EPR spectroscopy using a newly synthesized spin label for oligonucleotides, *Nucleic Acids Res.* **30**, 5328–5337, 2002.

23. Piton, N., Mu, Y., Stock, G., Prisner, T. F., Schiemann, O., and Engels, J. W. Base-specific spin-labeling of RNA for structure determination, *Nucleic Acids Res.* **35**, 3128–3143, 2007.

24. Cekan, P., Smith, A. L., Barhate, N., Robinson, B. H., and Sigurdsson, S. T. Rigid spin-labeled nucleoside Ç: A nonperturbing EPR probe of nucleic acid conformation, *Nucleic Acids Res.* **36**, 5946–5954, 2008.

25. Schiemann, O., Piton, N., Plackmeyer, J., Bode, B. E., Prisner, T. F., and Engels, J. W. Spin-labeling of oligonucleotides with the nitroxide TPA and use of PELDOR, a pulse EPR method, to measure intramolecular distances, *Nat. Protoc.* **2**, 904–923, 2007.

26. Sicoli, G., Wachowius, F., Bennati, M., and Hobartner, C. Probing secondary structures of spin-labeled RNA by pulsed EPR spectroscopy, *Angew. Chem. Int. Ed.* **49**, 6443–6447, 2010.

27. Hara, H., Horiuchi, T., Saneyoshi, M., and Nishimura, S. 4-Thiouridine-specific spin-labeling of *E. coli* transfer RNA, *Biochem. Biophys. Res. Comm.* **38**, 305–311, 1970.

28. Ramos, A. and Varani, G. A new method to detect long-range protein-RNA contacts: NMR detection of electron-proton relaxation induced by nitroxide spin-labeled RNA, *J. Am. Chem. Soc.* **120**, 10992–10993, 1998.

29. Qin, P. Z., Hideg, K., Feigon, J., and Hubbell, W. L. Monitoring RNA base structure and dynamics using site-directed spin-labeling, *Biochemistry* **42**, 6772–6783, 2003.

30. Edwards, T. E., Okonogi, T. M., Robinson, B. H., and Sigurdsson, S. T. Site-specific incorporation of nitroxide spin labels into internal sites of the TAR RNA; Structure-dependent dynamics of RNA by EPR spectroscopy, *J. Am. Chem. Soc.* **123**, 1527–1528, 2001.

31. Kim, N. K., Murali, A. and DeRose, V. J. A distance ruler for RNA using EPR and site-directed spin labeling, *Chem. Biol.* **11**, 939–948, 2004.

32. Qin, P. Z., Butcher, S. E., Feigon, J., and Hubbell, W. L. Quantitative analysis of the GAAA tetraloop/receptor interaction in solution: A site-directed spin labeling study, *Biochemistry* **40**, 6929–6936, 2001.

33. Popova, A. M., Kalai, T., Hideg, K., and Qin, P. Z. Site-specific DNA structural and dynamic features revealed by nucleotide-independent nitroxide probes, *Biochemistry* **48**, 8540–8550, 2009.

34. Macosko, J. C., Pio, M. S., Tinoco Jr., I., and Shin, Y. K. A novel 5' displacement spin-labeling technique for electron paramagnetic resonance spectroscopy of RNA, *RNA* **5**, 1158–1166, 1999.

35. Grant, G. P. G. and Qin, P. Z. A facile method for attaching nitroxide spin labels at the 5' terminus of nucleic acids, *Nucleic Acids Res.* **35**, e77, 2007.

36. Grant, G. P. G., Boyd, N., Herschlag, D., and Qin, P. Z. Motions of the substrate recognition duplex in a group I intron assessed by site-directed spin labeling, *J. Am. Chem. Soc.* **131**, 3136–3137, 2009.

37. Caron, M. and Dugas, H. Specific spin-labeling of transfer ribonucleic acid molecules, *Nucleic Acids Res.* **3**, 19–34, 1976.

38. Schiemann, O. 2009. Mapping global folds of oligonucleotides by pulsed electron-electron double resonance, In: *Methods in Enzymology, Vol 469: Biophysical, Chemical, and Functional Probes of RNA Structure, Interactions and Folding, Pt B*, ed. D. Herschlag, 329–351. San Diego: Elsevier Academic Press Inc.

39. Freed, J. H. 1976. Theory of slow tumbling ESR spectra for nitroxides, In: *Spin Labeling: Theory and Application*, ed. L. J. Berliner, pp. 53–132. New York: Academic Press.

40. Budil, D. E., Lee, S., Saxena, S., and Freed, J. H. Nonlinear-least-squares analysis of slow-motion EPR spectra in one and two dimensions using a modified Levenberg-Marquardt algorithm, *J. Magn. Reson. Ser. A* **120**, 155–189, 1996.

41. Edwards, T. E. and Sigurdsson, S. T. EPR spectroscopic analysis of U7 hammerhead ribozyme dynamics during metal ion induced folding, *Biochemistry* **44**, 12870–12878, 2005.

42. Kim, N. K., Bowman, M. K., and DeRose, V. J. Precise mapping of RNA tertiary structure via nanometer distance measurements with double electron-electron resonance spectroscopy, *J. Am. Chem. Soc.* **132**, 8882–8883, 2010.

43. Cekan, P., Jonsson, E. O., and Sigurdsson, S. T. Folding of the cocaine aptamer studied by EPR and fluorescence spectroscopies using the bifunctional spectroscopic probe Ç, *Nucleic Acids Res.* **37**, 3990–3995, 2009.

44. Rabenstein, M. D. and Shin, Y. K. Determination of the distance between two spin labels attached to a macromolecule, *Proc. Natl. Acad. Sci.* **92**, 8239–8243, 1995.

45. Hustedt, E. J. and Beth, A. H. 2001. Structural information from CW-EPR spectra of dipolar-coupled nitroxide spin-labels, In: *Biological Magnetic Resonance* ed. S. S. Eaton, G. R. Eaton, and L. J. Berliner, pp. 155–184. New York: Kluwer Academic/Plenum Press.

46. Schiemann, O., Fritscher, J., Kisseleva, N., Sigurdsson, S. T., and Prisner, T. F. Structural investigation of a high-affinity Mn-II binding site in the hammerhead ribozyme by EPR spectroscopy and DFT calculations. Effects of neomycin B on metal-ion binding, *ChemBioChem* **4**, 1057–1065, 2003.

47. Kisseleva, N., Khvorova, A., Westhof, E., and Schiemann, O. Binding of manganese(II) to a tertiary stabilized hammerhead ribozyme as studied by electron paramagnetic resonance spectroscopy, *RNA* **11**, 1–6, 2005.

48. Vogt, M., Lahiri, S., Hoogstraten, C. G., Britt, R. D., and DeRose, V. J. Coordination environment of a site-bound metal ion in the hammerhead ribozyme determined by N-15 and H-2 ESEEM spectroscopy, *J. Am. Chem. Soc.* **128**, 16764–16770, 2006.

49. Kisseleva, N., Kraut, S., Jaschke, A., and Schiemann, O. Characterizing multiple metal ion binding sites within a ribozyme by cadmium-induced EPR silencing, *HFSPJ.* **1**, 127–136, 2007.

50. Schiemann, O., Carmieli, R., and Goldfarb, D. W-band P-31-ENDOR on the high-affinity Mn^{2+} binding site in the minimal and tertiary stabilized hammerhead ribozymes, *Appl. Magn. Res.* **31**, 543–552, 2007.

51. Schiemann, O., Weber, A., Edwards, T. E., Prisner, T. F., and Sigurdsson, S. T. Nanometer distance measurements on RNA using PELDOR, *J. Am. Chem. Soc.* **125**, 3434–3435, 2003.

52. Schiemann, O., Piton, N., Mu, Y., Stock, G., Engels, J. W., and Prisner, T. F. A PELDOR-based nanometer distance ruler for oligonucleotides, *J. Am. Chem. Soc.* **126**, 5722–5729, 2004.

53. Martin, R. E., Pannier, M., Diederich, F., Gramlich, V., Hubrich, M., and Spiess, H. W. Determination of end-to-end distances in a series of TEMPO diradicals of up to 2.8 nm length with a new four-pulse double electron electron resonance experiment, *Angew. Chem. Int. Ed.* **37**, 2834–2837, 1998.

54. Jeschke, G., Chechik, V., Ionita, P., Godt, A., Zimmermann, H., Banham, J., Timmel, C. R., Hilger, D., and Jung, H. DeerAnalysis2006—A comprehensive software package for analyzing pulsed ELDOR data, *Appl. Magn. Res.* **30**, 473–498, 2006.

55. Jeschke, G. and Polyhach, Y. Distance measurements on spin-labelled biomacromolecules by pulsed electron paramagnetic resonance, *Phys. Chem. Chem. Phys.* **9**, 1895–1910, 2007.

56. Schiemann, O., Cekan, P., Margraf, D., Prisner, T. F., and Sigurdsson, S. T. Relative orientation of rigid nitroxides by PELDOR: Beyond distance measurements in nucleic acids, *Angew. Chem. Int. Ed.* **48**, 3292–3295, 2009.

57. Frankel, A. D. and Young, J. A. T. HIV-1: Fifteen proteins and an RNA, *Annu. Rev. Biochem.* **67**, 1–25, 1998.

58. Edwards, T. E. and Sigurdsson, S. T. Electron paramagnetic resonance dynamic signatures of TAR RNA-small molecule complexes provide insight into RNA structure and recognition, *Biochemistry* **41**, 14843–14847, 2002.

59. Edwards, T. E., Okonogi, T. M., and Sigurdsson, S. T. Investigation of RNA-protein and RNA-metal ion interactions by electron paramagnetic resonance spectroscopy. The HIV TAR-Tat motif, *Chem. Biol.* **9**, 699–706, 2002.

60. Edwards, T. E., Robinson, B. H., and Sigurdsson, S. T. Identification of amino acids that promote specific and rigid TAR RNA-tat protein complex formation, *Chem. Biol.* **12**, 329–337, 2005.

61. Edwards, T. E. and Sigurdsson, S. T. EPR spectroscopic analysis of TAR RNA-metal ion interactions, *Biochem. Biophys. Res. Commun.* **303**, 721–725, 2003.

62. Casiano-Negroni, A., Sun, X. Y., and Al-Hashimi, H. M. Probing Na^+-induced changes in the HIV-1 TAR conformational dynamics using NMR residual dipolar couplings: New insights into the role of counterions and electrostatic interactions in adaptive recognition, *Biochemistry* **46**, 6525–6535, 2007.

63. Ippolito, J. A. and Steitz, T. A. A 1.3-Å resolution crystal structure of the HIV-1 trans-activation response region RNA stem reveals a metal ion-dependent bulge conformation, *Proc. Natl. Acad. Sci. U.S.A.* **95**, 9819–9824, 1998.

64. Froeyen, M. and Herdewijn, P. RNA as a target for drug design, the example of Tat-TAR interaction, *Curr. Op. Med. Chem.* **2**, 1123–1145, 2002.

65. Bailly, C., Colson, P., Houssier, C., and Hamy, F. The binding mode of drugs to the TAR RNA of HIV-1 studied by electric linear dichroism, *Nuc. Acids Res.* **24**, 1460–1464, 1996.

66. Dassonneville, L., Hamy, F., Colson, P., Houssier, C., and Bailly, C. Binding of Hoechst 33258 to the TAR RNA of HIV-1. Recognition of a pyrimidine bulge-dependent structure, *Nucleic Acids Res.* **25**, 4487–4492, 1997.

67. Hamy, F., Brondani, V., Florsheimer, A., Stark, W., Blommers, M. J. J., and Klimkait, T. A new class of HIV-1 Tat antagonist acting through Tat-TAR inhibition, *Biochemistry* **37**, 5086–8095, 1998.

68. Gelus, N., Hamy, F., and Bailly, C. Molecular basis of HIV-1 TAR RNA specific recognition by an acridine Tat-antagonist, *Bioorg. Med. Chem.* **7**, 1075–1079, 1999.

69. Faber, C., Sticht, H., Schweimer, K., and Rösch, P. Structural rearrangement of HIV-1 Tat-responsive RNA upon binding of neomycin B, *J. Biol. Chem.* **275**, 20660–20666, 2000.

70. Brodsky, A. S. and Williamson, J. R. Solution structure of the HIV-2 TAR-argininamide complex, *J. Mol. Biol.* **267**, 624–639, 1997.
71. Baker, T. J., Luedtke, N. W., Tor, Y., and Goodman, M. Synthesis and anti-HIV activity of guanidinoglycosides, *J. Org. Chem.* **65**, 9054–9058, 2000.
72. Luedtke, N. W., Baker, T. J., Goodman, M., and Tor, Y. Guanidinoglycosides: A novel family of RNA ligands, *J. Am. Chem. Soc.* **122**, 12035–12036, 2000.
73. Calnan, B. J., Biancalana, S., Hudson, D., and Frankel, A. D. Analysis of arginine-rich peptides from the HIV Tat protein reveals unusual features of RNA-protein recognition, *Genes Devel.* **5**, 201–210, 1991.
74. Tao, J. and Frankel, A. D. Electrostatic interactions modulate the RNA-binding and transactivation specificities of the human immunodeficiency virus and simian immunodeficiency virus Tat proteins, *Proc. Natl. Acad. Sci. U.S.A.* **90**, 1571–1575, 1993.
75. Long, K. S. and Crothers, D. M. Characterization of the solution conformations of unbound and Tat peptide-bound forms of HIV-1 TAR RNA, *Biochemistry* **38**, 10059–10069, 1999.
76. Bayer, P., Kraft, M., Ejchart, A., Westendorp, M., Frank, R., and Rosch, P. Structural studies of the HIV-1 Tat protein, *J. Mol. Biol.* **247**, 529–535, 1995.
77. Qin, P. Z., Feigon, J., and Hubbell, W. L. Site-directed spin labeling studies reveal solution conformational changes in a GAAA tetraloop receptor upon Mg(2+)-dependent docking of a GAAA tetraloop, *J. Mol. Biol.* **351**, 1–8, 2005.
78. Sicoli, G., Mathis, G., Delalande, O., Boulard, Y., Gasparutto, D., and Gambarelli, S. Double electron-electron resonance (DEER): A convenient method to probe DNA conformational changes, *Angew. Chem. Int. Ed.* **47**, 735–737, 2008.
79. Sicoli, G., Mathis, G., Aci-Seche, S., Saint-Pierre, C., Boulard, Y., Gasparutto, D., and Gambarelli, S. Lesion-induced DNA weak structural changes detected by pulsed EPR spectroscopy combined with site-directed spin labelling, *Nucleic Acids Res.* **37**, 3165–3176, 2009.
80. Krstic, I., Frolow, O., Sezer, D., Endeward, B., Weigand, J. E., Suess, B., Engels, J. W., and Prisner, T. F. PELDOR spectroscopy reveals preorganization of the neomycin-responsive riboswitch tertiary structure, *J. Am. Chem. Soc.* **132**, 1454–1455, 2010.
81. Shelke, S. A. and Sigurdsson, S. T. Noncovalent and site-directed spin labeling of nucleic acids, *Angew. Chem. Int. Ed.* **49**, 7984–7986, 2010.

11 Electrochemical Approaches to the Study of DNA/Drug Interactions

Shinobu Sato and Shigeori Takenaka

CONTENTS

In this chapter, electrochemistry-based analysis of DNA/drug interactions is described. Electrodes modified with DNA allow quantitative analysis of DNA/drug interactions using minute amounts in homogenous solution. The detection principle is based on changes in the redox potential of the drug, as well as decreases in the drug's apparent diffusion coefficient at the electrode upon binding to DNA.

11.1 INTRODUCTION

Analysis of DNA/drug interactions is important for clarifying mechanisms of action that drugs have on DNA, as well as for developing future drugs that can interact with a diverse set of DNA structures [1]. This chapter discusses electrochemistry-based approaches to studying DNA/drug interactions. The technique exploits the fact that the redox reactions of drugs at metal electrodes change upon interaction with DNA [2,3]. The electrochemical approach is particularly useful in cases where the spectroscopic properties of the drug molecules do not appreciably change when they bind to DNA.

The nucleobases of DNA are themselves redox active, although electrochemical processes only occur at high voltages [4]. Generally, double-stranded DNA (dsDNA) is hardly amenable to redox reaction, as its nucleobases form interstrand hydrogen bonds. In contrast, single-stranded DNA (ssDNA) is more susceptible to redox reactions, as its nucleobases are more exposed to the solvent and electrodes, facilitating electron transfer. Such single-stranded regions are generated upon DNA damage in biological systems and in fact they were successfully analyzed by electrochemical means [5]. Since the redox potentials of the nucleobases are relatively high, proper choice of electrodes is essential for successful electrochemical analysis of DNA [5]. By properly choosing electrodes, electrochemistry of DNA can also be used to discriminate among methylated and unmethylated cytosines, important epigenetic modifications that regulate gene transcription [6].

As described above, under moderate potentials, DNA is electrochemically inactive, and drugs that interact with DNA may be selectively detected by proper choice of electrodes and applied potentials. The redox properties of antibiotics have been previously studied electrochemically, as many of them have redox active groups such as benzoquinone (Figure 11.1). Electrochemical analysis of drug/DNA interactions is a topic that has been summarized in several excellent reviews [2,3] and monographs [7]. In addition to these static aspects of drug/DNA interactions, the reaction process of DNA with drugs can be studied electrochemically. For example, adriamycin-induced oxidative damage to DNA that results in the formation of toxic 8-oxoguanine has been studied electrochemically [8].

First, the interactions of drugs with DNA in solution are discussed. When a potential is applied at an electrode immersed in a solution of DNA and drug, free DNA and drug-bound DNA molecules are concentrated near the electrode, where these molecules undergo electron-transfer redox reactions. Analysis of changes in the redox behavior of

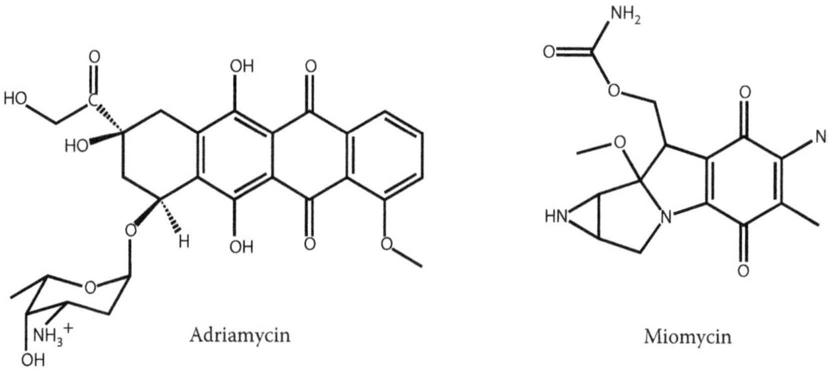

FIGURE 11.1   Chemical structure of some anticancer drugs.

drugs before and after binding to DNA yields useful information. Detailed studies of a DNA-binding metal complex with dsDNA were reported [9], and the interactions of dsDNA immobilized on a glassy carbon electrode (GCE) with the DNA-binding metal complex $Co(bpy)_3^{3+}$ were recently studied [10].

Immobilization of thiolated oligonucleotides (ssDNAs) on a gold electrode has been thoroughly explored for studying drug/DNA interactions. The relationship of the length of immobilized oligonucleotide with the surface density and the conformation of the oligonucleotides on the electrode was studied extensively [11]. More specifically, the orientation of oligonucleotides immobilized through thiol or disulfide groups on a single crystal Au(111) and a polycrystalline Au surface was explored by *in situ* scanning tunneling microscopy [12]. Finally, the mobility of electrode-immobilized thiolated oligonucleotides baring ferrocene moieties at opposite termini was analyzed from the electrochemical behavior of the ferrocene. It was found that 20-mer oligonucleotides are disposed with a 40° tilt from the electrode surface, and that their motion can be delineated by an elastic-bending model [13].

The amount of DNA immobilized on the electrode was evaluated on the basis of the fact that hexaaminoruthenium (III) $(Ru(NH_3)^{3+})$ binds with the phosphate anions of DNA at a 1:3 ratio [14]. It was also shown that methylene blue (MB) binds in the distal part of dsDNA when the dsDNA is immobilized densely on the electrode [15]. Also, when immobilized on the electrode at high density, dsDNAs are bound perpendicularly to the electrode surface. This in turn implies that nucleobases are disposed parallel to the electrode, through which electron transfer to and from the electrode may be enabled [16]. Those drugs that bear high affinity for dsDNA and are electrochemically active can be used as a sensor of DNA/drug interactions. Drug types include metal complexes, intercalators, and groove binders, as shown in Figure 11.2 [17]. Intercalators are especially prominent nucleic acid-binding drugs, and polyintercalators [18], including polymeric [19], threading [20], and anionic intercalators [21], have been studied extensively.

11.2 BASICS

11.2.1 Analysis in Solution

It is important to select electrodes for collecting data on the electrochemical behavior of drugs to be analyzed. Hanging mercury drop electrodes have been electrodes of choice as they can retain the active surface for a long period of time [4,5]. However, GCE is becoming more popular, as it is easier to handle [22]. Reviewed briefly below is one of such examples in which DNA/drug interactions in solution were analyzed.

Interactions of several DNA-binding metal complexes with calf thymus DNA (native dsDNA) were analyzed in detail by cyclic voltammetry (CV) [9]. Equilibrium between the DNA-bound and DNA-free forms is depicted in Scheme 11.1. The Nernst equation is transformed into Equation 11.1. Upon binding to calf thymus DNA, $Co(bpy)_3^{3+}$ underwent a shift in E_b toward the higher potential

$$E_b^{o'} - E_f^{o'} = 0.059 \log \frac{K_{2+}}{K_{3+}} \tag{11.1}$$

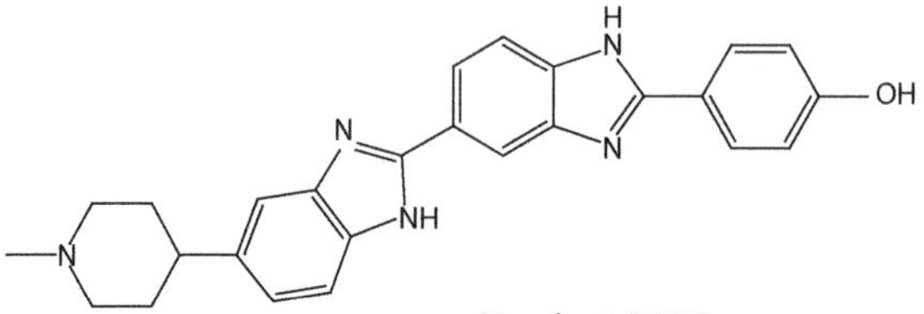

Hoechest 33258

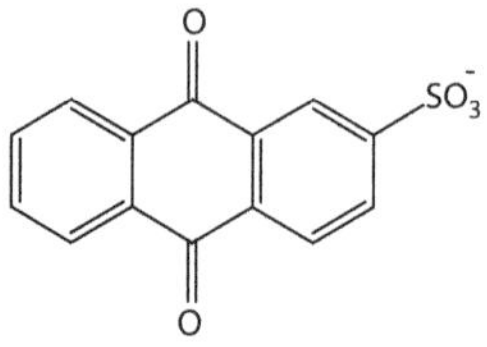

AQMS

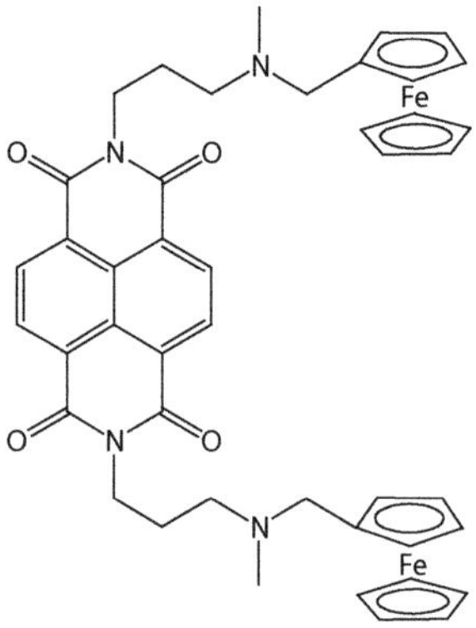

FND

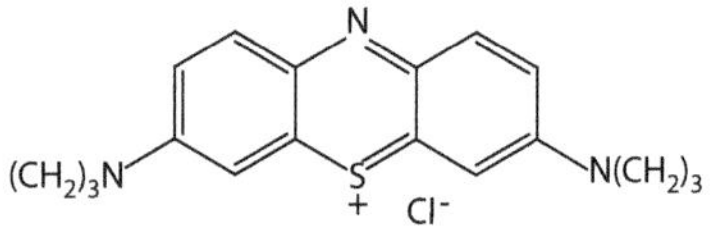

Methylene blue (MB)

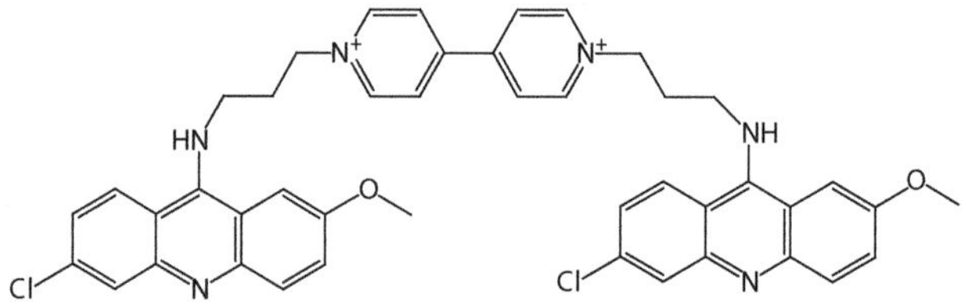

Bisacridinyl viologen

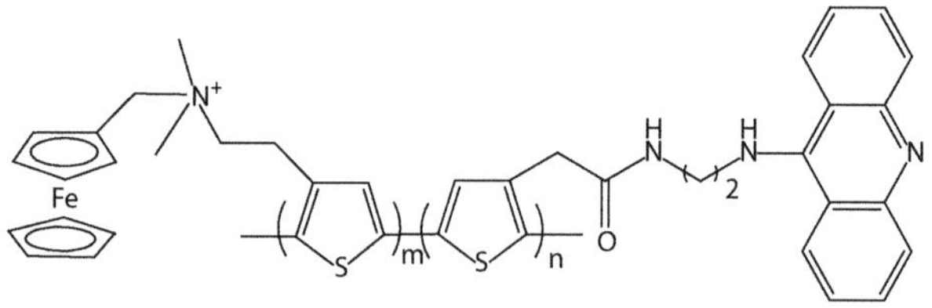

Multifunctional
conjugated polythiophene

FIGURE 11.2   Example of DNA hybridization indicators.

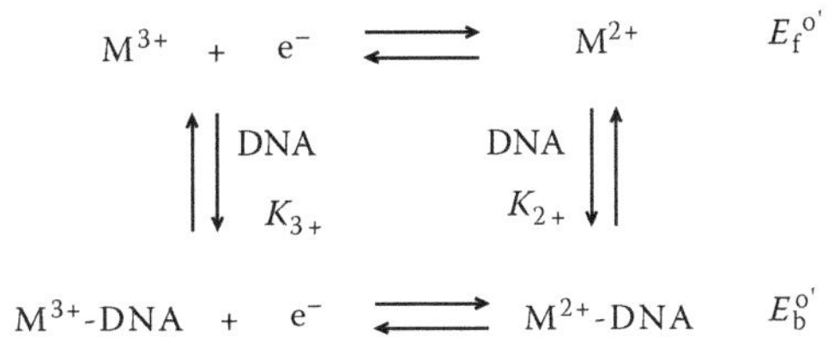

$$M^{3+} + e^- \rightleftharpoons M^{2+} \qquad E_f^{o'}$$

$$\updownarrow \text{DNA} \qquad \text{DNA} \updownarrow$$

$$K_{3+} \qquad K_{2+}$$

$$M^{3+}\text{-DNA} + e^- \rightleftharpoons M^{2+}\text{-DNA} \qquad E_b^{o'}$$

SCHEME 11.1

with a decrease in the current. The latter may be ascribed to a decrease in the diffusion rate due to binding to high-molecular-weight DNA. The Randles–Sevcik equation (Equation 11.2) holds here

$$i_p = 269An^{3/2}D^{1/2}Cv^{1/2} \tag{11.2}$$

where A represents the electrode area (cm^2), n the number of electrons involved in the reaction, D the diffusion coefficient (cm^2/s), C the analytical concentration of the complex (mol/L), and v the scan rate (V/s). The D_f for the metal complex alone was 5.0×10^{-6} cm^2/s at 25°C, whereas D_b for the DNA complex was 3.2×10^{-7} cm^2/s, that is, a tenth of the former. In addition, the negative potential shift associated with DNA binding is larger for K_{3+} than K_{2+}, suggesting that electrostatic interactions predominate over hydrophobic ones in this system.

The binding constant of the complex for DNA is determined from the peak current i_o at various DNA or complex concentrations. The current arises from diffusion of the DNA-bound and DNA-free forms toward the electrode. In the former, analytical methods vary whether the exchange of the DNA-bound and DNA-free forms occurs fast enough on the time scale of electrochemical measurements (mobile) or not (static). The i is expressed by $BC_t(D_fX_f + D_bX_b)^{1/2}$ and $B(D_f^{1/2}C_f + D_b^{1/2}C_b)$, respectively, where $X_f = C_f/C_t$, $X_b = C_b/C_t$, and $B = 269n^{2/3}Av^{1/2}$. When the DNA concentration [NP] is varied at a constant complex concentration (C_t) in the mobile system, the binding constant K and coordination number s are derived by plotting i against [NP] and fitting the data into the following equation. In one case, $K = 9400$ M^{-1} and $s \sim 3$ were obtained. Given the peak current in the absence of DNA as i_o and the one at saturation as i_{sat}, the fraction of the complex is expressed by Equation 11.3, which is rearranged into Equation 11.4 [9,23].

$$X_b = \frac{i^2 - i_o^2}{i_{sat}^2 - i_o^2} \tag{11.3}$$

$$X_b = \frac{C_b}{C_t} = \frac{b - \left(b^2 - 2K^2C_t[\text{NP}]/s\right)^{\frac{1}{2}}}{2KC_t} \tag{11.4}$$

$$b = 1 + KC_t + \frac{K[\text{NP}]}{2s} \tag{11.5}$$

Meanwhile, Mikkelsen and coworkers [10] assumed that i_o at saturation with respect to DNA represents that for the DNA/ligand complex, estimated the fraction of the complex from the half-maximal current, and obtained K and s on the basis of Scatchard plots and the McGhee–von Hippel equation (Equation 11.6) [24], which is one form of the Langmuir isotherms and is often used for the analysis of DNA/drug interactions.

$$\frac{r}{C_f} = K(1-sr)\left(\frac{1-sr}{1-(s-1)r}\right)^{s-1} \tag{11.6}$$

Specifically, $K = 3900$ M^{-1} and $s \sim 2.5$ were obtained for Co(bpy)$_3^{3+}$ and calf thymus DNA. These values are of the same order, but considerably smaller than those described above for the same system. It is noted that the McGhee–von Hippel equation is based on the binding of small molecules with one-dimensional polymers like DNA and may be more suitable for the analysis of these systems.

11.2.2   DNA-Immobilized Electrodes

Several methods are known to immobilize DNA on the electrode. Thus, GCE is oxidized with 2.5% K$_2$Cr$_2$O$_7$ and 10% HNO$_3$ at +1.5 V for 15 s to generate carboxylic groups on the surface, which are then converted to active esters of N-hydroxysuccinimide with water-soluble carbodiimide [25]. DNA is allowed to couple with the product to achieve immobilization [25]. It is supposed that the DNA is bound at amino groups of its guanine and cytosine bases in the single-stranded region through amide bonds. Hence, the structure of dsDNA may be perturbed more or less at the immobilization site.

In contrast, dsDNA may retain integrity in the immobilization of thiolated oligonucleotides on the gold electrode, when run under as mildest conditions as possible. Thus, 15-meric double-stranded oligonucleotides were immobilized to 2.46×10^{13} molecules/cm^2 [15]. Given the diameter of dsDNA at 20 Å, 55% of the electrode surface is covered with dsDNA. This is an order of magnitude smaller than a typical immobilization density of 5×10^{14} molecules/cm^2 for the self-assembled membrane (SAM) with n-alkanethiol. Nonetheless, virtually the entire surface is covered with dsDNA molecules. It was found that MB binds only at the distal sites of dsDNA immobilized on the electrode this way. In other words, dsDNA proximal to the electrode is so crowded, that even such small molecules as MB can hardly gain access. Incidentally, interactions of drugs with immobilized dsDNA can be analyzed by the same Langmuir-type isotherms described above [15].

Immobilization of thiolated oligonucleotides on a gold electrode was studied in great detail [11]. The random coil properties of the oligonucleotides to be immobilized were found to govern the resulting immobilization density. Thus, oligonucleotides shorter than 24-meric assume extended conformations and hence are immobilized at a relatively high density. In contrast, longer oligonucleotides assume a random coil and the immobilization density is considerably lower. Incidentally, reducing the concentration and/or shortening the coupling reaction time can lower the immobilization density for shorter oligonucleotides. Nearly complete hybridization with short complementary oligonucleotides to prepare dsDNA-immobilized electrode is possible at 10^{12} molecules/cm^2 immobilization density [14]. Some bare gold surface remains exposed at immobilization densities like these, but it can be masked by treatment with 6-mercaptohexanol (MCH), which forms a SAM in the bare gold regions and prevents nonspecific adsorption [26], as depicted in Figure 11.3. The immobilized DNA with an MCH SAM remains stable at temperatures as high as 75°C.

The amount of DNA immobilized on the electrode was determined by chronocoulometry (CC) in the presence of $Ru(NH_3)_3^{3+}$ [14]. In the Cottrell equation (Equation 11.7) for the chronocoulogram in the absence of $Ru(NH_3)_3^{3+}$, the y-intercept at zero time represents Q_{dl} or the charge in the electric bilayer, and $nFA\Gamma_0$ is obtained by subtracting Q_{dl} from the y-intercept of the chronocoulogram in the presence of $Ru(NH_3)_3^{3+}$. Accordingly, the amount of ssDNA present on the electrode is given by Equation 11.8, where N_A represents the Avogadro number, z the charge of $Ru(NH_3)_3^{3+}$, and m the number of bases of DNA.

$$Q = \frac{2nFAD^{1/2}C}{\pi^{1/2}}t^{1/2} + Q_{dl} + nFA\Gamma_0 \tag{11.7}$$

$$\Gamma_{DNA^{ss}} = \Gamma_0 \frac{z}{m} N_A \tag{11.8}$$

The amount of double-stranded DNA Γ_{DNA}^{ds} is estimated by CC after hybridization with the complementary DNA, and the efficiency of hybridization is expressed as follows:

$$\text{Hybridization efficiency (\%)} = \left\{ (\Gamma_{DNA}^{ds} - \Gamma_{DNA}^{ss}) / \Gamma_{DNA}^{ss} \right\} \times 100\%.$$

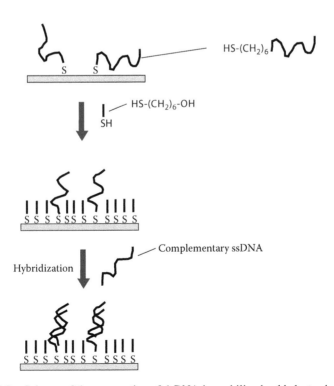

FIGURE 11.3   Scheme of the preparation of dsDNA-immobilized gold electrodes.

11.3 CASE STUDIES

Shown below is an analytical example using ferrocenylnaphthalenediimide (FND) as threading intercalator, which was developed as an indicator of hybridization [27]. Electrochemical measurements were made with a three-electrode configuration consisting of an Ag/AgCl reference electrode, a Pt counter electrode, and an ssDNA- or dsDNA-immobilized gold electrode masked with MCH as the working electrode, as shown in Figure 11.4.

11.3.1 DNA/Drug Interactions in Homogeneous Solution

As FND is easily adsorbed on the GCE and bare gold electrodes, MCH-masked electrodes were used in the experiment. Pretreatment of gold electrodes was carried out according to the literature [28]. The surface of a gold electrode (diameter 1.6 mm, area 0.020 cm^2) was physically polished (polishing with 6 μm diamond slurry for 30 min, 1 μm diamond slurry for 30 min, washed with Milli-Q water, 0.05 μm alumina slurry for 30 min, and sonication in Milli-Q water for 5 min three times) and subsequently electrochemically polished (1000–2000 scans over the range of −0.35 to −1.35 V versus Ag/AgCl at 2 V/s in 0.5 M NaOH until a stable voltammogram is obtained). After washing with Milli-Q water, the electrode was oxidized and reduced by applying 2 V for 5 s, −0.35 V for 10 s in 0.5 M H$_2$SO$_4$, and subsequently scanned 40 times over the range of −0.35–1.5 V versus Ag/AgCl at 4 V/s and finally scanned four times over the same voltage range at 100 mV/s. After washing with Milli-Q water, the electrode was dipped in 100 μL of 1 mM MCH aqueous solution and kept for 1 h at 45°C. The MCH-masked electrode was ready after washing with Milli-Q water.

Cyclic voltammograms of the MCH-masked electrode were taken in 0.10 M AcOH–AcOK (pH 5.5) containing 0.10 M KCl and 50 μM FND in the presence of various amounts of sonicated calf thymus DNA. Cyclic voltammograms of FND in

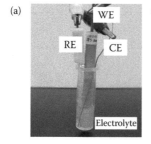
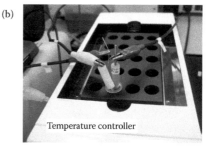

FIGURE 11.4 (a) Photograph of a microcentrifuge tube (2 mL, Thermo Fisher Scientific, USA) as the measuring cell equipped with a working electrode (WE, MCH-masked, ssDNA- or dsDNA-modified gold electrode, diameter 6 mm), a counter electrode (CE), and a reference electrode (RE, Ag/AgCl, ESA Biosciences, Inc.). (b) Temperature control unit for the cell during measurements using a temperature controller (Cool Mini Block Bath, CB-105, Scinics, Co., Japan).

the absence and presence of 5 mM calf thymus DNA at a scan rate of 100 mV/s are shown in Figure 11.5a. The $E_{1/2}$ was observed at 414 mV in the absence of DNA and it was shifted toward the negative potential by −36 mV and the current decreased in the presence of DNA. The scan rate dependence of i_{pa} in the absence and presence of DNA is shown in Figure 11.5b. The obtained ΔE_p of 72 or 77 mV in either case was larger than the theoretical value of 59 mV based on the diffusion-controlled process. Nonetheless, i_{pa}s were proportional to the root of scan rate in both cases, suggesting a pseudodiffusion-controlled process. The diffusion coefficient of FND in the absence and presence of DNA was obtained as $D_f = 6.58 \times 10^{-6}$ m²/s at 25°C, and $D_b = 1.30 \times 10^{-6}$ cm²/s using Equation 11.2. The latter value was larger than that for Co(bpy)$_3^{3+}$, presumably because of a difference in the treatment of calf thymus DNA, sonicated versus untreated. Commercial calf thymus DNA (untreated) has an average molecular weight of 10^7 Da, but this is reduced to 10^5–10^6 Da by sonication.

The i_{pa} decreased with an increase in the amount of DNA added and then leveled off. Binding parameters of $K = 5.0 \times 10^4$ M^{-1} and $s = 3.4$ were obtained by fitting the data with Equation 11.4, as shown in Figure 11.6a. A Scatchard plot shown in Figure 11.6b was also fitted with McGhee–von Hipple equation 11.6 to obtain $K = 0.7 \times 10^4$ M^{-1} and $s = 3.7$. The subtle difference in K between the two treatments is reminiscent of that for Co(bpy)$_3^{3+}$ (see above). Scatchard analysis based on the hypochromic and red shifts in absorption spectra upon addition of calf thymus DNA yield $K = 10^5$ M^{-1}, $s = 3$ in 10 mM 2-(N-morpholino)ethanesulfonic acid (MES) (pH 6.2) and 1 mM ethylenediaminetetraacetic acid (EDTA) containing 0.10 M NaCl [27]. These values are larger than those obtained electrochemically, but may be reasonable given the difference in the salt concentration. Observed binding constant,

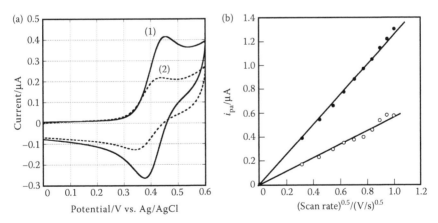

FIGURE 11.5 (a) Cyclic voltammograms of 50 μM FND in the absence (1) and presence (2) of 5 mM-bp calf thymus DNA in 0.10 M AcOH–AcOK (pH 5.5) and 0.10 M KCl, recorded with a sweep rate of 100 mV/s at 25°C. (b) Plot of i_{pa} of FND in the absence (closed circles) and presence (open circles) of calf thymus DNA versus scan rate. E_{pa} and E_{pc} in the absence or presence of DNA were 0.450 and 0.427 or 0.378 and 0.350 V, respectively.

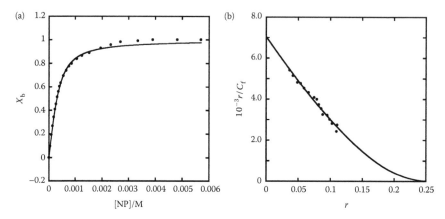

FIGURE 11.6 Binding analysis based on Equation 11.4 (a) or Equation 11.6 (b).

K_{obs}, of cationic drugs decreases with an increase in the salt concentration according to the following equation [29]:

$$\frac{\partial \ln K_{obs}}{\partial \ln\left[M^+\right]} = -m \tag{11.9}$$

where m' is number of ion pairs formed in the complex.

11.3.2 Interaction of Drugs with DNA Immobilized on an Electrode

11.3.2.1 Characterization of DNA-Immobilized Electrodes

First, a gold electrode (diameter 1.6 mm, area 0.020 cm²) was mechanically and electrochemically polished according to standard procedures. Cyclic voltammograms of the resulting electrode were taken in 0.05 M H_2SO_4. A large reduction peak with normalized area 390 ± 10 µC/cm² was observed for the polycrystalline gold electrode [30], which was assignable to oxygen desorption (the shaded portion in Figure 11.7). The surface area of this electrode was calculated as 1.56×10^{-5} C/390×10^{-6} C/cm² = 0.040 cm² from the area of the shaded portion (1.56×10^{-5} C) and its roughness factor was 2.0.

Commercially available thiol-modified oligonucleotides are protected with either a trityl or an $HO(CH_2)_6S$ group. The protecting group is removed by treatment with 0.10 M dithiothreitol (pH 8.3–8.5) at 25°C for 16 h and subsequently purified with a commercial simple cartridge or demineralization column for oligonucleotide purification. A solution of a certain concentration is prepared by dissolving in Milli-Q water after lyophilization of the DNA thus obtained. Since disulfide oligonucleotides carrying an $HO(CH_2)_6S$ protecting group are known to be immobilized on the gold electrode [12], a 24-meric disulfide oligonucleotide (5′-HO(CH₂)₆SS(CH₂)₆–ATG ATC GCG GGC GTC GGC GTG TTT-3′) was used without deprotection in our experiment. The gold electrode prepared above was dipped in Milli-Q water containing 0.10 µM oligonucleotide and 0.10 M NaCl and kept for 16 h at 37°C. The DNA-immobilized electrode was obtained by subsequent incubation in 100 µL of 1 mM MCH for 1 h at 45°C. After washing with Milli-Q water, CC was measured in the absence and presence of

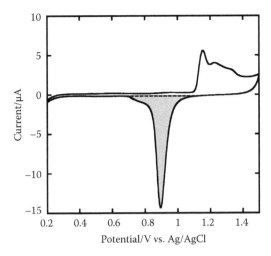

FIGURE 11.7 Cyclic voltammogram for a polycrystalline gold electrode after mechanical and electrochemical polishing in 0.05 M H_2SO_4 aqueous solution at 100 mV/s. The shaded area represents reduction of chemisorbed oxygen.

$Ru(NH_3)^{3+}$ (Figure 11.8). The modification density with this DNA was estimated as $\Gamma_{DNA}^{ss} = 1.2 \times 10^{12}$ molecules/cm^2 from Equations 11.7 and 11.8. A dsDNA-immobilized electrode was prepared by hybridization with a complementary oligonucleotide in $2 \times$ SSC (30 mM sodium citrate (pH 7.4) and 0.30 M NaCl) for 2 h at 15°C. The density of dsDNA on the electrode was found to be 2.5×10^{12} molecules/cm^2 by analogous CC measurements, implying quantitative hybridization.

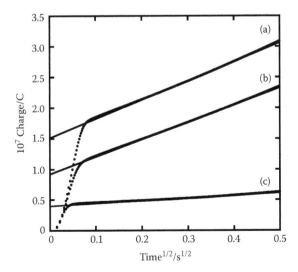

FIGURE 11.8 Chronocoulometric response curves for an ssDNA-immobilized electrode in the absence (a) and presence of 50 μM $Ru(NH_3)^{3+}$ (b) in 10 mM Tris–HCl, pH 7.4. Chronocoulometric response curves after hybridization with a complementary DNA were also measured in the presence of 50 μM $Ru(NH_3)^{3+}$ (c).

11.3.2.2 Interaction of Drugs with dsDNA-Immobilized Electrode

The CV of the dsDNA-immobilized electrode in 0.10 M AcOH–AcOK (pH 5.5) containing 5 μM FND and 0.10 M KCl is shown in Figure 11.9a. A redox peak based on FND was observed in the ssDNA and dsDNA-immobilized electrodes. But, the current was larger in the latter electrode, showing that FND has a preference for dsDNA and that a larger amount of FND was concentrated on the latter electrode. The scan rate dependence of the oxidation peak current for the dsDNA-immobilized electrode is shown in Figure 11.9b. The species responsible for this redox is FND concentrated on the electrode. A pure cyclic voltammogram for the dsDNA-immobilized electrode was obtained by subtracting the CV of an MCH-masked electrode to give $E_{1/2} = 412$ mV with $\Delta E_p = 24$ mV. Needless to say, all of the peak charge for dsDNA was larger than that for MCH-masked electrodes. The former value was in agreement with that in homogenous solution, but the latter, which ought to be $\Delta E_p = 0$ theoretically, was not, suggesting that FND did not absorb on the gold surface directly but was concentrated on the dsDNA on the electrode. This in turn implies that FND bound on the electrode has some freedom of motion.

The net charge per area of the dsDNA-immobilized electrode was plotted against the FND concentration (Figure 11.10). Since the dsDNA density on the electrode was estimated to be one tenth that of the dsDNA-immobilized one [15], a large portion of the surface is available for MCH masking in the dsDNA-immobilized electrode and the current based on the diffusion of bulk FND for the dsDNA-immobilized electrode was assumed to be similar to the MCH-masked one. Filled circles of Figure 11.10 represent the net charge for the dsDNA-immobilized electrode after subtracting the MCH-masked one. Two binding modes were suggested from this plot for the interaction of FND. Ten molecules of FND were to bind to

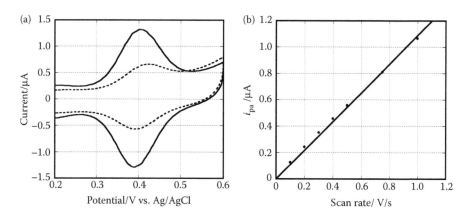

FIGURE 11.9 (a) Cyclic voltammograms of ssDNA- (dotted line) and dsDNA-immobilized electrodes in 0.10 M AcOH–AcOK (pH 5.5) containing 0.10 M KCl and 5 μM FND at 1 V/s and 15°C. (b) Plot of i_{pa} versus scan rate for the dsDNA-immobilized electrode. The E_{pa} and E_{pc} of the ssDNA and dsDNA-immobilized electrodes were 0.416 (0.73) and 0.432 (0.424) or 0.396 (0.455) and 0.403 (0.400) V, respectively. The values in brackets were obtained by subtracting those for the MCH-masked electrode from those for the ssDNA- or dsDNA-immobilized one.

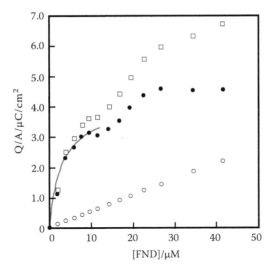

FIGURE 11.10 Plots of the charge per unit area for MCH-masked (open circles) and dsDNA-immobilized electrodes (open squares) calculated from the peak area of the CV curves against the FND concentration. The net increase in charge per unit area for the dsDNA-immobilized electrode is shown by the closed circles.

the 24-meric dsDNA on the electrode from the charge of the first step (3.05×10^{-6} C/cm^2), that is, concentrated FND $3.05 \times 10^{-6}/2/96,500 = 1.58 \times 10^{-11}$ mol/cm^2, ratio of FND to dsDNA $1.58 \times 10^{-11}/1.66 \times 10^{-11} = 9.5 \approx 10$. This estimate is reasonable given that the binding of FND with dsDNA conforms to the neighbor exclusion model [29]. Four more molecules of FND were to bind in the second step, suggesting the additional binding to the terminal part of dsDNA. From the analysis of the first step by a Langmuir-type isotherm [31], $K = 1.0 \times 10^5$ M^{-1} and $s = 15$ were obtained.

11.4 CONCLUSIONS

The analysis of DNA/drug interactions using solution electrochemistry and electrochemistry on DNA-immobilized electrodes was reviewed in this chapter. As electrochemical methods exploit redox reactions of drugs and their DNA complexes, they give information on the microenvironment where the drugs reside. Moreover, forces involved in the interaction of drugs with DNA may be identified. Recent advent of technology has exploited gold/sulfur interactions to immobilize various molecules and biomolecules on a metal electrode surface in an organized fashion, facilitating the reproducible construction of DNA-immobilized electrodes. This in turn enables detailed analyses of interactions of drugs with DNA. Notably, only a small amount (ca. a few pg) of synthetic DNA is required in such studies. This is advantageous, as one can design DNA sequences at will and optimize them for specific studies. Ultimately, electrochemical identification of a specific drug/DNA sequence combination may find medicinal applications.

11.5 PROSPECTS AND OUTLOOK

Nuclear magnetic resonance (NMR) is one of the methods commonly used for the analysis of DNA/drug interactions [32], although a relatively large amount of synthetic oligonucleotides is required for taking NMR spectra. Another common method is surface plasmon resonance (SPR) [33] in which only a few ng of samples suffice, but the technique is not well suited to provide detailed molecular information on the nature of the interactions. Electrochemical methods based on DNA-immobilized electrodes enable the analysis of DNA/drug interactions with a few pg of samples and give information on the microenvironment in which such interactions occur.

In addition, single-stranded DNA-immobilized electrodes can be used as a tool for gene analysis. As soon as high-performance, electrochemically active ligands specific for double-stranded DNA are developed, simple and quick diagnosis of genes based on electrochemistry may be realized. Drugs exemplified in Figure 11.2 may be promising lead compounds for future development of electrochemical platforms.

The spontaneous formation of a DNA duplex according to the intrinsic complementarity of its nucleobases has potential for the construction of DNA-based computers and nanoscale architectures based on DNA building blocks. Furthermore, since DNA is a one-dimensional polymer with nucleobases arranged in a sequence, DNA may be used as a programmable nanowire. Drugs that specifically bind to DNA could therefore enable its site-selective modification, thereby broadening the scope of DNA structures and usability as a material. Electrochemical approaches for studying these systems therefore play a crucial role in these undertakings.

11.6 APPENDIX

11.6.1 DERIVATION OF EQUATION 11.1

Equations 11.10 and 11.11 hold from the relationship of Scheme 11.1

$$K_{3+} = \frac{[M^{3+} - DNA]}{[M^{3+}][DNA]} \tag{11.10}$$

$$K_{2+} = \frac{[M^{2+} - DNA]}{[M^{2+}][DNA]} \tag{11.11}$$

$$E = E_f^{o'} + \frac{RT}{nF} \ln \frac{[M^{3+}]}{[M^{2+}]} \tag{11.12}$$

$$E = E_b^{o'} + \frac{RT}{nF} \ln \frac{[M^{3+} - DNA]}{[M^{2+} - DNA]} \tag{11.13}$$

As $E = 0$ at equilibrium, Equations 11.12 and 11.13 are transformed as follows:

$$E_f^{o'} = -\frac{RT}{nF} \ln \frac{[M^{3+}]}{[M^{2+}]} \tag{11.14}$$

$$E_b^{o'} = -\frac{RT}{nF} \ln \frac{[M^{3+} - DNA]}{[M^{2+} - DNA]} \tag{11.15}$$

Therefore, the following equation results:

$$E_b^{o'} - E_f^{o'} = -\frac{RT}{nF} \ln \frac{[M^{3+} - DNA]}{[M^{2+} - DNA]} + \frac{RT}{nF} \ln \frac{[M^{3+}]}{[M^{2+}]}$$

$$= \frac{RT}{nF} \ln \frac{[M^{2+} - DNA][M^{3+}][DNA]}{[M^{3+} - DNA][M^{3+}][DNA]}$$

When Equations 11.10 and 11.11 are combined with this equation, Equation 11.1 is obtained at 25°C with $n = 1$:

$$E_b^{o'} - E_f^{o'} = 0.059 \log \frac{K_{2+}}{K_{3+}}$$

11.6.2   DERIVATION OF EQUATION 11.3

Where a mobile condition holds, the observed diffusion coefficient D_{eff} in a certain ratio of a metal complex to DNA is expressed by the following equation:

$$D_{eff} = D_b X_b + D_f X_f$$

Under the experimental conditions, $X_f = 1 - X_b$ holds [34].

$$D_{eff} = D_b X_b + D_f \left(1 - X_b\right)$$

Then,

$$X_b = \frac{D_{eff} - D_f}{D_{sat} - D_f}$$

is obtained. As i_p is proportional to $D^{1/2}$ (Equation 11.2), Equation 11.3

$$X_b = \frac{i^2 - i_0^2}{i_{sat}^2 - i_0^2}$$

is obtained.

11.6.3   DERIVATION OF EQUATION 11.4

The interaction of a DNA-binding metal complex, M, with DNA composed of s base pairs with binding sites S to form a DNA complex, M–S, is expressed as follows: M + S = M–S, where the binding constant is defined as

$$K = \frac{C_b}{C_f C_s} \tag{11.16}$$

where C_b, C_f, and C_s represent the concentration of bound ligand, free ligand, and free binding site of DNA, respectively. When the total concentration of the ligand is set as C_t,

$$C_t = C_b + C_f \tag{11.17}$$

is obtained. When the average base number and DNA concentration are expressed as L and C_{DNA}, respectively,

$$C_{DNA} = C_b + C_s \tag{11.18}$$

is obtained, where $X = L/s$ and

$$C_{DNA} = [NP]/2L \tag{11.19}$$

From Equations 11.18 and 11.19,

$$x = [NP]/(2sC_{DNA}) \tag{11.20}$$

is obtained.

Equation 11.20 is combined with Equation 11.18 to give

$$[NP]/2s = C_b + C_s \tag{11.21}$$

The following equation is obtained by combining Equations 11.16 and 11.17.

$$C_s = \frac{C_b}{C_t - C_b} \frac{1}{K} \tag{11.22}$$

Equation 11.22 is then combined with Equation 11.21.

$$\frac{[NP]}{2s} = C_b + \frac{C_b}{C_t - C_b} \frac{1}{K}$$

This equation is rearranged with C_b as the variable.

$$KC_b^2 - \left(1 + KC_t + \frac{K[NP]}{2s}\right)C_b + \frac{K[NP]C_t}{2s} = 0$$

Here,

$$b = 1 + KC_t + \frac{K[NP]}{2s}$$

is introduced.

$$KC_b^2 - bC_b + \frac{K[NP]C_t}{2s} = 0$$

This equation is solved for C_b.

$$C_b = \frac{b \pm \left(b^2 - \frac{2K^2C_t[NP]}{s}\right)^{\frac{1}{2}}}{2K}$$

Where $X_b < 1$, Equation 11.4 is reduced as follows:

$$X_b = \frac{C_b}{C_t} = \frac{b - \left(b^2 - \frac{2K^2 C_t [\text{NP}]}{s} \right)^{\frac{1}{2}}}{2KC_t}$$

ACKNOWLEDGMENTS

Special thanks are due to Emeritus Professor Hiroki Kondo of Kyushu Institute of Technology for advice and helpful discussions.

REFERENCES

1. Haq. I. 2006. Reversible small molecule–nucleic acid interactions. In: *Nucleic Acids in Chemistry and Biology*, 3rd Edition, eds. G. M. Blackburn, M. J. Gait, D. Loakes, and D. M. Williams, pp. 341–379. Dorset: RSC Publishing.
2. Erdem, A. and Ozsoz, M. Electrochemical DNA biosensors based on DNA–drug interaction. *Electroanalysis*, **14**, 965–974, 2002.
3. Rauf, S., Gooding, J. J., Akhtar, K., Ghauri, M. A., Rahman, M., Anwar, M. A., and Khalid, A. M. Electrochemical approach of anticancer drugs–DNA interaction. *J. Pharm. Biomed. Anal.*, **37**, 205–217, 2005.
4. Palecek, E. and Jelen, F. Electrochemistry of nucleic acids. In: *Electrochemistry of Nucleic Acids and Proteins. Towards Electrochemical Sensors for Genomics and Proteomics*, eds. E. Palecek, F. Scheller, and J. Wang, pp. 73–173. New York: Elsevier B. V.
5. Fojta, M. Electrochemical sensors for DNA interactions and damage. *Electroanalysis*, **14**, 1449–1462, 2002.
6. Goto, K., Kato, D., Sekioka, N., Ueda, A., Hirono, S., and Niwa, O. Direct electrochemical detection of DNA methylation for retinoblastoma and CpG fragments using a nanocarbon film. *Anal. Biochem.*, **405**, 59–66, 2010.
7. Takenaka, S. 2003. Electrochemical detection of DNA with small molecules. In: *DNA and RNA Binders. From Small Molecules to Drugs*, eds. M. Demeunynck, C. Bailly, and W. D. Wilson, pp. 224–246. Weinheim: Wiley-VCH Verlag GmbH & Co. KGaA.
8. Oliveira-Brett, A. M., Vivan, M., Fernandes, I. R., and Piedade, J. A. P. Electrochemical detection of *in situ* adriamycin oxidative damage to DNA. *Talanta*, **56**, 959–970, 2002.
9. Carter, M. T., Rodriguez, M., and Bard, A. J. Voltammetric studies of the interaction of metal chelates with DNA. 2. Tris-chelated complexes of cobalt(III) and iron(II) with 1,10-phenanthroline and 2,2′-bipyridine. *J. Am. Chem. Soc.*, **111**, 8901–8911, 1989.
10. Millan, K. M., Spurmanis, A. J., and Mikkelsen, S. R. Covalent immobilization of DNA onto glassy carbon electrodes. *Electroanalysis*, **4**, 929–932, 1992.
11. Steel, A. B., Levicky, R. L., Herne, T. M., and Tarlov, M. J. Immobilization of nucleic acids at solid surface: Effect of oligonucleotide length on layer assembly. *Biophys. J.*, **79**, 975–981, 2000.
12. Wackerbarth, H., Marie, R., Grubb, M., Zhang, J. J., Hansen, A. G., Chorkendorff, I., Christensen, B. V., Boisen, A., and Ulstrup, J. *J. Solid State Electrochem.*, **8**, 474–481, 2004.
13. Anne, A. and Demaille, C. Dynamics of electron transport by elastic bending of short DNA duplexes. Experimental study and quantitative modeling of the cyclic voltammetric behavior of 3′-ferrocenyl DNA end-grafted on gold, *J. Am. Chem. Soc.*, **128**, 542–557, 2006.

14. Steel, A. B., Nerne, T. M., and Tarlov, M. J. Electrochemical quantitation of DNA immobilized on gold. *Anal. Chem.*, **70**, 4670–4677, 1998.
15. Kelley, S. O. and Barton, J. K. Electrochemistry of methylene blue bound to a DNA-modified electrode. *Bioconjugate Chem.*, **8**, 31–37, 1997.
16. Gorodetsky, A. A., Buzzeo, M. C., and Barton, J. K. DNA-mediated electrochemistry, *Bioconjugate Chem.*, 19, 2285–2296, 2008.
17. Hvastkovs, E. G. and Buttry, D. A. Recent advances in electrochemical DNA hybridization sensors. *Analyst*, **135**, 1817–1829, 2010.
18. Takenaka, S., Ihara, T., and Takagi, M. Bis-9-acridinyl derivative containing a viologen linker chain: Electrochemically active intercalator for reversible labeling of DNA. *Chem. Commun.*, 21, 1485–1487, 1990.
19. Zhang, L., Sun, H., Li, D., Song, S., Fan, C., and Wang, S. A conjugated polymer-based electrochemical DNA sensor: Design and application of a multi-functional and water-soluble conjugated polymer. *Macromol. Rapid Commun.* 29, 1489–1494, 2008.
20. Takenaka, S., Yamashita, K. Takagi, M., Uto, Y., and Kondo, H. DNA sensing on a DNA probe-modified electrode using ferrocenylnaphthalene diimide as the electrochemically active ligand. *Anal. Chem.*, **72**, 1334–1341, 2000.
21. Wong, E. L. S., Erohkin, P., and Gooding, J. J. A comparison of cationic and anionic intercalators for the electrochemical transduction of DNA hybridization via ling range electron transfer. *Electrochem. Commun.*, **6**, 648–654, 2004.
22. Lucarelli, F., Marraza, G., Turner, A. P. F., and Mascini, M. Carbon and gold electrodes as electrochemical transducers for DNA hybridization sensors. *Biosens. Bioelectron.*, **19**, 515–530, 2004.
23. Johnston, D. H. and Thorp, H. H. Cyclic voltammetry studies of polynucleotide binding and oxidation by metal complexes: Homogeneous electron-transfer kinetics. *J. Phys. Chem.*, **100**, 13837–13843, 1996.
24. McGhee, J. D. and von Hippel, P. H. Theoretical aspects of DNA-protein interactions: Co-operative and non-co-operative binding of large ligands to a one-dimensional homogeneous lattice. *J. Mol. Biol.*, **86**, 469–480, 1974.
25. Millan, K. M. and Mikkelsen, S. R. Sequence-selective biosensor for DNA based on electroactive hybridization indicators. *Anal. Chem.*, **65**, 2317–2323, 1993.
26. Peterlinz, K. A., Geogiadis, Herne, T. M., and Tarlov, M. J. Observation of hybridization and rehybridization of thiol-tethered DNA using two-color surface plasmon resonance spectroscopy. *J. Am. Chem. Soc.*, **119**, 3401–3402, 1997.
27. Sato, S., Nojima, T., Waki, M., and Takenaka, S. Supramolecular complex formation by β-cyclodextrin and ferrocenylnaphthalene diimide-intercalated double stranded DNA and improved electrochemical gene detection. *Molecules*, **10**, 693–707, 2005.
28. Xiao, Y., Lai, R. Y., and Plaxco, K. W. Preparation of electrode-immobilized, redox-modified oligonucleotides for electrochemical DNA and aptamer-based sensing. *Nat. Protocols*, **2**, 2875–2880, 2007.
29. Wilson, W. D. 1982. Intercalation in biological systems. In: *Intercalation Chemistry*, eds. M. S. Whittingham, A. J. Jacobson, pp. 445–501. New York: Academic Press, Inc.
30. Carvalhal, R. F., Freire, R. S., and Kubota, L. T. Polycrystalline gold electrodes: A comparative study of pretreatment procedures used for cleaning and thiol self-assembly monolayer formation. *Electroanalysis*, **17**, 1251–1259, 2005.
31. Carrasco, C., Facompre, M., Chisholm, J. D., Van Vranken, D. L., Wilson, W. D., and Bailly, C. DNA sequence recognition by the indolocarbazole antitumor antibiotic AT2433-B1 and its diastereoisomer. *Nucl. Acids Res.*, **30**, 1774–1781, 2002.
32. Nakamoto, K. Tsuboi, M., and Strahan, G, D. 2008. Drug–DNA interactions. *Structures and Spectra*. Hoboken, NJ: John Wiley & Sons, Inc.

33. Lin, L.-P., Huang, L.-S., Lin, C.-W., Lee, C.-K., Chen, J.-L., Hsu, S.-M., and Lin, S. Determination of binding constant of DNA-binding drug to target DNA by surface plasmon resonance biosensor technology. *Curr. Drug Targets Immune Endocr. Metabol. Disord.*, **5**, 61–72, 2005.

34. Welch, T. W., Corbett, A. H., and Thorp, H. H. Electrochemical detection of nucleic acid diffusion coefficients through noncovalent association of redox-active probe. *J. Phys. Chem.*, **99**, 11757–11763, 1995.

12 Nanopore Ion Microscope for Detecting Nucleic Acid/Drug Interactions

Meni Wanunu

CONTENTS

In this chapter, the use of ion microscopes for detecting interactions of nucleic acids with small molecules is described. The ion microscope consists of a nanoscale pore in an ultrathin membrane, in which ion current through the nanopore under an electric field is measured. Passage of individual nucleic acid molecules through very small pores gives rise to electronic signals that report on the fraction of small molecules bound to the nucleic acid, affording binding affinity measurements as well as spatial resolution.

12.1 INTRODUCTION

In recent years, nucleic acids have emerged as potential drug targets for numerous diseases, including RNA-mediated diseases, bacterial/viral infections, and other genetic diseases [1–5]. Further strengthening the need for nucleic acid targeting drugs comes from recent breakthroughs in understanding the immense role of nucleic acids in cellular components such as the ribosome [6], the discovery of epigenetic effects on gene regulation [7,8], as well as the major roles small RNA molecules play in posttranscriptional gene regulation [9]. Since the vast majority of drug development efforts have been focused on protein targets, there is a clear need for experimental and computational tools for surveying drugs that target genomic DNA or other nucleic acids with adequate specificity [1]. This is particularly true for antibacterial drugs, as many bacterial strains have evolved to be resistant to contemporary antibiotics, rendering them obsolete [10].

Nucleic acids are in general linear biopolymers composed of a uniform phosphate–sugar backbone and base side chains (four for DNA, namely A, G, C, and T). The order of the bases, that is, sequence of the nucleic acids, affects both their genetic code and global structure. In DNA molecules, different combinations of sequences form grooves that are binding sites for regulatory and other DNA-binding proteins, and therefore there is an immense set of potential drug targets for site-specific intercalating and groove-binding drugs. RNA molecules adopt even more elaborate secondary structure domains that in some cases resemble proteins in structure and function, and several RNA domains have already been identified and demonstrated as drug targets (i.e., the A-site of the prokaryotic ribosome).

Whether the nucleic acid target is a specific DNA sequence or an RNA domain, new methods for drug discovery and screening should ideally be able to probe binding thermodynamics/kinetics, the binding mode, as well as to address the binding site specificity and stoichiometry. Further, since many RNA molecules are chemically modified *in vivo* following transcription, the method of choice should also be capable of investigating native structures as extracted from cellular tissues. While many existing bulk techniques (e.g., calorimetry, crystallography, and footprinting) provide one parameter or another, rapid methods that facilitate multiparameter analysis for a test drug and target in solution using small amounts of precious nucleic acid can be extremely attractive in industrial and research environments alike.

A recent revolution in single-molecule biophysics has allowed scientists to develop tools that manipulate and characterize *individual* biomolecules in solution. The various techniques that afford this remarkable sensitivity were enabled by a few key technological achievements: (1) ability to introduce addressable chemical functionality to biomolecules, (2) incorporation of molecules and nanoscale objects that respond optically, conformationally, and/or electronically to biomolecular changes, and (3) ability to control and/or measure the relative position of nano- and microscale objects with subnanometer resolution. Virtually any technique that investigates individual biomolecules (atomic force microscopy, Forster resonance energy transfer, optical tweezers, magnetic tweezers, fluorescence correlation spectroscopy, etc.) utilizes at least two of these abilities.

This chapter compiles a body of recent results that hopefully communicate the feasibility of nanopores as tools for studying nucleic acid/ligand interactions. In the context of this chapter, nanopores are nanoscale ion microscopes that can analyze nucleic acid/small molecule interactions at the single-molecule level and with high throughput. Nanopores were inspired by Coulter counters, apertures that are hundreds of micrometers in diameters that have been used to analyze cell populations since the 1950s [11]. A nanopore, which is a molecular-scale version of the Coulter counter (sub-5 nm pore diameter), is very useful for detecting small molecules, polymers, and biopolymers in solution with high throughput.

The first use of nanopores for single-molecule analysis employed a small number of lipid-embedded protein channels [12]. Since then, the α-hemolysin toxin has been extensively used for investigating the structure of DNA and RNA oligonucleotides, for discriminating among different mononucleotides in solution, as well as for detecting nucleic acid complexes [13–17]. Some of these works have laid foundations and inspired sophisticated approaches to future-generation DNA sequencing [18], which can revolutionize personalized genomics by reducing the cost of sequencing a complete genome for <$1000. These commercial prospects of nanopores, as well as the need for robust and tunable pores, have stimulated the development of synthetic nanopores. Of these, nanopores in thin solid-state membranes have enabled a diversified set of nucleic acid structures to be studied, including double-stranded DNA [19–22], DNA complexes with peptide nucleic acids [23], circular viral DNA molecules [24], and other structures. The reference list here is not in any way exhaustive; more thorough surveys of the nanopore literature are found elsewhere [25–27].

This chapter will focus on sub-5 nm diameter nanopores in ultrathin silicon nitride membranes as tools for detecting various complexes of nucleic acids and small molecules. The goal of this research is to develop a method that quantitatively profiles small-molecule binding to various nucleic acid systems. Such a method could potentially quantify detection of DNA/drug complexes, recognize binding sites along DNA, and resolve binding and structural changes that result from binding to various RNA targets. After the basics of the operation of nanopore ion microscopes are described, several examples are shown in which the interactions of nucleic acids with small molecules were studied.

12.2 BASICS OF NANOPORE SENSING

12.2.1 RESISTIVE SENSING PRINCIPLE

Nanopore ion microscopes operate on the Coulter principle, or the principle of resistive sensing. As the name implies, resistive sensing employs as a signal the magnitude of electrolyte flow through the nanopore as a function of time. Discrete fluctuations in the resistance through the pore are attributed to individual nucleic acid molecules being driven across the pore. Therefore, the signal from a nanopore is an electrical current, as described below.

A scheme of a nanopore is shown in Figure 12.1a. In a typical experiment, a pair of Ag/AgCl electrodes is used to apply an electrical potential ΔV far away from the membrane separating the two chambers and containing the pore. The nanopore and electrodes are hydrated in a high ionic strength electrolyte solution (typically >0.2 M KCl), and the application of voltage results in ion migration toward the membrane. At the positive electrode, the electrolyte process that occurs is $Ag(s) + Cl^- \rightarrow AgCl$ $(s) + e^-$, and the reverse reaction occurs at the negative electrode, that is, Cl^- is released into the solution. These two reactions occur with fast kinetics at both electrodes, resulting in no potential drop across the electrode/electrolyte interface. Since the

(a)

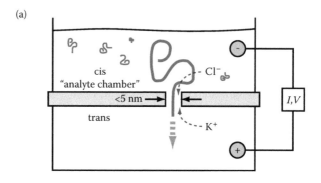

(b)

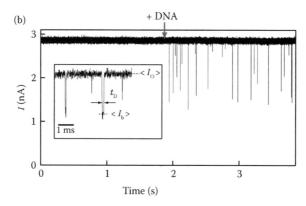

FIGURE 12.1 Nanopore ion microscopes. (a) Scheme of nucleic acid analysis using a nano-pore. The nanopore is a nanoscale hole (typically <5 nm in diameter) that penetrates through an insulating membrane (shown in gray). High ionic-strength electrolyte solution fills both sides of the membrane such that an electrolyte junction forms at the nanopore. A pair of Ag/AgCl electrodes is used to apply electrical potential across the membrane, which results in a steady-state ion current that is measured at high bandwidths using a patch-clamp amplifier. When a nucleic acid sample is placed in the analyte chamber (or *cis* chamber), individual molecules are stochastically threaded into and driven through the pore. (b) Typical current dataset of a 4 nm diameter nanopore, before and after the addition of 5 nM 400 base pair DNA into the *cis* chamber ($V = 300$ mV, $T = 21°C$, 1 M KCl). (Trace taken from Wanunu, M. et al. *Biophys. J.* **2008,** 95 (10), 4716–4725. With permission.) Shallow and deep spikes in the traces correspond to collisions and translocations of individual nucleic acid molecules. The inset shows a magnified portion of the current trace, where individual spikes can be seen. Computer-based analysis of the spikes yields for each molecule a characteristic transport time (t_d) and current amplitude ΔI, defined as $\Delta I = I_o - I_b$.

membrane is insulating, no ion transport can occur except across the nanopore, and the flux of ions across the pore is proportional to the electric current measured at the electrodes. During transport of a biomolecule, the ion current across the nanopore is diminished because the biomolecule impedes ion flow. Complete passage of the molecule to the other side is marked by return of the ion current to its original value. A current trace collected over a given time period contains a set of stochastic transient pulses. The pulses have a characteristic rate, depth, and duration that are to a first approximation influenced by the biomolecules' concentration, cross-sectional diameter, and length, respectively.

Typically, the current traces are digitally sampled to a computer and a two-state machine model is used to identify and analyze the events. Statistical analysis is then performed on event populations in order to characterize the distributions of current amplitudes (ΔI) and event durations (t_d). Sample current trace before and after the addition of DNA is shown in Figure 12.1b, and the inset shows expanded events and defines the measured quantities.

As Figure 12.1b shows, in a typical experiment, several nanoamps of steady-state current flow across the pore. Using modern patch-clamp amplifiers, this current signal can be recorded at a bandwidth of up to 100 kHz with noise levels that are well over an order of magnitude lower than the signal. Low-pass filtering the data to 100 kHz bandwidth yields maximum time resolutions of ~3–4 µs. Several groups have extensively studied the noise characteristics of nanopores [28–30]. A major source of electric noise comes from the capacitance of the membrane material. Since the membrane is ion impermeable, when voltage is applied, ion flow toward the membrane results in a transmembrane capacitance across it. Since capacitive noise dominates the high-frequency component of the current signal, practical bandwidths are often determined by the system's capacitance. Capacitance can be reduced by minimizing the area of the membrane that is in contact with the electrolyte. Optimizing the amplifier's input parasitic capacitance, as well as making improvements to the electrical shielding of the nanopore setup, further reduces the electrical noise.

The ultimate application of nanopores is to sequence DNA molecules. The original idea, proposed by Deamer and Branton in the early 1990s, was to pass single DNA molecules through an atomically thin nanopore constriction and read out the ion current as the sequence translocates across the pore. This goal has two grand challenges: (a) producing atomically thin nanopores and (b) resolving the current signal as each base translocates the pore. Since the very first paper in which nucleic acid translocation through a nanopore was demonstrated, reducing the mean translocation velocity of the nucleic acid molecules has been identified as an intriguing challenge [13]. Typical mean translocation rates are 10 ns to 1 µs per base for the various pores described in the literature, too fast to be recorded using conventional amplifiers due to bandwidth limitations. Approaches taken to reduce the speed of nucleic acid transport through the nanopore include increasing the solvent viscosity, reducing the applied voltage, and reducing the pore dimensions [31]. While the first two approaches have been successful, both decrease the electrolyte current and the current amplitude of biomolecular signals, thereby lowering the signal-to-noise of the measurement. The flexibility of solid-state nanopores enables studies of the effect of nanopore diameters on the translocation rates. Reducing the diameter of the nanopore to the sub-5 nm regime was found to have an enormous impact on slowing down DNA transport by at least an order of magnitude, as compared with transport through a 10 nm pore [22]. In addition, reduction of pore size forces the DNA to translocate head-to-tail (i.e., in a linear fashion). Using sub-5 nm pores, mean DNA translocation speeds are in the ~0.1–1 µs/base range (the word "mean" is used because there is no experimental evidence that translocation proceeds with uniform velocity). For this reason, all of the studies in this chapter were conducted with ultrathin solid-state nanopores of sub-5 nm diameters. More specifically, the nanopores used in this study were fabricated in ultrathin free-standing silicon nitride windows. Practical experimental details on setting up nanopore measurements are given elsewhere [32].

12.2.2   SOLID-STATE NANOPORES

Synthetic nanopores offer several features that lipid-embedded protein channels lack, namely size controllability, physical robustness, and chemical stability. Over the past two decades, many different approaches have been taken for making synthetic pores [27]. The most precise synthetic nanopores are nanopores fabricated in thin (<50 nm) solid-state membranes using a tightly focused electron beam. The fabrication process, which has been well studied over the past decade, utilizes the electron beam of a field-emission transmission electron microscope (TEM) [20]. When several hundred picoamps of 200 kV electrons are focused onto an nm-scale area on the silicon nitride membrane, the nitride begins to ablate locally, resulting in pore formation. Following this process, a slightly defocused electron beam can be used to locally heat and fine-tune the final diameter and shape of the nanopore with sub-nm resolution. The size of the nanopore is then conveniently measured using the TEM, and final adjustments can be made to tune the nanopore size to the desired specifications. Several groups have studied the process of nanopore formation. Because this process proceeds with local melting of the silicon nitride edge, minimization of surface energy during solidification results in an hourglass nanopore shape [33]. Under typical TEM-based pore fabrication conditions, the effective thickness of the membrane is several times smaller than the actual bulk membrane thickness, with control afforded by the electron beam diameter [34].

12.2.3   NUCLEIC ACID CAPTURE INTO NANOPORES

The mechanistic process of nucleic acid transport through nanopores consists of two fundamental steps: First, the biomolecule has to be threaded into the pore in a process called *capture*. Second, the biomolecule has to be transported through the pore via a process called *translocation*. As in any other single-molecule technique, characterization of a sample requires a statistical analysis based on measurements of hundreds to thousands of molecules. Therefore, for practical reasons, the rate of biomolecular capture by the nanopore needs to be moderate, and a minimum sample concentration is needed for this to occur. This may not be an issue for standard nucleic acids prepared using molecular biology techniques, for example, by polymerase chain reaction or by transfection. However, this may present a technical challenge for analyzing native nucleic acids extracted from biological sources, modified nucleic acids, as well as samples that need to be massively aliquoted for high-throughput drug screening. Therefore, there is a need to optimize the capture of DNA molecules into pores while preserving similar signal-to-noise values and translocation velocities.

When the dimensions of the pore are slightly larger than the cross-sectional diameter of a nucleic acid, biomolecules have to be properly threaded into the pore before being driven across it by the strong "pulling" force. As a result of this geometric constraint, not every biomolecular approach toward the pore results in successful transport, and collisions with the pore can often be observed as fast events with shallow blockade amplitudes. Similar behavior has been observed in experiments with single-stranded DNA using the lipid-embedded α-hemolysin protein channel [13].

To further explain the capture process, it is important to depict the electrical potential landscape around a nanopore under the conditions employed in nanopore experiments. A typical voltage applied to the electrodes during nanopore experiments is ~0.3 V. Although the voltage is applied to electrodes that are very far away from the pore (orders of magnitude further than the pore dimensions), it is important to keep in mind that the flux of ions in the solution dictates the electric field profile. This can easily be seen from Ohm's law, that is, $J = \sigma E$, where the ion flux J is proportional to the electric field E, with the solution conductance σ serving as the proportionality constant. The resulting electric field is strong near the pore (~10^5 V/cm) and decays with increasing distance from the pore entrance. Figure 12.2a shows a cross-sectional view of an unbiased and biased pore, where the colors illustrate the voltage profile (in arbitrary units). For electrode distances in the μm to mm range, the field profile across the nanopore does not change appreciably. Figure 12.2b shows a semilog plot of the electric field E that develops across the pore when +300 mV are applied to the *trans* chamber (values obtained from a finite-element simulation performed using COMSOL) [35]. The magnitude of E inside the pore (i.e., the "driving" field) is sufficient to irreversibly drive nucleic acid molecules across the pore in the direction of the *trans* chamber. In addition, interaction of the weak field outside the pore (the "capturing" field) with the highly charged nucleic acid molecule governs the capture process.

A recent study describes DNA capture into pores and suggests means for enhancing biomolecular capture. Under standard experimental conditions (300 mV, 4 nm pore, 21°C), normalized capture rates for DNA into 4 nm pores are ~1 molecule/s/nM. Without increasing the applied voltage, which would in turn speed up molecular transport and reduce the readout resolution, these capture rates can be enhanced by nearly two orders of magnitude by applying salt gradients across the pore (Figure 12.3) [35]. The capture enhancements enable the analysis of ~1 pM DNA concentrations with good throughput (at rates of 1 molecule/s), crucial for the analysis of trace samples such as cellular RNA and epigenetically modified DNA. For chamber volumes of 1 μL, easily achievable in the laboratory, attomole-level DNA amounts were rapidly detected using the salt gradient focusing method. Moreover, increasing the salt concentration in the *trans* chamber also increases the signal-to-noise of the measurement. Finally, the salt gradient approach allows experiments under various conditions and pH values, including physiological conditions, important for evaluating interactions of nucleic acids with potential drugs and proteins under the appropriate conditions (e.g., testing the pH selectivity of drug binding for cancer therapy).

12.3 CASE STUDIES

12.3.1 DETECTING INTERCALATION TO DOUBLE-STRANDED DNA

In this study, the presence of intercalators is studied using nanopore ion microscopes [36]. Many polycyclic aromatic molecules bind to double-stranded DNA via

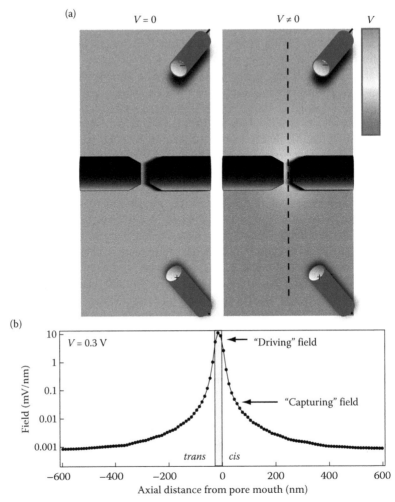

FIGURE 12.2 **(See color insert.)** Voltage drop across a nanopore. (a) Schematic map of the voltage drop across a membrane with a nanoscale pore when a voltage is applied across the pore using a pair of electrodes that are far away from the pore (depicted for high electrolyte concentration). Application of voltage results in ion flow toward the membrane, where ion flow is then impeded by the membrane and restricted to flow through the nanopore. (b) Electric field at various axial positions from the mouth of a 4 nm hourglass-shaped pore fabricated in a 25 nm thick silicon nitride membrane. (Adapted from Wanunu, M. et al. *Nat. Nanotechnol.* **2010,** 5 (2), 160–165.) The axis of the nanopore is defined by the dashed line in (a). For the simulation, 1 M KCl buffer was used as the electrolyte and a 300 mV voltage was applied at the two ends of the simulation box (see Ref. 35 for more details). The semilog plot shows that the field is very strongly concentrated at the pore (membrane height outlined by gray area in plot), with a very steep decay that is symmetric about the membrane center. Although weak, the interaction of the "capturing" field with the highly charged nucleic acid molecules captures nucleic acids toward the pore, where they are threaded through the pore by the strong "driving" field.

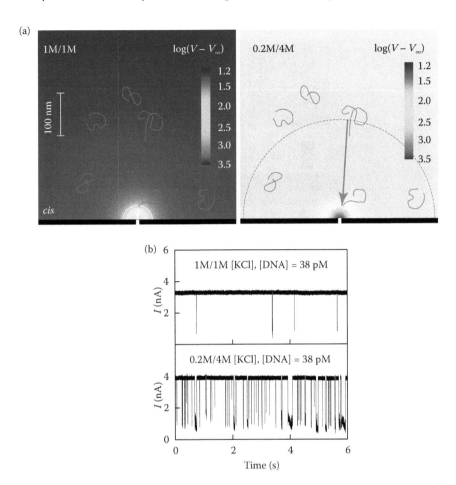

FIGURE 12.3 **(See color insert.)** Electrostatic focusing of nucleic acids into a nanopore: (a) Numerical simulations of the electrical potential in the *cis* side of a 4 nm nanopore for both symmetric (left) and asymmetric (right) KCl concentrations, where the *cis/trans* KCl concentrations are indicated in each image (simulation taken from Ref. 35). The pore mouth is positioned at the bottom center of each image (*trans* chamber not shown). The potential outside the pore increases with salt asymmetry, thereby funneling molecules to be captured. (b) Experimental time traces for a 38 pM 8000 bp solution under 1 M/1 M (left, rate = 0.4 s^{-1}) and 0.2 M/4 M (left, rate = 11.20 s^{-1}) *cis/trans* KCl concentrations. The 30-fold enhancement in the capture rate is a result of the electrical field outside the pore. (From Wanunu, M. et al. *Nat. Nanotechnol.* **2010,** 5 (2), 160–165. With permission.)

intercalation, a binding mode in which the molecule slides itself into the hydrophobic environment between two adjacent base pairs [37]. Affinities of intercalation can vary widely from the mM to nM range, depending on the small molecule and the binding site sequence. Because intercalation can result in errors in DNA replication, many intercalating agents are mutagenic. Ethidium bromide (EtBr) is an intercalating molecule used to stain DNA in gel electrophoresis, because the fluorescence of the DNA/ethidium complex can conveniently report DNA presence. Binding of

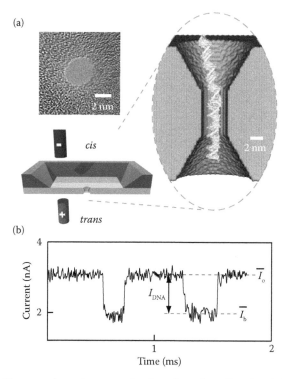

FIGURE 12.4 Nanopore ion microscope for detection of small-molecule binding to DNA. (a) A TEM image of a 3.5 nm pore fabricated in an ultrathin silicon nitride membrane. Magnified oval: 3D rendering of a 3.5 nm nanopore with ethidium-intercalated double-stranded DNA. (b) Ion current signal of two representative DNA translocations (the time between events was reduced for clarity). (From Wanunu, M., Sutin, J., and Meller, A. *Nano Lett.* **2009**, 9 (10), 3498–3502. With permission.)

ethidium (Et^+) to DNA is known to widen the normal B-DNA cross section by about 15% [38], as well as to elongate the fragment.

Nucleic acid-intercalating fluorophores such as ethidium are good candidates for validating nanopore data with bulk fluorescence measurements. Figure 12.4a shows a scheme of a 3.5 nm diameter nanopore into which an ethidium-intercalated DNA duplex has threaded. A TEM image of a 3.5 nm diameter nanopore is shown on top of the figure. When a DNA sample is placed in the *cis* chamber and a positive voltage is applied to the *trans* chamber, DNA molecules will thread and translocate through the nanopore. In Figure 12.4b, representative ion current traces are shown for the translocation of two DNA molecules. Each recorded ion current transient is characterized by its average current amplitude $\Delta I_b = \overline{I_o} - \overline{I_b}$, where $\overline{I_b}$ and $\overline{I_o}$ are average blocked-state and open-state currents for each translocation event, respectively. For each experiment, ~10^3 single-molecule translocations are analyzed to obtain ΔI_b distributions, the average capture rate, and translocation time distributions.

Figure 12.5a shows three different intercalating dyes used in this study. Ethidium differs from propidium in charge, while propidium differs from ethidium homodimer *I*

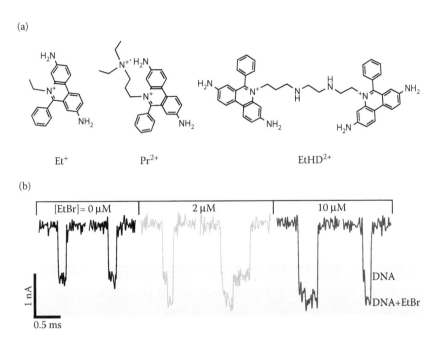

FIGURE 12.5 (a) Chemical structures of ethidium (Et), propidium (Pr), and ethidium homodimer *I* (EtHD). The counterions, which are usually the halides Br⁻ and I⁻, are not shown in the structures. (b) Representative single-molecule 400 bp DNA translocation events through a 3.5 nm pore at different EtBr concentrations, showing a deeper blocking current amplitude for the DNA/Et complex ($V = 300$ mV, $T = 21°C$ in all experiments). (From Wanunu, M., Sutin, J., and Meller, A. *Nano Lett.* **2009**, 9 (10), 3498–3502. With permission.)

in terms of the number of intercalating rings (but the charge is the same). In Figure 12.5b, representative ion current events are shown for a 400 bp DNA fragment translocating through a 3.5 nm pore, where the DNA was exposed to the indicated concentrations of EtBr. The free 400 bp DNA sample exhibits a single characteristic blocked current level of amplitude ~1 nA, whereas for increasing EtBr concentrations, deeper blockades of amplitude ~1.5 nA are observed. Interestingly, the mean event blockade amplitude is deeper with increasing EtBr levels, suggesting a correlation with the level of intercalation. This larger blockade amplitude may be related to the EtBr loading fraction by steric blockage of ion transport through the pore by the wider B-DNA. Another possibility is the partial charge neutralization of the DNA by the positively charged EtBr, although this is less likely because the experiments are carried out at high ionic strength.

The correlation between ΔI_b and EtBr loading is clearly observed by nanopore titration experiments, where measurements are made after aliquots of EtBr are added to the DNA-containing *cis* chamber. The surface plot in Figure 12.6a shows histograms of the event blockade amplitude as a function of EtBr concentration, obtained from the analysis of ~10³ molecules at each EtBr concentration. The major and minor populations in each ΔI_b histogram have Gaussian shapes, and their identity

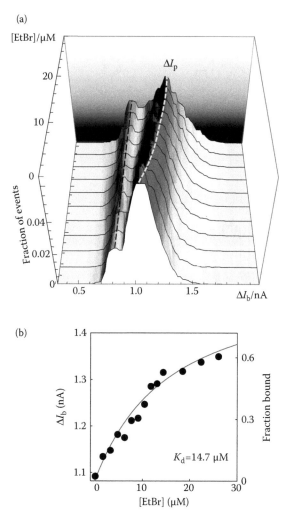

FIGURE 12.6 Nanopore titration of a DNA/EtBr complex. (a) Current amplitude histograms for a 400 bp DNA fragment as a function of EtBr concentration. A shift to increasing amplitudes is seen for the translocation peak (major population), whereas the collision peak (minor population), remains unchanged. (b) The peak current amplitude, ΔI_p, for ethidium/DNA as a function of EtBr concentration. The overlaid curve is a best fit to the fraction-bound ethidium from fluorescence measurements (see right axis), both measurements coinciding well with a K_d value of 14.7 μM. (From Wanunu, M., Sutin, J., and Meller, A. *Nano Lett.* **2009**, 9 (10), 3498–3502. With permission.)

corresponds to translocations and collisions, respectively [22]. As the EtBr concentration increases, the position of the translocation peak (red, defined here as ΔI_p, see Figure 12.6a) shifts to larger values, while the minor collision peak does not shift appreciably (purple). Plotting the dependence of ΔI_p (see Figure 12.6b) results in binding curves for EtBr to DNA. This plot is superimposed with a binding curve obtained from fluorescence, yielding with excellent agreement a dissociation constant $K_d = 14.7\ \mu M$.

Figure 12.7 shows nanopore-based binding curves for two other intercalators, namely propidium iodide (PrI) and ethidium homodimer *I* (EtHD). The affinities of these molecules to B-DNA are larger than ethidium because of their double charge. As described above for EtBr, the peak event amplitudes ΔI_p were measured for each of these dyes. Figure 12.4 displays the dependences of ΔI_p on dye concentrations for propidium (Figure 12.4a) and for ethidium homodimer *I* (Figure 12.4b). Bulk fluorescence binding assays for the same DNA molecules are overlaid, displaying good agreement with the nanopore measurements.

It is well known that a DNA stained with EtBr displays retarded mobility in gel electrophoresis as compared to unstained DNA due to the decreased effective charge and increased stiffness of the polymer [39]. These molecular changes to DNA

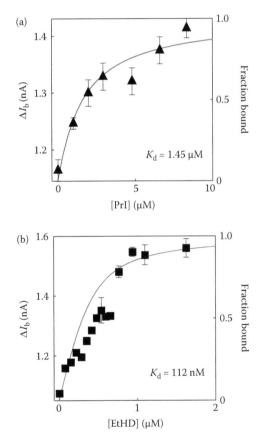

FIGURE 12.7 Nanopore titration of 400 bp DNA using propidium (a) and ethidium homodimer *I* (b). The peak current amplitude value (ΔI_p) is plotted as a function of dye concentrations (symbols). The overlaid curves are best fit to the fraction-bound dye obtained from fluorescence measurements. (From Wanunu, M., Sutin, J., and Meller, A. *Nano Lett.* **2009,** 9 (10), 3498–3502. With permission.)

TABLE 12.1

Average Capture Rates and Translocation Times for Free and Intercalator-Bound DNA

Molecule	Capture Rate (s^{-1} nM^{-1})		Translocation Time (ms)	
	Free DNA	Drug-Bound	Free DNA	Drug-Bound
EtBr	0.57 ± 0.02	0.13 ± 0.01	0.11 ± 0.01	0.44 ± 0.04
PrI	0.65 ± 0.02	0.12 ± 0.01	0.09 ± 0.02	0.55 ± 0.06
EtHD	0.52 ± 0.02	0.15 ± 0.01	0.10 ± 0.01	0.52 ± 0.04
EtBr	0.57 ± 0.02	0.13 ± 0.01	0.11 ± 0.01	0.44 ± 0.04
PrI	0.65 ± 0.02	0.12 ± 0.01	0.09 ± 0.02	0.55 ± 0.06

Source: From Wanunu, M., Sutin, J., and Meller, A. *Nano Lett.* **2009,** 9 (10), 3498–3502. With permission.

structure upon binding of cationic intercalators also affect transport through nanopores. Table 12.1 shows average normalized capture rates and translocation times for the free and drug-loaded DNA. It is interesting to note that average capture rates decreased by a factor of 3–5 upon intercalation of Et, Pr, or EtHD, and average translocation times are more retarded for both divalent intercalators EtHD and PrI (factor of 5–6) than for EtBr (factor of 4). This is expected for a DNA with a greater charge reduction. These findings are consistent with the reduced gel electrophoretic mobility of the intercalator/DNA complex (see Supplementary Information of Ref. 36), manifested in the case of nanopores by suppression of DNA capture and slowed transport through the pore.

12.3.2 SPATIALLY RESOLVING LIGAND BINDING ALONG DNA

Nanopores can also spatially resolve free DNA regions from regions that have bound ligands. To demonstrate this, a 2.2 nm pore (see Figure 12.4a) was used to probe the interaction of *single-stranded DNA* molecules with the cyanine dye SYBR Green II (SGII), an RNA-selective stain that intercalates with high-affinity single-stranded nucleic acids (Invitrogen Corp., Carlsbad, CA). As an exception, SGII has particularly low affinity toward a deoxyadenine homopolymer because the purine bases of the homopolymer undergo extensive stacking interactions that prevent intercalation [40,41]. Since SGII binds with high affinity to a random single-stranded DNA sequence, a two-segment $dA_{50}dN_{50}$ molecule was designed in order to create two synthetic fragments along the same molecule: a fragment that binds SGII (dN_{50}) and a region that does not (dA_{50}). As a negative control, the response of the $dA_{50}dN_{50}$ to SGII was compared to a dA_{60} homopolymer, which SGII does not bind to.

Figure 12.8 displays representative translocation events for dA_{60} before and after the addition of SGII to the *cis* chamber. As expected, poly(dA) translocation remains unaffected by SGII, as seen by the similar event shapes before and after SGII addition. In contrast, typical translocation events for the $dA_{50}dN_{50}$ sample with SGII

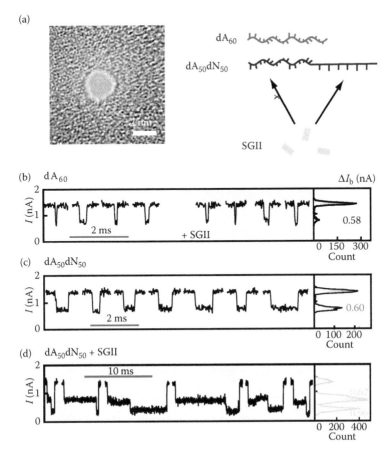

FIGURE 12.8 Detection of small-molecule binding to single-stranded DNA. (a) TEM image of a 2.2 nm pore and the molecules used in this study. Since SYBR Green II (SGII) binds to poly(dA) with very low affinity, dA_{60} was used as a negative control and $dA_{50}dN_{50}$ as a positive control. (b) Typical translocation events for dA_{60}, before and after the addition of SGII to the *cis* chamber (side histogram is an all-point histogram of the current data). Typical translocation events for $dA_{50}dN_{50}$, before and after the addition of SGII, are shown in (c) and (d), respectively. The two-step events prevalent after SGII binding were characterized by amplitudes of 0.58 and 0.96 nA, corresponding to dA_{50} and SGII-bound dN_{50}, respectively. (From Wanunu, M., Sutin, J., and Meller, A. *Nano Lett.* **2009**, 9 (10), 3498–3502. With permission.)

(Figure 12.8d) display strikingly longer translocation times than for $dA_{50}dN_{50}$ alone (Figure 12.8c), and the structures of the translocations exhibit a second, deeper blocked current level. The all-point histograms for these events clearly show that the deeper blockade is a discrete level with amplitude 0.96 nA, larger than the amplitude for the free DNA (0.60–0.62 nA). Translocation time distributions for the dA_{60} and $dA_{50}dN_{50}$ samples, shown in Figure 12.9a and b, respectively, show that SGII binding considerably slows the translocation process (histograms shown in log time units), suggesting that SGII binding induces secondary structure that stalls the entry of the

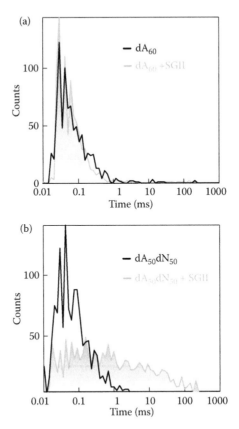

FIGURE 12.9   Translocation–time distributions for dA_{60} (a) and $dA_{50}dN_{50}$ (b), both before and after the addition of excess SGII. (From Wanunu, M., Sutin, J., and Meller, A. *Nano Lett.* **2009,** 9 (10), 3498–3502. With permission.)

dN_{50} portion of the molecule into the pore. Noteworthy, for the vast majority of events with $dA_{50}dN_{50}$ (>95%), the shallow blockade level precedes the deeper blockade level. This indicates that the entry of the dA_{50} portion of the molecule is favored over the entry of the bulkier SGII-bound dN_{50} portion. This favored entry can also be rationalized by considering the positive charge of SGII, which may decrease the overall charge of the dN_{50} region upon binding SGII, and/or condense the DNA structure to a coil. This conclusion is consistent with the literature, as certain DNA-binding molecules, including clinical drugs, have been shown to induce secondary structure in single-stranded DNA molecules [42].

It is important to mention that nanopore ion microscopes do not require the use of fluorogenic/radioactive probes or surface immobilization in order to operate. In addition, the method is quite fast, and can produce results extremely rapidly (seconds to minutes) by reading only ~10^3–10^4 DNA copies. Finally, the experiments can be carried out on nucleic acids of varying lengths and secondary structures, for example, single- and double-stranded DNA.

12.3.3   DETECTION OF A 1:1 NUCLEIC ACID:LIGAND COMPLEX

Although the above results indicate that nanopore ion microscopes are sensitive to the presence of intercalators on a DNA molecule and offer information about their location, it is unclear whether binding of individual small-molecule drugs can be detected. Figure 12.10 shows preliminary results that a peptide drug can indeed be detected (M. Wanunu, unpublished data). Actinomycin D (AD), a natural chemotherapeutic/antibiotic from the Lincosamide class, binds to GC-rich DNA, as well as to ssDNA [43]. In these experiments, AD was titrated in 2 nM intervals into 300 nM 106-mer random sequence ssDNA at 5°C. Statistical analysis of average ΔI values from 2000 molecules reveals the emergence of a new peak, with a shift of +300 pA. Single-molecule traces before AD addition (red traces) and after partial titration (green traces) show distinct pulses with $\Delta I = 300$ pA that are predominantly absent when AD is not added, perhaps indicating that nanopores can detect the binding of AD molecules to the ssDNA fragment.

Further evidence that nanopores are useful in detecting nucleic acid/ligand systems comes from a drug target that is of clinical importance for treating bacterial infections. A canonical RNA target for drugs is the A-site of prokaryotic ribosomal RNA (rRNA) [44–47]. Aminoglycoside antibiotics are very effective in targeting this key functional site, disrupting protein synthesis by inhibition of transfer-RNA (tRNA) binding and/or reducing the codon–anticodon recognition fidelity [48–50]. The A-site is a small loop within the 16S unit of rRNA that contains two key adenine residues, A1492 and A1493. It was previously shown that a truncated 27-nucleotide-long RNA construct that contains the A-site region mimics the function and antibiotics recognition features of the 16S rRNA [45,51]. Various methods are used to test binding of aminoglycosides to this A-site, including the use of fluorescent nucleotide analogs (see Chapter 7).

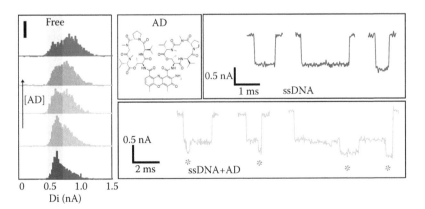

FIGURE 12.10   Detection of individual Actinomycin D (AD) molecules bound to single-stranded DNA. Histograms of the mean current amplitudes (ΔI) for 106-mer DNA at each increasing concentration of AD show a gradual emergence of a deeper blockade level (see structure to the right). Typical translocation traces for the 106-mer before (grey traces) and after (black traces) AD addition show a deeper level (marked with an asterisk) that may correspond to transport of a single AD/DNA complex through the pore.

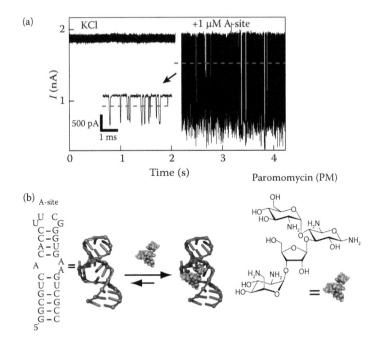

FIGURE 12.11   Counting individual RNA molecules and RNA/paromomycin (PM) complexes using a nanopore. (a) Continuous ion current traces are shown before and after the addition of 1 μM A-site RNA to the negatively biased chamber (1 M KCl, 3 nm pore, 8 nm thick membrane, $V = 500$ mV, $T = 0°C$). Inset shows an expanded portion of the current trace, with the appropriate scale bars. Dashed red lines indicate the threshold level used to detect the events. (b) The sequence and hairpin structure of the truncated A-site model used in this study, as well as the chemical structure of PM.

In order to facilitate the detection of short nucleic acids, such as the truncated A-site, very thin pores are required (to maximize the signal-to-noise and resolution of the nanopore). A recent systematic study has concluded that decreasing the thickness of the membrane increases the signal-to-noise of nanopore measurements [52]. In particular, 3 nm diameter nanopores fabricated in 7 nm thick membranes were found to be useful for detecting small nucleic acids such as microRNAs, short DNA sequences (as small as 10 bp), and tRNAs [52].

Figure 12.11a shows continuous two-second traces of the ion current through a 3 nm pore, both before and after the addition of ~1 μM A-site RNA to the *cis* chamber. The structures of the A-site, the model drug paromomycin (PM), as well as their complex, are shown in Figure 12.11b. The steady-state open pore current is similar after RNA addition, although stochastic occupancy of RNA molecules is indicated by high-frequency current blockade spikes in the current trace (500–2000 events/s).

Figure 12.12a shows continuous ~200 ms ion current traces for 1 μM of A-site under increasing concentrations of PM. The shaded areas at the troughs of the traces highlight the changing characteristic amplitudes of the events with increasing PM concentrations. Current amplitude histograms from the experiments shown in Figure 12.2a are shown in Figure 12.2b. Each histogram is based on the analysis

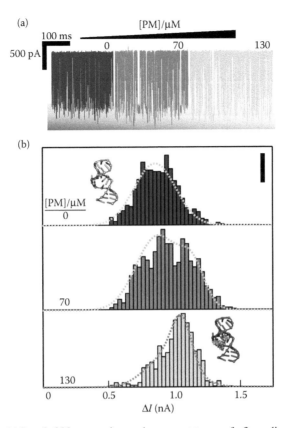

FIGURE 12.12   (a) Set of ~200 ms continuous ion current traces of a 3 nm diameter pore where the negative chamber contains 1 μM A-site and 0, 70, or 130 μM PM are added. The gray- and pink-shaded areas at the troughs of the traces are for visualization of the changing current amplitude level of the events upon increasing the PM/A-site ratio. (b) Histograms of the mean current amplitudes from ~1000 events of each experiment shown in (a). As the PM concentration increases, complexation to the A-site molecule increases the mean current amplitude of the events (the vertical bar represents 50 counts).

of the mean current amplitude of many events in a given experiment. Prior to the addition of PM, the current distribution fits a single Gaussian function with a peak at 0.86 nA, respectively (green-dashed curve in Figure 12.2b, $\psi^2_{red} = 0.88$). As more PM is added, the distribution of the current blockade amplitudes broadens, and a new current level emerges at 1.05 nA (see brown histogram in Figure 12.12b). The yellow- and orange-dashed curves are two-Gaussian fits to the histograms, which convey the emergence of a new blockade level at 1.05 nA upon binding of PM to the A-site.

Knowing the relative occupancy of the A-site for each PM concentration is sufficient to obtain a quantitative affinity curve, or binding isotherm. For simple 1:1 binding, that is, $PM + A \rightleftharpoons PM \cdot A$, the dissociation constant is $K_d = [PM][A]/[PM \cdot A]$, where [A] is the concentration of A-site RNA. Therefore, the fraction of complex F_C in solution as a function of [PM] can be expressed in terms of K_d by

$$F_C = \frac{[PM]}{[PM] + K_d}$$

Since the current populations in Figure 12.12 overlap, it is not straightforward to assign an exact value of F_C for each experiment (i.e., for a set of events collected from a sample). Instead, plotting the mean change in current amplitude for all of the events in each experiment can effectively quantify the fraction of bound molecules in the population. Figure 12.13 shows the mean current amplitude change as a function of PM concentration for two different bath salt concentrations. Each point in the graph corresponds to the mean ΔI value of >1000 events at each concentration of PM, subtracted from the mean amplitude for pure A-site in the absence of PM. The systematic trend in the amplitude change as the PM concentration increases can be fit to the equation for F_C, yielding a K_d value of 90 ± 34 µM at 1 M KCl and 44 ± 17 µM at 0.5 M KCl.

These results suggest that affinity can be extracted for a 1:1 nucleic acid:ligand complex using a nanopore ion microscope. The K_d values are similar with values obtained by other techniques, and the increased affinities of the A-site constructs to PM at lower salt concentrations are known to be due to electrostatic contributions to the interactions with the A-site [53]. Notably, statistically meaningful current blockage data (~500 events) are collected in under a minute for ~1 pmol levels of nucleic

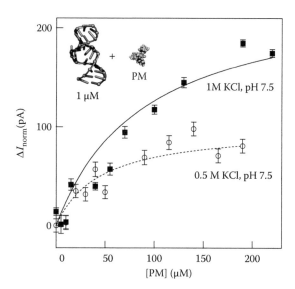

FIGURE 12.13   Mean current amplitude changes (ΔI_{norm}) for titrations of 1 µM A-site RNA with PM in 1 M KCl (solid squares) and 0.5 M KCl (open triangles) at pH 7.5, 0°C. The ΔI_{norm} values were obtained by subtracting the ΔI value at a PM/A-site ratio of 0 (i.e., A-site only). Fitting the data points to a simple binding model yields a K_d value of 90 ± 34 µM at 1 M KCl and 44 ± 17 µM at 0.5 M KCl.

acid, making this technique attractive for studying real-time binding kinetics of precious samples and for high-throughput screening of dilute nucleic acid samples.

12.4   CONCLUSIONS

In conclusion, nanopore ion microscopes are simple yet useful tools for studying the interactions between nucleic acids and small-molecule ligands. Specifically, solid-state nanopores are convenient tools for studying these systems because the dimensions of the nanopore can be tuned to the molecule of study. For example, in this chapter, double-stranded nucleic acids were threaded using 3.5 nm pores, single-stranded nucleic acids were threaded using ~2 nm pores, and nucleic acids as short as the 27 oligonucleotide A-site model were reliably detected using nanopores in ultrathin (6–7 nm thick) membranes with excellent signal-to-noise. The ion current is conveniently measured at high temporal bandwidths to provide a signal that is sensitive to the nucleic acid cross-sectional diameter. Therefore, as a ligand molecule interacts with the nucleic acid, the ion current drops to a level that is lower than that of the free nucleic acid. Using such a scheme, nucleic acid/ligand complexes formed by intercalation and groove binding were detected.

12.5   PROSPECTS AND OUTLOOK

Many avenues and research directions in nanotechnology have been inspired by larger-scale machines and objects from nature, such as cantilever springs for atomic force microscopy, nanoscale motors, and miniature transistors. The nanopore borrows a simple idea from a very bright scientist, Wallace Coulter, who started his research on cell counting using a hole he punctured in a cigarette wrapping film using a hot metal pin. Despite the idea's triviality and apparent simplicity, Coulter's approach has sparked a revolution in hematology. With less than two decades of research in its trail, the nanopore field is still in its infancy. Commencing with lipid-embedded protein channels and diverging to synthetic nanopores, the nanopore field is certain to deliver many years of fruitful research.

Nowadays, it is harder than ever to decouple progress in science from progress in technology. The diverse portfolio of emerging nanoscale technologies has enabled various approaches for nanopore fabrication. Of these, electron beam fabrication of nanopores in solid-state membranes appears the most promising for the analysis of biomolecules. In five decades of progress, electron beams have replaced the hot metal pin that Coulter used for making microscale pores in cigarette wrapping, and electrons have replaced the "eyes" with which pores can be seen and evaluated. Contrary to this progress, ions remain the reporters of nanopore activity. This is not necessarily a negative comment: ions respond to electric field much faster than biomolecules, and therefore ion activity can monitor biomolecular activity with high temporal resolution, as evident from *in vitro* nanopore experiments and from the field of electrophysiology. However, it would be a shame to not develop alternative means of biomolecular sensing, and several groups have taken up this challenge. Current developments in the field include the introduction of nanopores in other materials (e.g., graphene) [54–56], approaches for chemical modification of nanopores, insertion of protein channels to generate hybrid inorganic/biological pore devices [57], and the introduction of various carbonaceous materials and metal patterns near the pore as a means of gating transport. Apart from technological difficulties, the Achilles heel in all of these approaches is the challenge in producing devices with suitable surface properties. The nanoscale world is dominated by surface, and it is therefore critical to understand and be able to manipulate surfaces in order to interface with biomolecules in a reproducible fashion. By combining these developments with biological sensing in the appropriate context, nanopores may just grow to be the method of choice for investigating biomolecular systems.

Practical methods for screening the interactions of nucleic acids with ligand libraries require a highly reproducible and parallelizable nanopore platform. This presents production challenges (i.e., fabrication of $\sim 10^3$ nanopores with similar characteristics) as well as technological challenges (simultaneous electrical readout from $\sim 10^3$ nanopores at the required bandwidth). These challenges will undoubtedly be facilitated by progress in electron beam manipulation and computing power, respectively.

ACKNOWLEDGMENTS

My special thanks to Dr. Meller and the Meller group, specifically Jason Sutin, Gautam Soni, and Ben McNally, for experimental support and valuable discussions. I also thank Dr. Drndic and her group for supporting some of these projects. Finally, I thank my colleagues and collaborators Y. Tor, A. Aksimentiev, M. F. Kamenetskii, and their groups for having influenced and supported this work.

REFERENCES

1. Borman, S. *Chem. Eng. News* **2009,** 87 (39), 63–66.
2. Osborne, R. J. and Thornton, C. A. *Hum. Mol. Genet.* **2006,** 15, R162–R169.
3. Cooper, T. A., Wan, L. L., and Dreyfuss, G. *Cell* **2009,** 136 (4), 777–793.
4. O'Rourke, J. R., and Swanson, M. S. *J. Biol. Chem.* **2009,** 284 (12), 7419–7423.
5. Vicens, Q. *J. Incl. Phenom. Macrocycl. Chem.* **2009,** 65 (1–2), 171–188.
6. Yonath, A. *Annu. Rev. Biochem.* **2005,** 74, 649–679.
7. Gommersampt, J. H. and Borst, P. *Faseb J.* **1995,** 9 (11), 1034–1042.
8. Bird, A. *Genes Dev.* **2002,** 16 (1), 6–21.
9. Fire, A., Xu, S. Q., Montgomery, M. K., Kostas, S. A., Driver, S. E., and Mello, C. C. *Nature* **1998,** 391 (6669), 806–811.
10. Higgins, C. F. *Nature* **2007,** 446 (7137), 749–757.
11. Coulter, W. H. US Patent # 2,656,508, 1953.
12. Bezrukov, S. M., Vodyanoy, I., and Parsegian, V. A. *Nature* **1994,** 370 (6487), 279–281.
13. Kasianowicz, J. J., Brandin, E., Branton, D., and Deamer, D. W. *Proc. Natl. Acad. Sci. U.S.A.* **1996,** 93 (24), 13770–13773.
14. Gu, L. Q., Braha, O., Conlan, S., Cheley, S., and Bayley, H. *Nature* **1999,** 398 (6729), 686–690.
15. Akeson, M., Branton, D., Kasianowicz, J. J., Brandin, E., and Deamer, D. W. *Biophys. J.* **1999,** 77 (6), 3227–3233.
16. Meller, A., Nivon, L., Brandin, E., Golovchenko, J., and Branton, D. *Proc. Natl. Acad. Sci. U.S.A.* **2000,** 97 (3), 1079–1084.

17. Mathe, J., Aksimentiev, A., Nelson, D. R., Schulten, K., and Meller, A. *Proc. Natl. Acad. Sci. U.S.A.* **2005,** 102 (35), 12377–12382.
18. Branton, D., Deamer, D. W., Marziali, A., Bayley, H., Benner, S. A., Butler, T., Di Ventra, M., et al. *Nat. Biotechnol.* **2008,** 26 (10), 1146–1153.
19. Li, J., Stein, D., McMullan, C., Branton, D., Aziz, M. J., and Golovchenko, J. A. *Nature* **2001,** 412 (6843), 166–169.
20. Storm, A. J., Chen, J. H., Ling, X. S., Zandbergen, H. W., and Dekker, C. *Nat. Mater.* **2003,** 2 (8), 537–540.
21. Heng, J. B., Aksimentiev, A., Ho, C., Marks, P., Grinkova, Y. V., Sligar, S., Schulten, K., and Timp, G. *Nano Lett.* **2005,** 5 (10), 1883–1888.
22. Wanunu, M., Sutin, J., McNally, B., Chow, A., and Meller, A. *Biophys. J.* **2008,** 95 (10), 4716–4725.
23. Singer, A., Wanunu, M., Morrison, W., Kuhn, H., Frank-Kamenetskii, M., and Meller, A. *Nano Lett.* **2010,** 10 (2), 738–742.
24. Fologea, D., Brandin, E., Uplinger, J., Branton, D., and Li, J. *Electrophoresis* **2007,** 28 (18), 3186–3192.
25. Healy, K., Schiedt, B., and Morrison, A. P. *Nanomedicine* **2007,** 2 (6), 875–897.
26. Dekker, C. *Nat. Nanotechnol.* **2007,** 2 (4), 209–215.
27. Howorka, S. and Siwy, Z. *Chem. Soc. Rev.* **2009,** 38 (8), 2360–2384.
28. Tabard-Cossa, V., Trivedi, D., Wiggin, M., Jetha, N. N., and Marziali, A. *Nanotechnology* **2007,** 18 (30), 6.
29. Smeets, R. M. M., Keyser, U. F., Dekker, N. H., and Dekker, C. *Proc. Natl. Acad. Sci. U.S.A.* **2008,** 105 (2), 417–421.
30. Uram, J. D., Ke, K., and Mayer, M. *ACS Nano* **2008,** 2 (5), 857–872.
31. Healy, K. *Nanomedicine* **2007,** 2 (4), 459–481.
32. Wanunu, M. and Meller, A. Single-molecule analysis of nucleic acids and DNA-protein interactions using nanopores. In *Single-Molecule Techniques: A Laboratory Manual,* Selvin, P. and Ha, T. J., Eds. Cold Spring Harbor Laboratory Press: Cold Spring Harbor, New York, 2008, pp 395–420.
33. Kim, M. J., Wanunu, M., Bell, D. C., and Meller, A. *Adv. Mater.* **2006,** 18 (23), 3149–3155.
34. van den Hout, M., Hall, A. R., Wu, M. Y., Zandbergen, H. W., Dekker, C., and Dekker, N. H. *Nanotechnology* **2010,** 21, 115304.
35. Wanunu, M., Morrison, W., Rabin, Y., Grosberg, A. Y., and Meller, A. *Nat. Nanotechnol.* **2010,** 5 (2), 160–165.
36. Wanunu, M., Sutin, J., and Meller, A. *Nano Lett.* **2009,** 9 (10), 3498–3502.
37. Waring, M. J. and Chaires, J. B., *DNA Binders and Related Subjects,* In: Topics in Current Chemistry. Springer: Berlin, 2005, Vol. 253.
38. Sobell, H. M., Tsai, C., Jain, S. C., and Gilbert, S. G. *J. Mol. Biol.* **1977,** 114 (3), 333–365.
39. Nielsen, P. E., Zhen, W. P., Henriksen, U., and Buchardt, O. *Biochemistry* **1988,** 27 (1), 67–73.
40. Holcomb, D. N. and Tinoco, I. *Biopolymers* **1965,** 3 (2), 121.
41. Saenger, W., Riecke, J., and Suck, D. *J. Mol. Biol.* **1975,** 93 (4), 529–534.
42. Adamcik, J., Valle, F., Witz, G., Rechendorff, K., and Dietler, G. *Nanotechnology* **2008,** 19 (38), 7.
43. Wadkins, R. M. and Jovin, T. M. *Biochemistry* **1991,** 30 (39), 9469–9478.
44. Vicens, Q. and Westhof, E. *Structure* **2001,** 9 (8), 647–658.
45. Hermann, T. *Biochimie* **2006,** 88 (8), 1021–1026.
46. Tor, Y. *Biochimie* **2006,** 88 (8), 1045–1051.
47. Tor, Y., Exploring RNA-ligand interactions, *Pure and Applied Chem.* **2009,** 263–272.
48. Moazed, D. and Noller, H. F. *Nature* **1987,** 327 (6121), 389–394.

49. Lynch, S. R., Gonzalez, R. L., and Puglisi, J. D. *Structure* **2003,** 11 (1), 43–53.
50. Llano-Sotelo, B., Klepacki, D., and Mankin, A. S. *J. Mol. Biol.* **2009,** 391 (5), 813–819.
51. Purohit, P. and Stern, S. *Nature* **1994,** 370 (6491), 659–662.
52. Wanunu, M., Dadosh, T., Ray, V., Jin, J. M., McReynolds, L., and Drndic, M. *Nat. Nanotechnol.* **2010,** 5 (11), 807–814.
53. Pilch, D. S., Kaul, M., and Barbieri, C. M. Ribosomal RNA recognition by aminoglycoside antibiotics. In *DNA Binders and Related Subjects*, Springer-Verlag: Berlin, 2005, Vol. 253, pp. 179–204.
54. Schneider, G. F., Kowalczyk, S. W., Calado, V. E., Pandraud, G., Zandbergen, H. W., Vandersypen, L. M. K., and Dekker, C. *Nano Lett.* **2010,** 10 (8), 3163–3167.
55. Merchant, C. A., Healy, K., Wanunu, M., Ray, V., Peterman, N., Bartel, J., Fischbein, M. D., et al. *Nano Lett.* **2010,** 10 (8), 2915–2921.
56. Garaj, S., Hubbard, W., Reina, A., Kong, J., Branton, D., and Golovchenko, J. A. *Nature* **2010,** 467 (7312), 190-U73.
57. Hall, A. R., Scott, A., Rotem, D., Mehta, K. K., Bayley, H., and Dekker, C. *Nat. Nanotechnol.* **2010,** 5 (12), 874–877.

13 A Primer for Relaxation Kinetic Measurements

Ulai Noomnarm and Robert M. Clegg

CONTENTS

The theory of chemical relaxation kinetics is reviewed in this chapter. Relaxation lifetimes are derived using Castellan's method. Several kinetic simulations demonstrate the method. Relaxation amplitudes are introduced, along with a brief description of the time and frequency-domain methods of data acquisition. Aspects pertaining to dye binding to nucleic acid structures are discussed.

13.1 INTRODUCTION

Binding of small molecules to macromolecules can exhibit very complex behavior, especially for binding to polyelectrolytes such as nucleic acids. Equilibrium titrations to determine binding constants of dyes to DNA have been extensively investigated. To interpret equilibrium titrations, models of binding must be proposed, which take into account excluded-site binding, sequence dependence, and polyelectrolyte effects; these complexities pertain especially to binding to DNA, and binding theories have been well developed. However, a more thorough understanding of mechanisms, sequence specificity, and contributions of multiple-binding conformers is available from kinetics. Intercalation between base pairs and interaction of small molecules in the minor or major grooves (or both simultaneously) are the major modes of binding. In general, multistep mechanisms are involved in the binding of small molecules to DNA; unfortunately, equilibrium titrations cannot reliably distinguish (or often even detect) intermediate steps and intermediate conformations, unless the spectra of the different conformers are spectrally distinct. Even then, it is difficult or unfeasible to distinguish many mechanisms of binding, and of course impossible to detect fast reaction steps that are transient.

The best way for gaining more insight is to disperse the binding event onto the time scale, by following the binding events in time [1]. It is often found that the dynamics of different steps of binding are well separated temporally, which aides greatly in deciphering kinetic pathways. By spreading the process out onto an extra dimension (the time axis), individual steps can be detected that would otherwise remain hidden in an overall binding signal. Relaxation kinetic experiments (commonly known as T-jump, P-jump, E-jump, etc.) have been indispensable in elucidating kinetic steps of binding to DNA and RNA. In addition, the lessons learned from dye–DNA kinetic experiments, as well as coupling dye–DNA binding to measurements of proteins and enzymes interactions with nucleic acids, have been key for understanding many processes involving nucleic acid structures.

In this chapter, we give an overview of the basic theory of fast relaxation kinetic methods and show some examples. This is not a review of the literature but a pedagogical tutorial of some of the basics pertaining to the general theory of relaxation kinetics, and to point out issues that are important and specific when studying the binding of dyes to polyelectrolytes such as DNA. We have included more detailed discussion of the assumptions and mathematical procedures than is usual. Our goal is to provide uninitiated readers an introduction to the original and more recent literature, and to provide the background necessary to carry out the procedures for analyzing mechanisms pertaining to their own experiments, and straightforwardly extend the analysis to more complicated mechanisms. For this reason, we present the methodology for the most common simpler multistep reactions. We consider only thermodynamically independent-coupled reactions. It is straightforward to extend the methodology to thermodynamically coupled reactions, and for this because it is beyond the purview of this tutorial, and for additional details and examples, we refer the reader to the literature [2–17].

It is important to emphasize that as in all kinetic studies it is easier to show which kinetic reaction schemes are not consistent with kinetic data than to prove that a particular mechanism is the true one. Nevertheless, relaxation methods have provided valuable insight into many kinetic events revealing separate kinetic steps that are central for understanding biological mechanisms. In a sense, one can say that relaxation kinetic methods disperse interdependent kinetic steps of complex reaction mechanisms onto the time axis, so that individual steps can be observed. The perturbations of the chemical reaction system are small so that the kinetic response remains within the confines of linear response theory; thus, kinetic systems can be studied that are difficult or impossible to observe from equilibrium measurements, and which cannot easily be studied with nonlinear methods, such as stopped-flow.

13.1.1 Perturbations and Normal Modes of Relaxation Kinetics

A chemical relaxation experiment is typically carried out by perturbing the equilibrium state of a chemical reaction system by a small amount. The usual perturbation is a time-dependent change in an internal thermodynamic parameter, such as temperature, pressure, electric field, or sometimes the concentration of a reaction component. The perturbation brings the reaction system temporarily out of its equilibrium state, and the reaction system responds by relaxing from the original equilibrium state (before the perturbation) to the new equilibrium state (after the perturbation). "Small" means that the free energy change of the system, which defines the equilibrium state, can be expanded in a Taylor series, retaining only the first-order linear terms [2,6,9–11,18]. The simplest time-dependent perturbation in chemical relaxation experiments is a sudden abrupt change in temperature, pressure, or electric field. Provided that the free energy departure of each participating chemical reaction from equilibrium is sufficiently small, the system relaxes exponentially (not necessarily one exponential, but also possibly a series of exponentials). That is, the relaxation of the chemical system, which is often a complex system of several interdependent and independent chemical reactions, can be described as a series of exponentials, where each exponential component represents a "normal mode" of the overall relaxing reaction system. If the chemical system consists of a single chemical reaction, then the normal mode is the extent of reaction of just this single reaction. The kinetic reaction system is described by a set of linear differential equations. The solution of these equations can be expressed in terms of normal modes; each normal mode is a solution that relaxes in time exponentially, and independently, from the other normal modes. Many of the chemical species participate in several of the reactions, and many of the chemical species from different independent reactions contribute to each normal mode. That is, depending on the relative reaction velocities of the different elementary reactions, those chemical species that participate in multiple elementary reactions are involved in several normal modes of relaxation. Thus, even if a spectroscopic signal from only one of the chemical species is used to follow the progress of kinetics, the exponential relaxation of several normal modes can be observed just by following changes in concentration of this one reaction species. This is described in detail in the next sections.

Castellan's scheme [2] is the most general and straightforward way to generate the equations from which relaxation times (and amplitudes) corresponding to different mechanisms can be derived. This method also clearly associates the thermodynamic perturbations to the extents of the independent, coupled reactions. Often one must consider several possible disparate kinetic mechanisms to describe kinetic data. Therefore, it is necessary to derive expressions that can represent kinetic data corresponding to different mechanisms. With the method of Castellan reviewed below, it is relatively easy to obtain mathematical expressions that describe various mechanisms. Thereby diverse mechanisms can be analyzed and hypothesized, conditions can be varied, and experiments carried out that can discriminate between different mechanisms.

13.2 BASIC THEORY

The theory presented in this section can be followed in more detail in Castellan's original paper [2]; we follow the presentation in Castellan's paper (with a few clarifying comments), and retain Castellan's notation, so the reader can easily consult that publication for more detail. This section introduces the relationship between the rate of the reactions in terms of the deviation of the free energy of each reaction from its equilibrium state caused by the perturbation. If the reader accepts Equation 13.18, they can proceed to the next section. The rate equations are described in terms of deviations of the extents of the reactions from their equilibrium values.

The chemical reaction system is defined as a collection of elementary chemical reactions.

$$\sum_{i=1}^{N} v_{i\alpha} B_i = 0; \quad \alpha = 1, 2, ..., R \tag{13.1}$$

where R is the number of independent elementary reactions and B_i refers to the ith chemical species (not every B_i belongs to every Rth reaction). For a reaction written according to Equation 13.1, $v_{i\alpha}$, which is a stoichiometric coefficient, is negative for reactants and is positive for products.

The advancement of the αth reaction, ξ_α, is given in molar units. Thus the moles of the ith chemical species at any time are defined as

$$n_i = n_i^0 + \sum_{\alpha}^{R} v_{i\alpha} \xi_\alpha \tag{13.2}$$

where n_i^0 is the total number of moles of the ith species; note that each chemical species can belong to multiple different reactions. n_i can also be defined similarly in terms of the number of moles of the ith species at equilibrium, $\bar{n}_i$ (the line over the character refers to equilibrium values)

$$n_i = \bar{n}_i + \sum_{\alpha}^{R} v_{i\alpha} \Delta\xi_\alpha \tag{13.3}$$

$\Delta\xi_\alpha = \xi_\alpha - \bar{\xi}_\alpha$ is then the deviation of the αth reaction (at equilibrium, $\Delta\xi_\alpha = 0$, for all reactions). The concentrations (per volume, V) of the different chemical species, c_i, are then defined as

$$c_i = \bar{c}_i + \frac{1}{V} \sum_{\alpha}^{R} v_{i\alpha} \Delta\xi_\alpha = \bar{c}_i + \Delta c_i \tag{13.4}$$

The reason why the description of the state of the chemical reaction system is chosen in terms of the molar extents (advancements) of the different reactions, instead of directly in terms of only the chemical species concentrations, is because the changes in the extents of reactions can be defined directly in terms of the changes in the free energy change of each reaction (caused by the perturbation)—see below. Then the rate of the αth reaction can be written as

$$\frac{1}{V}\frac{\partial \xi_\alpha}{\partial t} = k_\alpha \prod_j a_j^{-v_{j\alpha}} - k_{-\alpha} \prod_{j'} a_{j'}^{v_{j'\alpha}} \tag{13.5}$$

Here, we have assumed that the activity coefficients of the different species in each reaction are not affected by the perturbations [2,11]. The term $a_j^{-v_{j\alpha}}$ is the activity of the jth reaction species raised to the stoichiometric coefficient of its participation as a reactant in the αth reaction (remember that the stoichiometric coefficient of reactants is defined to be negative). $a_{j'}^{v_{j'\alpha}}$ is the corresponding term for the j'th reactions species entering as a product in the αth reaction. k_α and $k_{-\alpha}$ are the forward and reverse rate constants of the αth chemical reaction.

At equilibrium, the forward and reverse rates of the αth reactions are equal (e.g., to r_α)

$$r_\alpha = k_\alpha \prod_j \bar{a}_j^{-v_{j\alpha}} = k_{-\alpha} \prod_{j'} \bar{a}_{j'}^{v_{j'\alpha}} \tag{13.6}$$

r_α is called the exchange rate of the αth reaction. The $\bar{a}_j$ are the chemical species activities at equilibrium. The rate of the αth reaction can therefore be written as

$$\frac{1}{V}\frac{\partial \xi_\alpha}{\partial t} = r_\alpha \prod_j \left(\frac{a_j}{\bar{a}_j}\right)^{-v_{j\alpha}}\left[1 - \prod_i^N \left(\frac{a_i}{\bar{a}_i}\right)^{v_{i\alpha}}\right] \tag{13.7}$$

Note that the term $\prod_i^N \left(a_i/\bar{a}_i\right)^{v_{i\alpha}}$ is a product over all species of the αth reaction (products and reactants). The product $\prod_j \left(a_j/\bar{a}_j\right)^{-v_{j\alpha}} \approx 1$ (for very small deviations from equilibrium) is a product only over the reactants.

At this point, the products of the activities for the αth reaction can be written as follows:

$$\prod_i \bar{a}_i^{v_{i\alpha}} = K_\alpha \text{ and } \prod_i a_i^{v_{i\alpha}} = Q_\alpha \tag{13.8}$$

where K_α is the equilibrium constant of the αth reaction. Q_α is the equivalent to K_α of the αth reaction, but for values representing the nonequilibrium species activities during the relaxation to equilibrium. Thus, the term in brackets in Equation 13.7 can be written as $[1 - Q_\alpha/K_\alpha]$.

Using the usual definition of the chemical potential of each reaction species, $\mu_i = \mu_i^0 + RT \ln a_i$ [19,20], for example, and its relation to the free energy change of a reaction, G_α, we have

$$G_\alpha = \left(\frac{\partial G}{\partial \xi_\alpha}\right)_{T,P,\xi_\beta} = \sum_i^N v_{i\alpha}\mu_i = -RT\left(\ln K_\alpha - \ln Q_\alpha\right) = RT \ln\left(\frac{Q_\alpha}{K_\alpha}\right) \tag{13.9}$$

Because near equilibrium $\left(G_\alpha/RT\right) \ll 1$, then $Q_\alpha/K_\alpha = \exp\left(G_\alpha/RT\right) \approx 1 + G_\alpha/RT$, and we have $[1 - Q_\alpha/K_\alpha] \approx -G_\alpha/RT$; therefore

$$\frac{1}{V}\frac{\partial \xi_\alpha}{\partial t} = -\frac{r_\alpha G_\alpha}{RT} = \frac{-r_\alpha \sum_i^N v_{i\alpha}\mu_i}{RT} \tag{13.10}$$

Equation 13.10 is for relaxation chemical kinetics the fundamental relationship between the free energy deviation from equilibrium of the αth reaction $G_\alpha = \Sigma_i^N v_{i\alpha}\mu_i$ and the rate of this reaction $(1/V)\partial\xi_\alpha/\partial t$. It also contains the exchange rates of the reaction at equilibrium, r_α. One sees clearly from this equation that the rate of the reaction depends on the exchange rate of the reaction exactly at equilibrium, and the amplitude of the reaction rate depends on the ratio of the free energy deviation (caused by the perturbation) and RT.

In general, the reactions of a multistep kinetic mechanism are coupled with each other through shared reaction species; this means that the free energy deviation of the αth reaction is coupled to the G_β of other participating reactions β. That is, $G_\alpha = \Sigma_{\beta=1}^R G_{\alpha\beta}\Delta\xi_\beta$, where $G_{\alpha\beta} = \left(\partial G_\alpha/\partial\xi_\beta\right)_{T,P}$. Using the definition of the chemical potentials, $G_{\alpha\beta}$ can be written as

$$G_{\alpha\beta} = \sum_{i=1}^N v_{i\alpha}\left(\frac{\partial\mu_i}{\partial\xi_\beta}\right) = \sum_{i=1}^N v_{i\alpha}\sum_{j=1}^N \left(\frac{\partial\mu_i}{\partial c_j}\right)\left(\frac{\partial c_j}{\partial\xi_\beta}\right) \tag{13.11}$$

In the last equation, the actual concentrations of the reaction species have been introduced, since observing the reaction species' concentrations is the way one follows the progress of the reaction. Recalling that the chemical potentials can be related to the concentrations and the activity coefficients, $\mu_i = \mu_i^0 + RT\ln a_i = \mu_i^0 + RT\ln c_i + RT\ln\gamma_i$, and if the assumption is made that the perturbation is small enough so that the activity coefficients of the species do not depend on the change in concentration of any of the reaction species, that is, $\dfrac{\partial\ln\gamma_i}{\partial c_j} = 0$, then one can write

$$G_\alpha = \sum_{\beta=1}^R G_{\alpha\beta}\Delta\xi_\beta = \frac{RT}{V}\sum_{\beta=1}^R\left[\sum_{i=1}^N \frac{v_{i\alpha}v_{i\beta}}{\overline{c}_i}\right] = \frac{RT}{V}\sum_{\beta=1}^R g_{\alpha\beta}\Delta\xi_\beta \tag{13.12}$$

where $\left[\Sigma_{i=1}^N v_{i\alpha}v_{i\beta}/\overline{c}_i\right] \equiv g_{\alpha\beta}$. The reader should consult Castellan's paper to see the complete expression, including the contributions to possible changes in activity coefficients. However, it is usually assumed that this is not relevant and that high concentration reaction components are buffered. (In this case, then the terms $g_{\alpha\beta}$ involving these chemical species are not included in the sum.)

Now one can write the very simple equation

$$\frac{1}{V}\frac{\partial\xi_\alpha}{\partial t} = -r_\alpha G_\alpha/RT = \left(\frac{-r_\alpha}{RT}\right)\frac{RT}{V}\sum_{\beta=1}^R g_{\alpha\beta}\Delta\xi_\beta \tag{13.13}$$

or more simply,

$$\frac{\partial\xi_\alpha}{\partial t} = -r_\alpha G_\alpha/RT = -r_\alpha\sum_{\beta=1}^R g_{\alpha\beta}\Delta\xi_\beta = -\sum_{\beta=1}^R r_\alpha\left[\sum_{i=1}^N v_{i\alpha}v_{i\beta}/\overline{c}_i\right]\Delta\xi_\beta \tag{13.14}$$

Equation 13.14 is the basic differential equation to be solved in order to calculate the relaxation time and the relaxation amplitudes. Note that $g_{\alpha\beta} = g_{\beta\alpha}$. The rate of change of the advancement of the αth reaction depends not only on its own rate of exchange, r_α, and on its own advancement deviation from equilibrium, $\Delta\xi_\alpha$, but also on the deviation of the advancements of all the other reactions $\Delta\xi_\beta$ through a term containing the concomitant common chemical species of reactions α and β, $v_{i\alpha}v_{i\beta}/\overline{c_i}$.

13.2.1 CALCULATING NORMAL MODE RELAXATION TIME CONSTANTS AND AMPLITUDES

Equation 13.14 is a linear coupled differential equation with the time-dependent variables $\Delta\xi_\beta$, and constant coefficients $-r_\alpha\left[\sum_{i=1}^{N} v_{i\alpha}v_{i\beta}/\overline{c_i}\right] = -r_\alpha g_{\alpha\beta}$. The equilibrium concentrations of the chemical species $\overline{c_i}$ must be known. The stoichiometric coefficients $v_{i\beta}$ are defined according to the reaction mechanism. The solution of such differential equations is well known. The impulse solution is given in terms of normal modes, which we represent by the subscript label ε. The time-dependent progress of the αth reaction is a sum of exponentials, Equation 13.15, where λ_ε are the eigenvalues (the rates of relaxation of εth normal modes, which are the reciprocals of the relaxation lifetimes $\lambda_\varepsilon = 1/\tau_\varepsilon$) and $m_{\alpha\varepsilon}$ are the coefficients representing the contributions of the different ε normal modes to the αth reaction. $m_{\alpha\varepsilon}$ contain contributions of all the separate reactions that are coupled to the αth reaction through shared reaction species.

$$\Delta\xi_\alpha = \sum_{\varepsilon=1}^{R} m_{\alpha\varepsilon}e^{-\lambda_\varepsilon t} \tag{13.15}$$

Substituting this solution into the differential equation, one arrives at the expression

$$\Delta\dot\xi_\alpha = \sum_{\varepsilon=1}^{R} m_{\alpha\varepsilon}\lambda_\varepsilon e^{-\lambda_\varepsilon t} = r_\alpha\sum_{\beta=1}^{R}\sum_{\varepsilon=1}^{R} g_{\alpha\beta}m_{\alpha\varepsilon}e^{-\lambda_\varepsilon t} \tag{13.16}$$

Or, because all the terms for ε are independent (normal modes), $m_{\alpha\varepsilon}\lambda_\varepsilon = r_\alpha\sum_{\beta=1}^{R} g_{\alpha\beta}m_{\alpha\varepsilon}$, and

$$\sum_{\beta=1}^{R}\left(r_\alpha g_{\alpha\beta} - \lambda_\varepsilon\delta_{\alpha\beta}\right)m_{\beta\varepsilon} = 0 \tag{13.17}$$

These linear algebraic equations with constant coefficients only have a solution provided that the following determinant is zero (see, e.g., [21]),

$$\left|r_\alpha g_{\alpha\beta} - \lambda_\varepsilon\delta_{\alpha\beta}\right| = 0 \tag{13.18}$$

If we know r_α and $g_{\alpha\beta}$, we can easily formally (and often numerically) solve for every normal mode eigenvalue $\lambda_\varepsilon = 1/\tau_\varepsilon$. Once we know the eigenvalues, we can solve for the eigenvectors of Equation 13.18. The eigenvectors are needed to calculate the measured amplitudes. The eigenvectors contain contributions from all the different reactions; that is, each normal mode, ε, which is relaxing in time with the rate $\lambda_\varepsilon = 1/\tau_\varepsilon$ has contributions from all the different reactions β, each represented by $m_{\beta\varepsilon}$.

The measured amplitudes of a relaxation experiment result from the changes in concentrations of particular reaction species. Often these observable reaction species are dyes; so the measured signal is due to observation of some optical signal, for instance, the change in concentration observed by absorption or fluorescence of one of the contributing species. However, any signal related to changes in species concentrations can be used (and it does not have to be an optical signal; for example, dielectric dispersion experiments use the conductance of the solution, and ultrasound chemical kinetics observe the absorption of acoustic energy).

Using the determinant of Equation 13.18, we can automatically simply write down the equations to be solved for calculating the relaxation lifetimes of any reaction mechanism. The simplicity of this approach is powerful, and also deceptively simple. For instance, one must know the equilibrium concentrations of every chemical species of the reaction mechanism, and the resulting exact analytical expressions for the relaxation lifetimes are often unwieldy (higher-order polynomials in λ_ε that can be solved explicitly). For multistep cases, this is a formidable problem, and exact analytical solutions are not very transparent. However, it is often the case that one can choose the reaction conditions in order to considerably simplify the situation. And if the exchange rates (e.g., rate constants) of some of the reactions are considerably different from each other, approximations can be made. We discuss these approximations in the next section. These simpler solutions are often germane to many problems, and the behavior of lifetimes using such approximations is easily understood. Nevertheless, one always has to keep in mind that the approximations of separate reaction rates may not hold, and the full expressions may be necessary. This is in principle not difficult if numerical methods are used.

13.2.2 EXAMPLES OF SOME SIMPLE RELAXATION MECHANISMS

In order to show how to use Castellan's method, it is best to consider several simple examples. We emphasize that the general determinant, Equation 13.18, describing the time course of the relaxation, from which the normal mode relaxation times are derived, applies to any mechanism where the elementary reactions can be expressed as in Equation 13.1. The difficulty is the calculation of the equilibrium concentrations of the chemical species. We demonstrate that this is especially so when we consider the actual experiments of dyes binding to nucleic acids. The solution for the relaxation lifetimes is given by solving the determinant in Equation 13.18 in terms of the equilibrium concentrations of the reaction species, and the rate constants and the corresponding equilibrium constants. The measured signal is related to one or more of these chemical species, and the equations representing the measured signal amplitudes must be obtained through a transformation of Equation 13.15 to a parameter system in terms of the species concentrations. We discuss this later. Now we consider some simple example reaction mechanisms. We use expression for the forward and reverse exchange rates in terms of concentrations:

$$r_\alpha = k_\alpha \prod_j \overline{a}_j^{-v_{j\alpha}} = k_\alpha \prod_j \overline{c}_j^{-v_{j\alpha}} = k_{-\alpha} \prod_{j'} \overline{a}_{j'}^{v_{j'\alpha}} = k_{-\alpha} \prod_{j'} \overline{c}_{j'}^{v_{j'\alpha}}$$

(i.e., we assume that the perturbation did not affect the activity coefficients), and $g_{\alpha\beta} = \left[\Sigma_{i=1}^{N} \, v_{i\alpha} v_{i\beta} / \bar{c}_i \right]$. Then we set up the determinant Equation 13.18.

The goal of this chapter is pedagogical. We shall consider six kinetic mechanisms in detail in order to introduce the reader to the Castellan analysis, and to demonstrate how to derive expressions for the relaxation time, especially when certain approximations can be made about the rates of the elementary reactions steps of a reaction mechanism. These mechanisms, and the assumption of well-separated rates of the elementary reaction steps, have been applied to many relaxation experiments in the literature, and are also useful when considering dye binding to nucleic acids. Of course, expressions can be derived for all these mechanisms without any such assumptions about the rates. This can be done by simply deriving expressions, either analytically or numerically, that are solutions of the full quadratic (or higher) algebraic equations, which result from expanding the determinant Equation 13.18 without any simplifications. However, the exact equations are cumbersome and not very illuminating unless assumptions of limiting conditions are subsequently made. Therefore, because our goal is to introduce the reader to the Castellan procedure, and since in many cases the assumption of considerable differences in the rates of the elementary reactions is valid, we will concentrate on the cases of well-separated rates. This does not mean that it may be necessary to deal with the general, more complex, cases, without these assumptions.

Mechanism 1

$$A \underset{k_{-1}}{\overset{k_1}{\rightleftharpoons}} B$$

This is trivial to write down from first principles, but let us use the general expression, and write down all the intermediate steps, just to demonstrate:

$$\left| r_1 g_{11} - \lambda_1 \right| = k_1 \bar{A} \left[\frac{1}{\bar{A}} + \frac{1}{\bar{B}} \right] - \lambda_1 = k_1 \left[1 + \frac{\bar{A}}{\bar{B}} \right] - \lambda_1 = k_1 \left[1 + \frac{k_{-1}}{k_1} \right] - \lambda_1 = \left[k_1 + k_{-1} \right] - \lambda_1$$

where the reverse equilibrium constant $K_1^{\text{diss}} = \dfrac{\bar{A}}{\bar{B}} = \dfrac{k_{-1}}{k_1}$. Therefore,

$$\lambda_1 = 1/\tau_1 = k_1 + k_{-1} \tag{13.19}$$

The rate of this reaction—and therefore the relaxation time—is concentration independent and is equal to the sum of the rate constants. But we see that we do not have to write down any differential equations, and the result arises straightforwardly from the determinant in Equation 13.18. The relaxation experiment can tell us the sum of the rate constants. To arrive at the individual rate constants k_1 and k_{-1}, we can use information of the equilibrium constant, $K_1^{\text{diss}} = \left(k_{-1}/k_1 \right)$, which has to be determined from an equilibrium experiment (for instance, if the absorption or fluorescence spectra are known from the two conformers, one can linearly decompose the spectrum into the two components).

Such a mechanism is also often observed, also in single-molecule experiments, where one is observing a simple one-step conformational change. However, there is a difference between ensemble relaxation experiments and single-molecule experiments. In single-molecule experiments, the individual rate constants, k_1 and k_{-1}, can be determined from the histograms of the dwell times in each conformation. In the ensemble experiment, one can only determine the sum of the constants, and one needs the additional information of the equilibrium constant to obtain the individual rate constants k_1 and k_{-1}.

Mechanism 2

$$A + B \underset{k_{-1}}{\overset{k_1}{\rightleftharpoons}} C$$

This is the simplest single-step binding mechanism.

$$\left| r_1 g_{11} - \lambda_1 \right| = k_1 \bar{A}\bar{B} \left[\frac{1}{\bar{A}} + \frac{1}{\bar{B}} + \frac{1}{\bar{C}} \right] - \lambda_1 = k_1 \left[\bar{B} + \bar{A} + \frac{\bar{A}\bar{B}}{\bar{C}} \right] - \lambda = k_1 \left[\bar{A} + \bar{B} \right] + k_{-1} - \lambda$$

where

$$\frac{\bar{A}\bar{B}}{\bar{C}} = K_1^{\text{diss}} = \frac{k_{-1}}{k_1}$$

Therefore, the relaxation time is

$$1/\tau_1 = k_1 \left[\bar{A} + \bar{B} \right] + k_{-1} \tag{13.20}$$

The two kinetic constants are determined by varying the equilibrium concentration of $\bar{A} + \bar{B}$, and making a plot of $1/\tau_1$ versus $\left[\bar{A} + \bar{B} \right]$ (see Figure 13.1). This results in a straight line with intercept k_{-1} and slope k_1. If one has previously measured the equilibrium constant $K_1^{\text{diss}} = \bar{A}\bar{B}/\bar{C}$, then one can calculate the concentrations $\bar{A}$ and $\bar{B}$ for different starting concentrations of starting concentrations of A_0 and B_0. Alternatively, if it is possible to start with just C_0, then $\bar{A} = \bar{B} = C_0 - \bar{C}$.

It is often better to use an excess of one reaction species; for instance, $A_0 \gg B_0$; then $1/\tau_1 \approx k_1 A_0 + k_{-1}$. In this case, the plot of $1/\tau_1$ versus A_0 (which one knows accurately) also has an intercept k_{-1} and slope k_1. For small molecules, each with just one binding site, it makes no difference which reactant is in excess. However, when dealing with dye binding to DNA (or proteins binding to DNA), where there are multiple sites on the DNA for dye binding, care must be taken. If it is assumed that there are multiple-binding sites on every DNA molecules, and that they all have identical

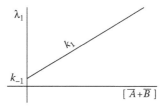

FIGURE 13.1   Plot of $1/\tau_1$ versus $\left[\bar{A} + \bar{B} \right]$ for Mechanism 2.

binding kinetics, then it is best to arrange conditions such that the concentration of DNA-binding sites, $[\text{DNA-binding sites}]_0$, is in large excess over the dye concentration $[\text{Dye}]_0$. Then $1/\tau_1 = k_1[\text{DNA-binding sites}]_0 + k_{-1}$. By varying the concentration of DNA, it is straightforward to determine the rate constants as above. However, when concentration of the ligand (dye) is in the range of the binding dissociation constant, and either $[\text{DNA-binding sites}]_0 \approx [\text{Dye}]_0$ or $[\text{DNA-binding sites}]_0 \approx [\text{Dye}]_c$ there is a problem of excluded site binding (see Section 13.3). This effect arises whenever sites on the DNA that are already bound exclude neighboring sites; the neighboring sites may be unoccupied, but they cannot bind ligand in the presence of a neighboring site that has already bound a ligand. When this happens, it is clear that calculating the concentration of free binding sites involves knowing the fraction of sites already bound. Special statistics is required to calculate the number of sites on the DNA that are still free for binding a ligand [22–24]. For instance, if a certain fraction of the original binding sites is occupied, the number of unoccupied sites that are free for binding per molecule will depend on the distribution of occupied sites on the DNA. This is examined in more detail in Section 13.3. For a discussion of chemical relaxation experiments with cooperative binding to a linear polymer such as DNA, see, for example, Refs. [24,25].

Mechanism 3

$$A + B_1 \underset{k_{-1}}{\overset{k_1}{\rightleftharpoons}} C_1, \; A + B_2 \underset{k_{-2}}{\overset{k_2}{\rightleftharpoons}} C_2$$

This mechanism is an extension of Mechanism 2; here there are two different conformations of the bound complex, and two binding species B_1 and B_2. This could represent, for instance, a possible mechanism for a dye (A) binding to two different sites on DNA, B_1 and B_2, where we assume in this mechanism there is no excluded site-binding effect between the B_1 and B_2 sites on the DNA.

$$\left| r_\alpha g_{\alpha\beta} - \lambda_\varepsilon \delta_{\alpha\beta} \right| = \begin{vmatrix} r_1 g_{11} - \lambda_\varepsilon & r_1 g_{12} \\ r_2 g_{21} & r_2 g_{22} - \lambda_\varepsilon \end{vmatrix} = \begin{vmatrix} k_1\left[\bar{A}+\bar{B_1}\right]+k_{-1}-\lambda_\varepsilon & k_1\bar{A}\bar{B_1}\left[\dfrac{1}{\bar{A}}\right] \\ k_2\bar{A}\bar{B_2}\left[\dfrac{1}{\bar{A}}\right] & k_2\left[\bar{A}+\bar{B_2}\right]+k_{-2}-\lambda_\varepsilon \end{vmatrix}$$

$$= \begin{vmatrix} k_1\left[\bar{A}+\bar{B_1}\right]+k_{-1}-\lambda_\varepsilon & k_1\bar{B_1} \\ k_2\bar{B_2} & k_2\left[\bar{A}+\bar{B_2}\right]+k_{-2}-\lambda_\varepsilon \end{vmatrix} = 0$$

If it is assumed that B_1 and B_2 are on a DNA molecule, then the fraction of B_1 and B_2 is constant for every DNA molecule. (The simplest case is just one B_1 and B_2 site on every DNA molecule.) Several approximations are relevant depending on the relative rates of the reaction rate constants, and the concentrations of the reaction species. See Figure 13.2a and b for plots of Case 1 and 2 for this mechanism.

Case 1: Assume we arrange conditions such that $B_{1,0}, B_{2,0} \gg A_0$ (DNA concentrations much higher than dye concentrations). Assume also that $k_1 > k_2$ and $k_{-1} > k_{-2}$. Then the determinant is $\cong \left(k_1\left[\bar{B_1}\right]+k_{-1}-\lambda_\varepsilon\right)\left(-\lambda_\varepsilon\right) = 0$; therefore

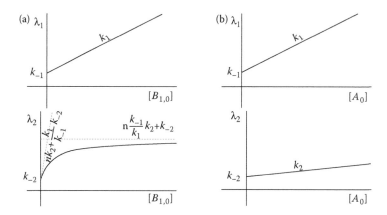

FIGURE 13.2 For Mechanism 3, plots of (a) $1/\tau$ versus $B_{1,0}$ for Case 1 ($B_{1,0}, B_{2,0} \gg A_0$; $k_1 > k_2$ and $k_{-1} > k_{-2}$) and of (b) $1/\tau_1$ versus $[A_0]$ for Case 2 ($A_0 \gg B_{1,0}, B_{2,0}$). For the slower time (τ_2) in Case 1, the dotted line indicates the initial slope and the dashed line the plateau.

$$\lambda_1 = 1/\tau_1 = k_1 \left[B_{1,0} \right] + k_{-1} \tag{13.21}$$

this is the fastest normal mode relaxation and consists only of the fastest reaction. If we can observe this relaxation normal mode (for instance, if an optical parameter of the dye molecule (A) changes upon binding), then by plotting, $1/\tau_1$ versus $\left[B_{1,0} \right]$ we can determine the rate constants k_1 from the slope of the straight line, and k_{-1} from the intercept.

How about the relaxation of the slow normal mode? In this case, $k_1 \left[\bar{A} + \bar{B}_1 \right] + k_{-1} \gg \lambda_2$, and the determinant equations result in approximately $\left(k_1 \left[\bar{A} + B_{1,0} \right] + k_{-1} \right)\left(k_2 \left[\bar{A} + B_{2,0} \right] + k_{-2} - \lambda_2 \right) - k_1 k_2 B_{1,0} B_{2,0} = 0$; therefore

$$
\begin{aligned}
\lambda_2 &\approx k_2 \left[B_{2,0} \right] + k_{-2} - \frac{k_1 k_2 B_{1,0} B_{2,0}}{\left(k_1 \left[B_{1,0} \right] + k_{-1} \right)} \\
&= \frac{\left(k_2 \left[B_{2,0} \right] + k_{-2} \right)\left(k_1 \left[B_{1,0} \right] + k_{-1} \right) - k_1 k_2 B_{1,0} B_{2,0}}{\left(k_1 \left[B_{1,0} \right] + k_{-1} \right)} \\
&= \frac{k_2 k_{-1} \left[B_{2,0} \right] + k_1 k_{-2} \left[B_{1,0} \right] + k_{-1} k_{-2}}{\left(k_1 \left[B_{1,0} \right] + k_{-1} \right)}
\end{aligned}
\tag{13.22a}
$$

If the ratio of the binding sites on the DNA is $B_{2,0}/B_{1,0} = n > 1$ (the ratio must be an integer), then

$$
\begin{aligned}
\lambda_2 &\approx \frac{k_2 k_{-1} n \left[B_{1,0} \right] + k_1 k_{-2} \left[B_{1,0} \right] + k_{-1} k_{-2}}{\left(k_1 \left[B_{1,0} \right] + k_{-1} \right)} \\
&= \frac{\left[n k_2 k_{-1} + k_1 k_{-2} \right]\left[B_{1,0} \right] + k_{-1} k_{-2}}{\left(k_1 \left[B_{1,0} \right] + k_{-1} \right)}
\end{aligned}
\tag{13.22b}
$$

Then, plotting $1/\tau_2$ versus $B_{1,0}$, the beginning slope is $nk_2 + (k_1/k_{-1})k_{-2}$, the intercept at very low $B_{1,0}$ is k_{-2}, and the plot plateaus at a value of $nk_2(k_{-1}/k_1) + k_{-2}$. Therefore, if we know the values of k_1 and k_{-1} from $1/\tau_1$, then from the intercept and slope or plateau of $1/\tau_2$ we can determine k_2 and k_{-2}.

Case 2: Assume we arrange conditions such that $A_0 \gg B_{1,0}, B_{2,0}$; this does not cause a problem with excluded site binding, because we have assumed that the sites B_1 and B_2 do not interfere. In this case

$$\lambda_1 = 1/\tau_1 = k_1[A_0] + k_{-1} \tag{13.23}$$

and just as before, we can easily determine k_1 and k_{-1}. By the same reasoning as for Case 1, we have for the slower normal mode relaxation

$$\left(k_2[A_0] + k_{-2} - \lambda_2\right) \approx \frac{k_1 k_2 \bar{B}_1 \bar{B}_2}{\left(k_1[A_0] + k_{-1}\right)}$$

thus

$$\lambda_2 \approx k_2[A_0] + k_{-2} - \frac{k_1 k_2 \bar{B}_1 \bar{B}_2}{\left(k_1[A_0] + k_{-1}\right)} \tag{13.24}$$

So, the slope is essentially k_2 and the intercept is $\sim k_{-2}$. That is, for this reaction mechanism (with multiple sites on the DNA, but no excluded binding, and if one binding reaction is much faster than the other), it is much easier to determine the rate constants if $B_{1,0}, B_{2,0} \gg A_0$, rather than $B_{1,0}, B_{2,0} \gg A_0$.

If both binding steps have similar rate constants, then the determinant equation will result in a quadratic equation, and then no approximations can be made. If the concentrations of the two binding sites B_1 and B_2 can be independently varied, instead of being present on one DNA molecule, then similar equations can be derived for the case that $B_{1,0} \gg A_0, B_{2,0}$ and/or $B_{2,0} \gg A_0, B_{1,0}$, and these conditions would allow the equations again to be simplified. Also, it is possible that the two DNA-binding sites could interact, and then one has to consider either excluded site binding or an extended mechanism whereby the binding tone site affects the binding affinity of the other. In this case, there would be two binding steps for each site—one when both sites are free, and one when the other site is already bound.

Mechanism 4

$$A + B \underset{k_{-1}}{\overset{k_1}{\rightleftharpoons}} C \underset{k_{-2}}{\overset{k_2}{\rightleftharpoons}} D$$

This mechanism is often proposed for dyes (A) binding to DNA (B), where one assumes that the dye first binds to the helix in an unspecific manner (for instance simple electrostatic binding to the outside of the helix), and then subsequently in the second step the dye fits into a specific binding conformation (for instance, intercalating between base pairs, or fitting snugly into the minor or major groove).

The determinant equation for this case is

$$\left| r_\alpha g_{\alpha\beta} - \lambda_\varepsilon \delta_{\alpha\beta} \right| = \begin{vmatrix} r_1 g_{11} - \lambda_\varepsilon & r_1 g_{12} \\ r_2 g_{21} & r_2 g_{22} - \lambda_\varepsilon \end{vmatrix}$$

$$= \begin{vmatrix} k_1 \overline{AB}\left[\dfrac{1}{\overline{A}} + \dfrac{1}{\overline{B}} + \dfrac{1}{\overline{C}} \right] - \lambda_\varepsilon & k_1 \overline{AB}\left[\dfrac{-1}{\overline{C}} \right] \\ k_2 \overline{C}\left[\dfrac{-1}{\overline{C}} \right] & k_2 \overline{C}\left[\dfrac{1}{\overline{C}} + \dfrac{1}{\overline{D}} \right] - \lambda_\varepsilon \end{vmatrix}$$

$$= \begin{vmatrix} k_1 \left[\overline{A} + \overline{B} \right] + k_{-1} - \lambda_\varepsilon & -k_{-1} \\ -k_2 & k_2 + k_{-2} - \lambda_\varepsilon \end{vmatrix} = 0$$

See Figure 13.3a and b for plots of Case 1 and 2 for this mechanism. The second step is sometimes considered to involve diffusion of the dye molecule along the outside of the helix, before finally settling into the binding site.

Case 1: First, we assume that $k_1 [\overline{A} + \overline{B}]$ and $k_{-1} \gg k_2$ and k_{-2}; in considering binding of a small molecule (dye) to DNA we define B as the binding sites on DNA and A as the dye molecule. We first find the faster relaxation, where $\lambda_1 \gg \lambda_2$. By the same reasoning in the last example, we have

$$\lambda_1 = 1/\tau_1 = k_1 \left[\overline{A} + \overline{B} \right] + k_{-1} \tag{13.25}$$

Thus, if the dye changes some optical parameter upon unspecific binding to the outside of the helix, we can determine the two rate constants k_1 and k_{-1} from a plot

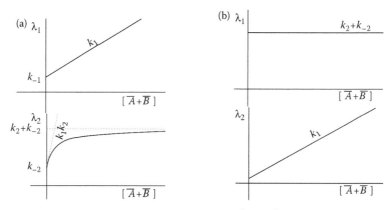

FIGURE 13.3 For Mechanism 4, plots of $1/\tau_1$ versus $\left[\overline{A} + \overline{B} \right]$ for (a) Case 1, in which the first reaction is faster than the second, and of (b) Case 2, in which the second reaction is faster than the first. For the slower time (τ_2) in Case 1, the dotted line indicates the initial slope and the dashed line the plateau.

of $1/\tau_1$ versus $\left[\bar{A}+\bar{B}\right]$. As before, if one of the reaction species is much larger than the other, then we do not have to determine the equilibrium concentrations. However, often as we mentioned in the case of the single-step bimolecular reaction, if one of the reaction species is DNA with many overlapping sites, then it is easier to use the conditions $\left[\text{DNA-binding sites}\right]_0 \gg \left[\text{Dye}\right]_0$. However, often it is not possible to observe the first bimolecular step (no signal change).

To find the slower relaxation time, λ_2, we divide the first row of the determinant by $k_{-1}\bar{C}$. We also take into consideration that $k_1\left[\bar{A}+\bar{B}\right]+k_{-1} \gg \lambda_2$. This gives the following approximation:

$$\begin{vmatrix} \dfrac{1}{\bar{A}}+\dfrac{1}{\bar{B}}+\dfrac{1}{\bar{C}} & -\dfrac{1}{\bar{C}} \\ -k_2 & k_2+k_{-2}-\lambda_2 \end{vmatrix} = 0$$

The expansion of the determinant leads to

$$\left(\dfrac{1}{\bar{A}}+\dfrac{1}{\bar{B}}+\dfrac{1}{\bar{C}}\right)\left(k_2+k_{-2}-\lambda_2\right)=\dfrac{k_2}{\bar{C}}; \quad \left(k_2+k_{-2}-\lambda_2\right)\left(K_1\left[\bar{A}+\bar{B}\right]+1\right)=k_2;$$

$$K_1=\dfrac{k_1}{k_{-1}}=\dfrac{\bar{C}}{\bar{A}\bar{B}}$$

Solving for λ_2,

$$\lambda_2 = k_{-2}+k_2\left[\dfrac{K_1\left[\bar{A}+\bar{B}\right]}{K_1\left[\bar{A}+\bar{B}\right]+1}\right] \tag{13.26}$$

Plotting $\lambda_2 = 1/\tau_2$ versus $\left[\bar{A}+\bar{B}\right]$, the intercept is k_{-2} and the beginning slope is K_1k_2. At high concentrations, the plateau is k_2+k_{-2}. The same precautions apply to this mechanism as we discuss in the simple one-step mechanism $A+B\underset{k_{-1}}{\overset{k_1}{\rightleftharpoons}}C$ when one of the reactant species is a polymer such as DNA that can have excluded site binding. This will be discussed in the Section 13.3 below.

Case 2: If the second step is faster than the binding step, then the determinant equations lead to the following two eigenvalues (provided the concentrations are low enough):

$$\lambda_1 = k_{-2}+k_2 \tag{13.27}$$

$$\begin{vmatrix} k_1\left[\bar{A}+\bar{B}\right]+k_{-1}-\lambda_2 & -k_{-1} \\ -k_2 & k_2+k_{-2} \end{vmatrix} = 0;$$

$$\left(k_1\left[\bar{A}+\bar{B}\right]+k_{-1}-\lambda_2\right)\left(k_2+k_{-2}\right)-k_{-1}k_2 = 0 \tag{13.28}$$

$$\lambda_2 = k_1\left[\bar{A}+\bar{B}\right]+k_{-1}\left(\dfrac{k_{-2}}{k_2+k_{-2}}\right)$$

If we know $K_2 = k_2/k_{-2} = \bar{D}/\bar{C}$, then we can determine k_2 and k_{-2} from λ_1. From a plot of λ_2 versus $\left[\bar{A} + \bar{B}\right]$, we can then get k_1 from the slope and k_{-1} from the intercept (if we know k_2 and k_{-2}).

Mechanism 5

$$A + B \underset{k_{-1}}{\overset{k_1}{\rightleftharpoons}} C$$

$$A + B \underset{k_{-2}}{\overset{k_2}{\rightleftharpoons}} D$$

This is different from Mechanism 3; the binding of A and B can form two different species of the bound state; each bound state excludes the formation of the other bound state. Such a binding can arise with dye binding to DNA (say to one of the grooves) where there are two sequences of DNA that are close and overlapping, such that binding to one of the sites excludes further dye from binding to the other site.

$$\left| r_\alpha g_{\alpha\beta} - \lambda_\varepsilon \delta_{\alpha\beta} \right| = \begin{vmatrix} r_1 g_{11} - \lambda_\varepsilon & r_1 g_{12} \\ r_2 g_{21} & r_2 g_{22} - \lambda_\varepsilon \end{vmatrix}$$

$$= \begin{vmatrix} k_1 \bar{A}\bar{B}\left[\dfrac{1}{\bar{A}} + \dfrac{1}{\bar{B}} + \dfrac{1}{\bar{C}}\right] - \lambda_\varepsilon & k_1 \bar{A}\bar{B}\left[\dfrac{1}{\bar{A}} + \dfrac{1}{\bar{B}}\right] \\ k_2 \bar{A}\bar{B}\left[\dfrac{1}{\bar{A}} + \dfrac{1}{\bar{B}}\right] & k_2 \bar{A}\bar{B}\left[\dfrac{1}{\bar{A}} + \dfrac{1}{\bar{B}} + \dfrac{1}{\bar{D}}\right] - \lambda_\varepsilon \end{vmatrix}$$

$$= \begin{vmatrix} k_1\left[\bar{A} + \bar{B}\right] + k_{-1} - \lambda_\varepsilon & k_1\left[\bar{A} + \bar{B}\right] \\ k_2\left[\bar{A} + \bar{B}\right] & k_2\left[\bar{A} + \bar{B}\right] + k_{-2} - \lambda_\varepsilon \end{vmatrix} = 0$$

If $k_1\left[\bar{A} + \bar{B}\right], k_{-1} \gg k_2\left[\bar{A} + \bar{B}\right], k_{-2}$, then (see Figure 13.4)

$$\lambda_1 = k_1\left[\bar{A} + \bar{B}\right] + k_{-1} \tag{13.29}$$

For the slower relaxation, we have

$$\left(k_1\left[\bar{A} + \bar{B}\right] + k_{-1}\right)\left(k_2\left[\bar{A} + \bar{B}\right] + \left(k_{-2} - \lambda_2\right)\right) - k_1 k_2 \left[\bar{A} + \bar{B}\right]^2 = 0$$

$$k_{-1}\left(k_2\left[\bar{A} + \bar{B}\right] + k_{-2} - \lambda_2\right) + \left(k_1\left[\bar{A} + \bar{B}\right]\right)\left(k_{-2} - \lambda_2\right) = 0$$

$$\left(-\lambda_2\right)\left(k_{-1} + k_1\left[\bar{A} + \bar{B}\right]\right) + k_{-1}\left(k_2\left[\bar{A} + \bar{B}\right] + k_{-2}\right) + k_{-2}\left(k_1\left[\bar{A} + \bar{B}\right]\right) = 0$$

$$\lambda_2 = \frac{k_{-1}\left(k_2\left[\bar{A} + \bar{B}\right] + k_{-2}\right) + k_{-2}\left(k_1\left[\bar{A} + \bar{B}\right]\right)}{\left(k_{-1} + k_1\left[\bar{A} + \bar{B}\right]\right)}$$

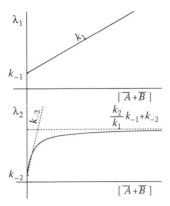

FIGURE 13.4 For Mechanism 5, plots of $1/\tau_1$ versus $\left[\overline{A}+\overline{B}\right]$. For the slower time ($\tau_2$), the dotted line indicates the initial slope and the dashed line the plateau.

$$\lambda_2 = \frac{\left(k_{-1}k_2 + k_{-2}k_1\right)\left[\overline{A}+\overline{B}\right] + k_{-1}k_{-2}}{\left(k_1\left[\overline{A}+\overline{B}\right] + k_{-1}\right)} \qquad (13.30)$$

A plot of $1/\tau_2$ versus $\left[\overline{A}+\overline{B}\right]$ has an intercept of k_{-2}, a beginning slope of $(k_{-1}k_2 + k_{-2}k_1)/k_{-1}$, and a plateau of $\left(\left(k_{-1}k_2 + k_{-2}k_1\right)/k_1\right) = \left(\left(k_2/k_1\right)k_{-1} + k_{-2}\right)$ (Figure 13.4). This model should be compared to model (3). We see that when the B species are in excess, then these two models (3) and (5) are identical; but when species A is in excess, then the models behave very differently. Model (5) plateaus at this concentrations and model (3) does not. This is due to the excluded site binding, where in model (5) the species B competes for overlapping sites that cannot be simultaneously occupied. This is the simplest model of excluded (overlapping) site binding.

Mechanism 6

$$B' \underset{k_{-1}}{\overset{k_1}{\rightleftharpoons}} B$$

$$A + B \underset{k_{-2}}{\overset{k_2}{\rightleftharpoons}} C$$

In this mechanism, there is a conformational change in species B, and only B (not B'), can bind to A. An example would be for B and B' to be two different local conformations of helical DNA, and A is a dye molecule that intercalates between base pairs (see, e.g., [23]). It is instructive to compare this mechanism to that of Mechanism 4 because as we will show, the expressions for the relaxation times are very similar, and it can be difficult to distinguish these two mechanisms.

$$\left| r_\alpha g_{\alpha\beta} - \lambda_\varepsilon \delta_{\alpha\beta} \right| = \begin{vmatrix} r_1 g_{11} - \lambda_\varepsilon & r_1 g_{12} \\ r_2 g_{21} & r_2 g_{22} - \lambda_\varepsilon \end{vmatrix}$$

$$= \begin{vmatrix} k_1 \bar{B}' \left[\dfrac{1}{\bar{B}'} + \dfrac{1}{\bar{B}} \right] - \lambda_\varepsilon & -k_1 \bar{B}' \left[\dfrac{1}{\bar{B}} \right] \\ -k_2 \bar{A}\bar{B} \left[\dfrac{1}{\bar{B}} \right] & k_2 \bar{A}\bar{B} \left[\dfrac{1}{\bar{A}} + \dfrac{1}{\bar{B}} + \dfrac{1}{\bar{C}} \right] - \lambda_\varepsilon \end{vmatrix}$$

$$= \begin{vmatrix} (k_1 + k_{-1}) - \lambda_\varepsilon & -k_{-1} \\ -k_2 \bar{A} & k_2 \left[\bar{A} + \bar{B} \right] + k_{-2} - \lambda_\varepsilon \end{vmatrix} = 0$$

See Figure 13.5a and b for plots of Case 1 and 2 for this mechanism.
    Case 1: Assume step 1 is much faster than step 2. In this case,

$$\lambda_1 = k_1 + k_{-1} \tag{13.31}$$

The following equations describe the slower relaxation time (when $\lambda_1 \gg \lambda_2$):

$$\left(k_2 \left[\bar{A} + \bar{B} \right] + k_{-2} - \lambda_2 \right) \left(k_1 + k_{-1} \right) = k_{-1} k_2 \bar{A}$$

$$\lambda_2 = -\dfrac{k_{-1} k_2 \bar{A}}{k_1 + k_{-1}} + k_2 \left[\bar{A} + \bar{B} \right] + k_{-2} = \left(1 - \dfrac{k_{-1}}{k_1 + k_{-1}} \right) k_2 \bar{A} + k_2 \left[\bar{B} \right] + k_{-2}$$

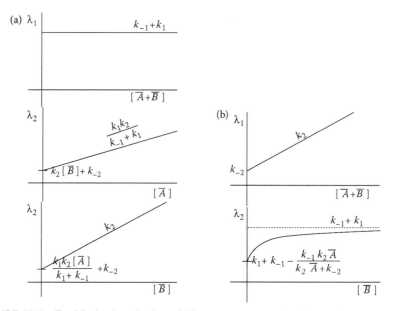

FIGURE 13.5 For Mechanism 6, plots of $1/t$ versus concentration for (a) Case 1, in which step 1 is faster than step 2, and (b) Case 2, in which step 2 is much faster than step 1. For the slower time (τ_2) in Case 2, the dashed line indicates the plateau.

$$\lambda_2 = \left(\frac{k_1}{k_1 + k_{-1}} \right) k_2 \bar{A} + k_2 \left[\bar{B} \right] + k_{-2} \qquad (13.32)$$

One sees by comparing the expressions for Case 2 of Mechanism 4 and the present Case 1 of Mechanism 6 that the equations for Mechanisms 4 and 6 are very similar, although not identical. It would be difficult to distinguish the two mechanisms from just lifetime data. Also, if species B is helical DNA and A is a dye molecule, in order to avoid analysis complications due to excluded site binding, one would have $\bar{B} \gg \bar{A}$. Then the expressions for the fast time are identical in terms of the corresponding rate constants, and the slower relaxation times for the two Mechanisms 4 and 6 are indistinguishable unless one has independently determined the equilibrium constant for the two steps of the reactions. Even then, this would require very accurate data.

Case 2: Assume step 2 is much faster than step 1. In this case,

$$\lambda_1 = k_2 \left[\bar{A} + \bar{B} \right] + k_{-2} \qquad (13.33)$$

$$\lambda_2 = \left(k_1 + k_{-1} \right) - \frac{k_{-1} k_2 \bar{A}}{\left(k_2 \left[\bar{A} + \bar{B} \right] + k_{-2} \right)} \qquad (13.34)$$

Again, this mechanism for this case is very difficult to distinguish from Case 1 of Mechanism 4. If species B is helical DNA, there is no relaxation signal for the dye involved in the conformational change of the DNA; however, as in Mechanism 4, the conformational change step includes the dye molecule because of the kinetic coupling, and there will be some signal resulting from dye binding even in the slow step.

13.2.3 CALCULATING THE CONCENTRATIONS

It is surely clear to the reader from the above expressions for the relaxation times that a general analysis of the relaxation lifetimes requires knowledge of the equilibrium concentrations of the reactants. This is a nontrivial exercise. A great simplification can be made if the concentration one of the reaction species (those involving a bimolecular-binding step) is in excess. Then, in the analysis of the relaxation times (not the amplitudes), one can ignore the concentrations of the other binding parameter. However, this is not always possible, and sometimes it is important to carry out the experiments in concentration ranges where this approximation cannot be made. In addition, the equilibrium constants may not be known before the relaxation experiment is carried out, and determining equilibrium constants becomes part of the overall analysis. Further, knowing the equilibrium concentrations of the reaction components can be difficult for dye binding to DNA because of excluded site effects and the intricate salt–polyelectrolyte interactions. There are many numerical algorithms that have been developed to calculate concentrations of more complex chemical-binding mechanisms. We refer the reader to a method of calculating concentrations that was developed especially for use in relaxation kinetics, which is related to the Castellan formalism [26]. We have used this in conjunction with the Castellan method, for the analysis of relaxation times as well as amplitudes [23,27,28]. Calculating concentrations is an important component analysis of relaxation kinetic data; however, a thorough discussion of this is outside of the scope of this chapter.

13.2.4 RELAXATION AMPLITUDES

We will only briefly consider amplitudes. An analysis of relaxation amplitude data is often mathematically complex, and is not always carried out. Nevertheless, an analysis of amplitudes is highly informative and in many cases crucial for the interpretation of some experiments; especially for distinguishing between different mechanisms that have similar behavior of relaxation times, and to derive the kinetic parameters. It is essential, if one wants to derive thermodynamic parameters from the kinetic data. This would require a complete separate chapter to cover adequately relaxation amplitudes. Here, for completeness, and to guide the reader into the literature, we will present a short description of relaxation amplitudes in the *modus operandi* of Castellan's method. We retain the notation of Castellan, as has also been carried through in a series of publications of Thusius [5,29–32]; we will follow the derivation of Thusius [5] closely in this section (see also [33]). A similar discussion of amplitudes within the Castellan formalism is by Jovin [8]. A series of articles by Hayman have presented a thorough formal description of chemical relaxation theory, also according to the Castellan formalism [14,34–37]. Of course, other publications can be consulted that treat the theoretical basis of chemical relaxation with a formalism different from that of Castellan [6,12,38–43].

In order to express the amplitudes concisely, it is necessary to express the equations in vector and matrix format.

The change in the concentration of chemical species "i," Δc_i, in terms of the net advancement of reaction α, $\Delta\xi_\alpha$, was given above in Equation 13.4, which we repeat here as

$$c_i = \overline{c}_i + \frac{1}{V}\sum_{\alpha}^{R} \nu_{i\alpha}\Delta\xi_\alpha = \overline{c}_i + \Delta c_i \tag{13.35}$$

This can be expressed in matrix format as

$$\Delta \mathbf{c} = \left(\frac{1}{V}\right)\underline{\underline{\nu}}\Delta\xi \tag{13.36}$$

where $\Delta \mathbf{c}$ is a vector of length N (the number of separate chemical species) representing the concentration deviations from equilibrium, $\Delta\xi$ is a vector of length R (the number of independent primary chemical reactions) representing the net advancements of the reactions, and $\underline{\underline{\nu}}$ is a matrix of dimension $N \times R$, where each element $\nu_{i\alpha}$ is the stoichiometric factor of the ith chemical species in the αth reaction. A single line under a bold character signifies a vector, and two lines under a bold character signify a matrix.

The deviation from equilibrium is caused by the thermodynamic perturbation of the equilibrium constant of each elementary reaction. We use the following thermodynamic functions for temperature and pressure perturbations:

$$\Delta \ln K_\alpha = \frac{\Delta K_\alpha}{K} = \left(\frac{\Delta H_\alpha}{RT}\right)\frac{\Delta T}{T}; \text{ where } \Delta H_\alpha = \sum_{i-1}^{N} v_{i\alpha} \bar{H}_i$$

$$\Delta \ln K_\alpha = \frac{\Delta K_\alpha}{K} = \left(\frac{\Delta V_\alpha}{RT}\right)\Delta P; \text{ where } \Delta V_\alpha = \sum_{i-1}^{N} v_{i\alpha} \bar{V}_i$$

$$K_\alpha = \prod_i^{N} \bar{c}_i^{v_{i\alpha}}$$

and

$$G_\alpha = \sum_{\beta=1}^{R} G_{\alpha\beta}\Delta\xi_\beta = \frac{RT}{V}\sum_{\beta=1}^{R}\left[\sum_{i=1}^{N}\frac{v_{i\alpha}v_{i\beta}}{\bar{c}_i}\right]\Delta\xi_\beta = \frac{RT}{V}\sum_{\beta=1}^{R}g_{\alpha\beta}\Delta\xi_\beta \qquad (13.37)$$

In matrix format therefore, using the equations in the Section 13.2

$$\frac{\Delta \mathbf{K}}{\mathbf{K}} = \left(\frac{1}{V}\right)\underline{\underline{g}}\Delta\underline{\xi} \qquad (13.38)$$

because $(\Delta K_\alpha/K)=(G_\alpha/RT)$ for every reaction, and remembering $g_{\alpha\beta} \equiv \left[\sum_{i=1}^{N} v_{i\alpha}v_{i\beta}/\bar{c}_i\right]$.

Therefore, using the relationship between $(\Delta\mathbf{K}/\mathbf{K})$ and $\Delta\underline{\xi}$ of Equation 13.38 we have

$$\Delta\mathbf{c} = \left(\frac{1}{V}\right)\underline{\underline{v}}\Delta\underline{\xi} = \underline{\underline{v}}\underline{\underline{g}}^{-1}\Delta\mathbf{K}/\mathbf{K} \qquad (13.39)$$

$\underline{\underline{g}}^{-1}$ is the inverse of the symmetric matrix $\underline{\underline{g}}$, and the $g_{\alpha\beta}^{-1}$ element of the inverse of a matrix $\underline{\underline{g}}$ is equal to $g_{\alpha\beta}^{-1} = g_{\beta\alpha}^{-1} = (-1)^{\alpha+\beta}|D_{\alpha\beta}|/|D|$, where $|D|$ is the determinant of $\underline{\underline{g}}$, and $|D_{\alpha\beta}|$ is the determinant of $\underline{\underline{g}}$ without the αth column and the βth row [21]. For larger reaction systems this looks cumbersome, but it is straightforward to carry out numerically.

The measured signal is related to the changes in the concentrations of the reaction species. The equation in the last paragraph tells us how to relate $\Delta\mathbf{c}$ to $\Delta\mathbf{K}/\mathbf{K}$, and consequently to either the change in molar enthalpies and ΔT or the change in the molar volumes and ΔP. The signal can be expressed as

$$\Delta P_{tot}^0 = \sum_{i=1}^{N} \phi_i\Delta\bar{c}_i \text{ where } \phi_i = \partial P/\partial\bar{c}_i$$

ΔP_{tot}^0 is the total change in the signal between the starting and end equilibrium state (before and after the perturbation).

This can also be written as

$$\Delta P_{tot}^0 = \phi^T\underline{\underline{v}}\underline{\underline{g}}^{-1}\Delta\mathbf{K}/\mathbf{K} \qquad (13.40)$$

the superscript "T" means transpose.

Because the concentrations are part of each normal mode of relaxation, the time dependent signal is distributed over the R normal modes as

$$\Delta P(t) = \sum_{n=1}^{R} \Delta P_n^0 \exp\left(-t/\tau_n\right) \tag{13.41}$$

ΔP_n^0 is the total equilibrium change in the measured signal of the nth normal mode, which contains all the chemical species participating in the nth normal mode relaxation. Thus, we see that knowing the general description of the reaction mechanism in terms of $\underline{\underline{v}}$ and $\underline{g}$, and the relation between concentrations of chemical species and the measured signal, it is possible formally to simply write down the amplitudes (albeit usually with the use of a computer) in terms of the thermodynamic perturbations.

13.2.5 ACQUIRING AND ANALYZING RELAXATION DATA—TIME AND FREQUENCY MODES

There are two ways to categorize the acquisition of relaxation kinetic data: (1) time domain and (2) frequency domain. Both have been extensively used to investigate models of ligand binding, protein–protein interactions, protein–nucleic acid interactions, and conformational changes. We do not discuss the instrumentation, but refer the reader to the literature [6,44].

13.2.5.1 Time Domain

The time domain has already been introduced. An internal thermodynamic condition (e.g., temperature or pressure) is changed abruptly; this perturbation is usually assumed, within the time constraints of equipment, to be a sudden square perturbation from one equilibrium condition to another. In this case, the perturbed system relaxes as a sum of exponentials. Each exponential time constant corresponds to the relaxation of a normal mode; for instance, the εth normal mode relaxes with the time constant $\tau_\varepsilon = 1/\lambda_\varepsilon$. For instance, the expression for the relaxation of the αth elementary reaction (given above, Equation 13.16, repeated here as Equation 13.42) is

$$\Delta\xi_\alpha = \sum_{\varepsilon=1}^{R} m_{\alpha\varepsilon}\lambda_\varepsilon e^{-\lambda_\varepsilon t} = r_\alpha \sum_{\beta=1}^{R} \sum_{\varepsilon=1}^{R} g_{\alpha\beta} m_{\alpha\varepsilon} e^{-\lambda_\varepsilon t} \tag{13.42}$$

The measured signal is related to concentration changes of the molecules that change some spectroscopic signal with each normal mode of relaxation (see Equations 13.40 and 13.41).

$$\Delta P(t) = \sum_{n=1}^{R} \Delta P_n^0 \exp\left(-t/\tau_n\right) \tag{13.43}$$

A major analysis undertaking is to "fit" the signal to a sum of exponentials, and to decide how many exponentials are necessary to account for the data (within some statistical criterion). This is a numerical problem that has occupied (plagued) many fields that rely on the analysis of exponentially decaying kinetic processes directly in the time domain (e.g., relaxation kinetics and also time-domain fluorescence lifetime measurements). This is in general carried out by iterative nonlinear regression methods [45]. The relaxation times and amplitudes are determined at various different equilibrium concentrations of selected chemical species and at differing solution conditions. The procedure of carrying out the relaxation measurements over a wide range of concentrations is a cardinal procedure in any relaxation kinetic experiment. The times and amplitudes are then analyzed in terms of kinetic models to establish a consistent kinetic mechanism and to determine the corresponding kinetic constants and thermodynamic parameters (molar reaction enthalpies and volume changes). There are various methods to carry out such an analysis. In the sequential data analysis method, which is the most common method, the recorded data is fitted with exponentials (This was earlier carried out by plotting the data recorded on photographs of oscilloscope traces on logarithmic graph paper.) the relaxation times and amplitudes are determined, and then the times and amplitudes are subsequently analyzed to determine the kinetic constants and thermodynamic parameters of various kinetic reaction schemes consistent with the data. This method is straightforward, but essentially requires one to decide on the number of exponentials before analyzing in terms of a kinetic mechanism; this sometimes requires the difficult task of separating exponentials with very similar lifetimes over certain concentration ranges. If the (proposed) kinetic model is fitted directly to the relaxation data, and if the data is fitted globally (simultaneously) over the whole data set, often kinetic models can be determined that are otherwise difficult to realize by the sequential analysis method of fitting first the relaxation times separately and then analyzing the fitted times according to different mechanisms [6,45].

13.2.5.2 Frequency Domain

In the frequency domain, the perturbation is carried out repetitively at a certain frequency. This is a very common way to perturb mechanical relaxing physical systems and in electrical circuits. The classical repetitive perturbations used in chemical relaxation kinetic methods are electric fields in dielectric dispersion [46–52] and acoustics and ultrasound [53–55]. Repetitive pressure jumps have also been used [56]. Temperature modulation is not common in this mode; however, recently repetitive temperature changes have also been employed for relaxation temperature jump in a biological cell [57].

We have shown above that the time-dependent relaxation signal of a chemical system in response to a sudden perturbation is a sum of exponentials. If instead of a sudden change in an internal thermodynamic variable, the perturbation is a repetitive modulation with a period T (radial frequency $\omega = 2\pi/T$). In the following discussion, for simplicity, we assume that the perturbation has the form of a perfect sinusoidal perturbation; this is not necessary, because any repetitive function can be expanded in a Fourier series, and then each frequency component is in the form of a sinusoidal function. The relationship between the frequency response of a chemical reaction system responding to a repetitive sinusoid application of a thermodynamic perturbation is given as a Fourier transformation (FT) of the time-dependent relaxation following a sudden perturbation, which, as shown above, consists of a sum of normal mode real decaying exponentials, $\Delta P(t) = \sum_{n=1}^{R} \Delta P_n^0 \exp(-t/\tau_n)$. The frequency-domain signal is then given at steady state as [43]

$$\text{FT}\left(\Delta P(t)\right) = \hat{P}(\omega) = \sum_{n=1}^{R} \frac{\Delta P_n^0 \tau_n}{\left(1 + i\omega\tau_n\right)}$$

$$\equiv \hat{P}(\omega) = \sum_{n=1}^{R} \frac{\alpha_n}{\left(1 + i\omega\tau_n\right)} = \sum_{n=1}^{R} \frac{\alpha_n}{\sqrt{1 + \left(\omega\tau_n\right)^2}} \exp\left(-\tan^{-1}\left(\omega\tau_n\right)\right) \qquad (13.44)$$

where $\phi_n = \tan^{-1}\left(\omega\tau_n\right)$ is the phase of the nth normal mode relaxation relative to the perturbation, and $\left(1/\sqrt{1 + \left(\omega\tau_n\right)^2}\right)$ is the demodulation factor of the nth normal mode. The usual analysis of this modulated signal $\hat{P}(\omega)$ can be made exactly as the analysis of frequency-domain fluorescence lifetime signals [58,59].

13.3 CASE STUDIES: BINDING TO NUCLEIC ACIDS AND STATISTICALLY EXCLUDED SITE BINDING

13.3.1 Equilibrium Binding to an Extended Lattice of Binding Sites

As already mentioned, in order to analyze the relaxation kinetic measurements involving binding steps, it is necessary to carry out the experiments at different concentrations of components. We have already indicated that this is not always trivial even for simple reactions (see Section 13.2.3). Often one chooses the concentrations of one reaction component to be in large excess; then one can vary the concentration of that component and analyze the lifetime data in that fashion. However, as we indicated in the section above, for samples that have either overlapping or interacting binding sites, one must carefully consider which reaction component one chooses to be in excess.

In the case of binding sites that interact with each other, it becomes more complicated. Excluded site binding to DNA molecules is an example of this type of binding, as we discussed shortly for Mechanisms 2 and 4. The problem is, to calculate the concentration of free sites. It is necessary to take into consideration in the analysis that ligands already bound to the DNA may affect the further binding to other sites, by either physically excluding neighboring sites or changing the binding affinities of nearby free sites. This is not trivial to take this into account; however, general formalisms have been developed. We give only a short summary of excluded site binding. The equilibrium aspects of ligand binding to a linear lattice of overlapping binding sites such as that presented by DNA have been treated theoretically in Refs. [22,60–63]. We do not consider any further complications, such as cooperativity effects.

The concentration of potential free binding sites, [sites], depends on the fraction of sites already occupied such that

$$\left[\text{sites}\right] = \left[\text{sites}\right]_0 f(r) \qquad (13.45)$$

where the subscript "0" refers to the total concentration of potential intercalation sites in the lattice, with no bound ligand. In the model as applied to double-stranded DNA, a site is an intercalation site between any two adjacent base pairs, and $[sites]_0$ is equal to the base pair concentration (provided that the number of base pairs per DNA molecule is large enough so that end effects can be ignored). Thus, the DNA concentrations are given in base pairs. In the model of McGhee and von Hippel [22], the parameter $f(r)$ accounts for the exclusion of potential binding sites arising from the statistical nature of the distribution of the ligands upon the binding lattice, and is given by Equation 13.46 (where the simpler case is considered, where all sites are equivalent),

$$f(r) \equiv (1 - nr)^n \left[1 - (n-1)r\right]^{1-n} \tag{13.46}$$

n is the number of intercalation sites per bound ligand in the presence of saturating ligand concentration. Therefore, the number of intercalation sites that are directly occupied or otherwise perturbed by an isolated bound ligand molecule (and which are removed from further binding interactions) equals $(2n-1)$. Because potential intercalation sites are multiply excluded by flanking bound ligands, the fractional lattice occupancy approaches the limit of $1/n$; that is, an average of n intercalation sites per bound ligand. The fraction of binding sites occupied, r, is defined as $r = [X]/[sites]_0$, where $[X]$ is the equilibrium concentration of the bound ligand species. The free ligand concentration is given by the usual conservation condition; that is, $[L] = [L]_0 - [X]$, where $[L]_0$ is the total analytical ligand concentration. These relationships can be used to obtain the following equation:

$$r/[L] = K_a (1 - nr) f(r) \tag{13.47}$$

where K_a is the association binding constant. For values of n greater than 1, the plot of $r/[L]$ versus r is similar to the conventional Scatchard plot, that is, $r/[L] = K_a (m - r)$. In the case of ethidium bromide (EB) binding to DNA, at low occupancy, m is the number of binding sites per intercalation site—where an intercalation site is located between two adjacent base pairs, and as the ligand concentration increases, the plot exhibits positive curvature. A line extrapolated from the initial slope at low r intersects the r axis at $1/(2n-1)$. In the case of EB–DNA interactions, before the theory of excluded site binding was developed, this intercept was found to be 0.33 [64–67]. This implies that $2n-1 = 3$ intercalation sites are excluded by each bound ligand at low-binding levels. It was also found that a saturated lattice has a ligand attached to every other ($n = 2$) intercalation site. Thus, the intercalation of a dye molecule perturbs the structure of the two sites directly adjacent to it (nearest-neighbor sites) so as to preclude intercalation at these sites. That the number of potential intercalation sites lost by binding one ligand to a free region of the lattice is equal to $2n-1$ and not n arises from the formalism of the model, which assumes a vectorial ligand lattice interaction [22]. This assumption is formally equivalent to excluding an equal number of lattice sites on both sides of the bound ligand. One sees that the calculations of the concentrations of free sites and free ligand are more involved than for simple noninteracting sites. This will be emulated in the chemical relaxation plots of the times and the amplitudes, as shown in the next section.

13.3.2   KINETICS OF BINDING TO AN EXTENDED LATTICE OF BINDING SITES

As an example, where it is necessary to take excluded site binding into account, it is the simple case of EB binding to double-stranded DNA. Assume that the mechanism is

$$\text{Sites} + \text{EB} \underset{k_{-1}}{\overset{k_1}{\rightleftharpoons}} \text{complex} \tag{13.48}$$

This is related to Mechanism 3 discussed above. However, now, because of excluded site binding, the relaxation time is [23,24,28]:

$$\tau^{-1} = k_1 \left(f(r) [\text{sites}]_0 - f'(r) [\text{EB}] \right) + k_{-1} \tag{13.49}$$

where $f'(r) = \partial f(r)/(\partial r)$, that is, $f'(r) = [nr(n-1) - 2n + 1](1 - nr)^{n-1}[1 - (n-1)r]^{n-1}$.

Note if $n = 1$, $f(r) = 1$, $f'(r) = -1$, and Equation 13.49 becomes the same as Equation 13.20. Original references can be consulted for detailed accounts and references for temperature jump [23] and pressure jump [23,28]. For an analysis of EB binding to DNA, including excluded site binding, and comparing the experimental results to Mechanisms 2, 4, and 6, see Refs. [23–25,28,68].

13.4   CONCLUSIONS

We have demonstrated a very general analysis method, originally due to Castellan and expanded by others, to derive relatively simple linear mathematical expressions for interpreting chemical relaxation experiments. Relaxation kinetic experiments have been crucial for understanding the mechanisms of binding of ligands (and proteins) to nucleic acids, as well as nucleic acid–nucleic acid interactions. Chemical relaxation kinetic experiments are especially advantageous when studying complex-binding mechanisms because free energy expressions defining reaction mechanisms can be expanded in a linear region. In effect, the reaction steps are dispersed on a temporal scale, such that coupled steps can be separated more easily. In addition, very rapid reaction steps can be measured that are otherwise too fast to observe. Kinetic rate constants can be obtained from the measured relaxation decay constants, and thermodynamic parameters (e.g., ΔH, ΔV) of the individual steps of reactions can be derived from the amplitudes. Using the analysis method of Castellan, a system of equations is easily derived, which can be analyzed readily with commonly available computer algorithms. However, a great benefit of relaxation methods is that insight can often be gleaned from inspection of data with simple approximations to the analytical expressions. After presenting how to set up the analysis of a general case, we show how to derive frequently used approximations for the most common mechanisms that provide valuable insight into reaction mechanisms according to simple plots.

REFERENCES

1. Eigen, M., Immeasurably fast reactions. *Nobel Lectures Chemistry*, 1963–1970. 1972, Amsterdam: Elsevier. pp. 163–205.

2. Castellan, G.W., Calculation of the spectrum of chemical relaxation times for a general reaction mechanism. *Berichte der Bunsengesellschaft für physikalische Chemie*, 1963. 67(9–10): 898–908.

3. Hammes, G.G. and P.R. Schimmel, Relaxation spectra of enzymatic reactions. *J. Phys. Chem.*, 1967. 71(4): 917–923.

4. Hammes, G.G. and P.R. Schimmel, Chemical relaxation spectra: Calculation of relaxation times for complex mechanisms. *J. Phys. Chem.*, 1966. 70(7): 2319–2324.

5. Thusius, D., *The Analysis of Relaxation Amplitudes, in Chemical and Biological Applications of Relaxation Spectrometry: Proceedings of the NATO Advanced Study Institute*, held at the University of Salford, Salford England, August 29–September 12, 1974, E. Wyn-Jones, Editor. 1975, D. Reidel Pub. Co.: Dordrecht; Boston. p. ix, 526 p.

6. Eigen, M. and L. De Maeyer, Theoretical basis of relaxation spectrometry, in G.G. Hammes and E.S. Lewis, Editors. *Investigation of Rates and Mechanisms of Reactions*, 1973, New York: Wiley-Interscience.

7. Hammes, G.G., Fast reactions in solution. *Annu. Rev. Phys. Chem.*, 1964. 15: 13–30.

8. Jovin, T.M., Fluorimetric kinetic techniques: Chemical relaxation and stopped-flow, in R.F. Chen and H. Edelhoch, Editors. *Biochemical Fluorescence: Concepts*, 1975, New York: Marcel Dekker, pp. 305–374.

9. Meixner, J., Thermodynamik und Relaxationserscheinungen. *Zeitschrift Naturforschung Teil A*, 1949. 4a: 594–600.

10. Meixner, J., Die thermodynamische Theorie der Relaxationserscheinungen und ihr Zusammenhang mit der Nachwirkungstheorie. Sonderausgabe der Kolloid-Zeitschrift. *Das Relaxationsverhalten der Materia*, 1953. 134(1): 3–20.

11. Meixner, J., New thermodynamic theory of relaxation phenomena. *Adv. Mol. Relax. Process.*, 1972. 3(1–4): 227–234.

12. Bernasconi, C.F., *Relaxation Kinetics*. 1976, New York: Academic Press. p. xi, 288 p.

13. Pecht, I. and R. Rigler, Chemical relaxation in molecular biology. *Molecular Biology, Biochemistry, and Biophysics*. 1977, Berlin; New York: Springer-Verlag. p. xvi, 418 p.

14. Hayman, H.J.G., Matrix methods and normalization procedures in chemical relaxation. *Isr. J. Chem.*, 1973. 11(2–3): 489–507.

15. Perlmutter-Hayman, B., Observability of normal modes of chemical relaxation. *Adv. Mol. Relax. Process.*, 1975. 7(2): 133–147.

16. Perlmutter-Hayman, B., Relaxation-times for chemical systems involving parallel reaction paths. *Adv. Mol. Relax. Interact. Process.*, 1977. 11(1–2): 1–8.

17. Ilgenfritz, G., Theory and simulation of chemical relaxation spectra. *Mol. Biol. Biochem. Biophys.*, 1977. 24: 1–42.

18. Kirkwood, J.G. and B.J. Crawford, The macroscopic equations of transport. *J. Phys. Chem.*, 1952. 56: 1048–1051.

19. McQuarrie, D.A. and J.D. *Simon, Physical Chemistry A Molecular Approach*. 1st ed. 1997, Sausalito: University Science Books. 1360 p.

20. Kuhn, H. and H.-D. *Foersterling, Principles of Physical Chemistry*. 2nd ed. 2009, Hoboken: John Wiley & Sons. 1036 p.

21. Kreyszig, E., *Advanced Engineering Mathematics*. 1968, New York: John Wiley and Sons, Inc. p. 898.

22. McGhee, J.D. and P.H. von Hippel, Theoretical aspects of DNA-protein interactions: Co-operative and non-co-operative binding of large ligands to a one-dimensional homogeneous lattice. *J. Mol. Biol.*, 1974. 86(2): 469–489.

23. Macgregor, R.B., Jr., R.M. Clegg, and T.M. Jovin, Pressure-jump study of the kinetics of ethidium bromide binding to DNA. *Biochemistry*, 1985. 24(20): 5503–5510.

24. Jovin, T.M. and G. Striker, *Chemical Relaxation Kinetic Studies of E. coli RNA Polymerase Binding to Poly [d(A−T)] Using Ethidium Bromide as a Fluorescence Probe, in Chemical Relaxation in Molecular Biology*, in I. Pecht and R. Rigler, Editors. 1977, Berlin: Springer-Verlag, pp. 245–281.

25. Schwarz, G., Chemical relaxation of cooperative conformational transitions of linear biopolymers. *J. Theor. Biol.*, 1972. 36(3): 569–580.

26. Avery, L., A numerical-method for finding the concentrations of chemicals in equilibrium. *J. Chem. Phys.*, 1982. 76(6): 3242–3248.

27. MacGregor, R.B., Jr. and R.M. Clegg, Ethidium-DNA binding kinetics in an agarose gel. *Biopolymers*, 1987. 26(12): 2103–2106.

28. Macgregor, R.B., Jr., R.M. Clegg, and T.M. Jovin, Viscosity dependence of ethidium-DNA intercalation kinetics. *Biochemistry*, 1987. 26(13): 4008–4016.

29. Thusius, D., Relaxation amplitudes for systems of 2 coupled equilibria. *J. Am. Chem. Soc.*, 1972. 94(2): 356–363.

30. Thusius, D., Simultaneous determination of equilibrium constants and thermodynamic functions by means of relaxation amplitude measurements. *Biochimie*, 1973. 55(3): 277–282.

31. Thusius, D., On the analysis of chemical relaxation amplitudes. *Biophys. Chem.*, 1977. 7(2): 87–93.

32. Thusius, D. and B. Vandenbunder, The temperature jump relaxation technique. *Biochimie*, 1975. 57(4): 437–443.

33. Schimmel, P., On the calculation of relaxation amplitudes. *J. Chem. Phys.*, 1971. 54: 4136–4137.

34. Hayman, H.J.G., Orthonormal chemical reactions and chemical relaxation. Part 1. Reactions in dilute solution. *Trans. Faraday Soc.*, 1969. 65: 2918–2929.

35. Hayman, H.J.G., Orthonormal chemical reactions and chemical relaxation. Part 2. Reactions in non-ideal solutions. *Trans. Faraday Soc.*, 1970. 66: 1402–1410.

36. Hayman, H.J.G., Orthonormal chemical reactions and chemical relaxation. Part 3. Reactions in solution under various thermodynamic conditions. *Trans. Faraday Soc.*, 1971. 67: 3240–3246.

37. Hayman, H.J.G., Orthonormal chemical reactions and chemical relaxation. Part 4. Reactions in solution with concomitant entropy and volume changes. *Isr. J. Chem.*, 1971. 9: 419–427.

38. Citi, M., F. Secco, and M. Venturini, Dynamic method of analysis—Evaluation of thermodynamic parameters from the amplitudes of chemical relaxation. *J. Phys. Chem.*, 1988. 92(22): 6399–6404.

39. Muirhead-Gould, J. and J.E. Stuehr, Temperature-jump relaxation amplitudes for single-step processes—Methyl-orange system. *J. Phys. Chem.*, 1975. 79(23): 2461–2464.

40. Schelly, Z.A. and M.M. Wong, Solvent-jump relaxation method using continuous-flow. *Int. J. Chem. Kinet.*, 1974. 6(5): 687–692.

41. Gutfreund, H., Transients and relaxation kinetics of enzyme reactions. *Ann. Rev. Biochem.*, 1971. 40: 315–344.

42. Tamura, K. and Z.A. Schelly, Analysis of the relaxation response induced by various forms of pulse perturbation of coupled chemical-equilibria. *Adv. Mol. Relax. Interact. Process.*, 1982. 24(4): 259–269.

43. Schwarz, G., Kinetic analysis by chemical relaxation methods. *Rev. Mod. Phys.*, 1968. 40(1): 206–218.

44. French, T.C. and G.G. Hammes, The temperature-jump method, in K. Kenneth, Editor. *Methods in Enzymology*, 1969, Waltham, MA: Academic Press, pp. 3–30.

45. Istratova, A.A. and O.F. Vyvenko, Exponential analysis in physical phenomena. *Rev. Sci. Instrum.*, 1999. 70(2): 1233–1257.

46. Cole, K.S. and R.H. Cole, Dispersion and absorption in dielectrics. *J. Chem. Phys.*, 1941. 9: 341–351.
47. Schwarz, G., Dielectric relaxation due to chemical rate processes. *J. Phys. Chem.*, 1967. 71(12): 4021–4030.
48. De Maeyer, L., M. Eigen, and J. Suarez, Dielectric dispersion and chemical relaxation. *J. Am. Chem. Soc.*, 1968. 90: 3157–3161.
49. Bottcher, C.J.F. and P. Bordewijk, Theory of electric polarization. 2nd ed. *Dielectrics in Time-Dependent Fields. Vol. II.* 1978, New York: Elsevier.
50. Hill, N.E., Vaughan, W.E., Price, A.H., and Davies, M. *Dielectric Properties and Molecular Behavior*. 1969, New York: van Nostrand Reinhold Company.
51. Jonscher, A.K., *Dielectric Relaxation in Solids*. 1983, London: Chelsea Dielectrics Press.
52. Bordi, F., C. Cametti, and R.H. Colby, Dielectric spectroscopy and conductivity of poly-electrolyte solutions. *J. Phys.: Condens. Matter*, 2004. 16: R1423–R1463.
53. Manes, M., Relationships between kinetics and acoustic phenomena in equilibrium systems. *J. Chem. Phys.*, 1953. 21: 1791–1796.
54. Markham, J.J., R.T. Beyer, and R.B. Lindsay, Absorption of sound in fluids. *Rev. Mod. Phys.*, 1951. 23: 353–411.
55. Feng, W. and A.I. Isayev, Continuous ultrasonic devulcanization of unfilled butyl rubber. *J. Appl. Polym. Sci.*, 2004. 94: 1316–1325.
56. Clegg, R.M., E.L. Elson, and B.W. Maxfield, New technique for optical observation of kinetics of chemical-reactions perturbed by small pressure changes. *Biopolymers*, 1975. 14(4): 883–887.
57. Ebbinghaus, S., Dhar, A., McDonald, D., and Gruebele, M. Protein folding stability and dynamics imaged in a living cell. *Nat. Meth.*, 2010. 7(4): 319–323.
58. Scheider, W., Real-time measurement of dielectric relaxation of biomolecules: Kinetics of a protein-ligand binding reaction. *Ann. N Y Acad. Sci.*, 1977. 303(1): 47–56.
59. Periasamy, A. and R.M. Clegg, *FLIM Microscopy in Biology and Medicine*. 2010, Boca Raton: Taylor & Francis. p. xxix, 407 p.
60. Crothers, D.M., Calculation of binding isotherms for heterogenous polymers. *Biopolymers*, 1968. 6(4): 575–584.
61. Zasedatelev, A.S., G.V. Gurskii, and M.V. Vol'kenshtein, Theory of one-dimensional adsorption. I. Adsorption of small molecules on a homopolymer. *Mol. Biol.*, 1971. 5(2): 194–198.
62. Schellman, J.A., Cooperative Multisite binding to DNA. *Isr. J. Chem.*, 1974. 12(1–2): 219–238.
63. Schellman, J.A., Macromolecular binding. *Biopolymers*, 1975. 14: 999–1018.
64. Lepecq, J.B., P. Yot, and C. Paoletti, Interaction of ethidium bromhydrate (Bet) with nucleic acids (Na). spectrofluorimetric study. *C R Hebd Seances Acad. Sci.*, 1964. 259: 1786–1789.
65. LePecq, J.B. and C. Paoletti, A fluorescent complex between ethidium bromide and nucleic acids. Physical-chemical characterization. *J. Mol. Biol.*, 1967. 27(1): 87–106.
66. Bauer, W. and J. Vinograd, Interaction of closed circular DNA with intercalative dyes. II. The free energy of superhelix formation in SV40 DNA. *J. Mol. Biol.*, 1970. 47(3): 419–435.
67. Bresloff, J.L. and D.M. Crothers, DNA-ethidium reaction kinetics: Demonstration of direct ligand transfer between DNA binding sites. *J. Mol. Biol.*, 1975. 95(1): 103–123.
68. Clegg, R.M., Qualitative exact diagnostic tests for two common relaxation kinetic models. *J. Chem. Phys.*, 1984. 81(12): 5546–5551.

14 DNA–Drug Interactions
A Theoretical Perspective

B. Jayaram, Tanya Singh, and Marcia O. Fenley

CONTENTS

The increasing availability of genomic information coupled with the advances in predictive computational tools for characterizing structure, electrostatics, and dynamics of DNA and for estimating DNA–ligand binding free energies ushers in an era of DNA-targeted computer-aided drug discovery to combat infectious diseases and noncommunicable disorders, with higher levels of reliability.

14.1 INTRODUCTION

DNA as a drug target—be it the DNA of humans, as in cancers, or that of infectious agents—proves attractive due to the availability of the well-studied three-dimensional (3D) DNA structures and the predictability of their accessible chemical functional groups. However, the number of known DNA-based drug targets is still very limited in comparison to the protein-based drug targets (Figure 14.1). The number of available structures of DNA–drug complexes is also small relative to protein–drug complexes deposited in the Research Collaboratory for Structural Bioinformatics (RCSB) database [1–5] (as shown in Figure 14.1 and Table 14.1), which indicates a heavy underrepresentation of DNA in the structural databases. Two concurrent developments, viz., increasing the availability of genomic sequences and advances in drug delivery systems, are expected to change this scenario drastically in the coming years.

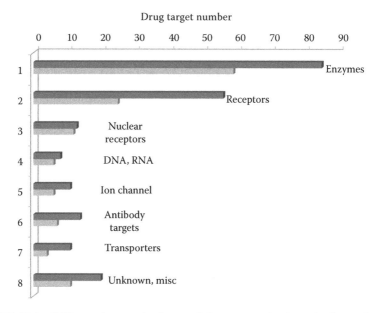

FIGURE 14.1 Different therapeutic classes of drug targets for drugs in the market [1–4]. The top bar indicates the total number of known targets for that class and the bottom bar indicates the number of three-dimensional structures available in the RCSB database [1] for that class. The targets correspond to approved drugs.

14.1.1 DRUG TARGETS FROM GENOMIC INFORMATION

Figure 14.2a and b helps explain how the knowledge of genomic sequences can aid in identifying the drug targets. For example, based on the knowledge of the genomic information of the malarial parasite *P. falciparum*, three DNA sequences, viz., (a) TGCATGCA, (b) GTGTGCACAC, and (c) GCACGCGTGC, have been identified as regulatory and essential for the functioning of the organism [6,7]. Naturally, if drugs

TABLE 14.1

Number of Three-Dimensional Structures of DNA and DNA–Drug Complexes Reported in the Nucleic Acid Database (NDB)

Number	Type of Complex/Binding	Total Number of PDB and NDB Entries
1	DNA	2689
2	DNA minor groove binders	109
3	DNA major groove binders	11
4	DNA intercalators	148
5	DNA–protein complexes	1692
6	Protein–ligand complexes	22,312

Note: The protein–ligand statistics was taken from the RCSB database [1].

are to be developed against these sequences, one wishes to know the frequency with which these sequences occur in humans in order to ensure that the drugs do not hamper the normal functioning of humans. The availability of genomic sequences, preferably with annotation, makes these choices feasible.

14.1.2   OPTIMUM SIZE OF DNA AS A TARGET AND THE DRUGS FOR SPECIFIC BINDING

The genomic sequences also bring to light other considerations. Figure 14.2a suggests that drugs have to bind to at least 16-mers to 18-mers or longer DNA sequences in order to ensure selectivity. Examination of the genomic sequences in humans (Figure 14.2b) indicates that the drugs have to cover at least 18 base pairs (bp) in order to uniquely bind to their targets with both high affinity and specificity. These numbers are reminiscent of the classical genetic switch [8] of the λ cro and λ repressor-operator systems wherein the operators are 17 bp long. Similarly, although each zinc-finger DNA-binding motif covers only 3 bp, multiple zinc-finger motifs bind in tandem to cover longer DNA sequences for transcriptional activation/regulation [9–15].

The above discussion brings to focus the optimal dimensions a drug molecule must possess for specific binding to its DNA target [16–20]. The axial distance of 20 bp (approximately two turns) of B-DNA is ~68 Å. The contour length along the grooves is longer (~90 Å). If the designed drug is a groove binder, its end-to-end distance has to be >90 Å. Typical lengths of some known DNA-binding ligands (drugs) are shown in Table 14.2 along with the target DNA base sequences. This data suggests the need to design drugs that are at least four times longer than netropsin, with overall molecular weights in the range of 1600–1800 Da, to ensure specific binding to unique DNA targets (Figure 14.3).

The length considerations of DNA-targeted drugs bring to fore nonconformity with Lipinski's rules [21,22] and drug delivery issues. A possible unexplored solution that has considerable promise is the design of monomeric drugs that bind to shorter sequences but could then form homo- or heterooligomeric drug(s) when binding to its DNA target. We propose that, for instance, four netropsin-like molecules with different sequence specificities and with an end-on coupling can easily cover any unique target sequence akin to the well-known binding of proteins to DNA as either dimers [8] or higher-order oligomers [23]. Recent advances in nano-biotechnology are expected to overcome hurdles that are still present in the delivery of drug molecules to their target site. These possibilities strongly affirm the potential for the success of novel drugs targeted to DNA.

Other design considerations, apart from length/size, include the chemical information content in the DNA grooves. The minor groove has a lower informational content than the major groove in the sense that both AT and TA base pairs project electronegative atoms into the minor groove. There exists a subtle polarity of the electrostatic potentials between GC and CG sequences in the minor grooves. Most proteins target the major groove. However, most known drugs target the minor groove enabling a snug fit between the binding partners. The wider major groove requires bulkier drugs with larger molecular weights. Cationic drugs have been the

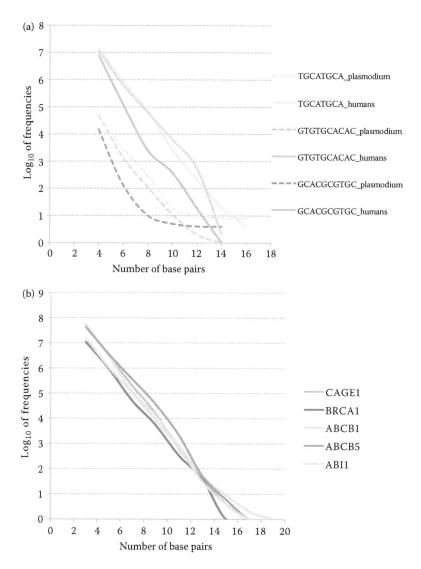

FIGURE 14.2 (a) Logarithm of the frequencies of the occurrence of base sequences of lengths 4–18 base pairs in *Plasmodium falciparum* and in humans embedding a regulatory sequence TGCATGCA, GTGTGCACAC, and GCACGCGTGC or parts thereof, of the plasmodium. The solid lines and the dashed lines correspond to humans and plasmodium, respectively. Curves lying between 0 and 1 on the log scale indicate occurrences in single digits. (b) Logarithm of the frequencies of occurrence of base sequences from 3 to 18 base pairs in humans embedding a regulatory sequence AAGCTGTCATTA or parts thereof of a cancer-causing CAGE1 gene [16], GACTGAGTCAA or parts thereof of a cancer-causing BRCA1 gene [17], CTCTAAGTCAT or parts thereof of a cancer-causing gene ABCB1 [18], GATATGTTAAAGC or parts thereof of a cancer-causing gene ABCB5 [19], and CTTCTGGGAA or parts thereof of a cancer-causing gene ABI1 [20].

TABLE 14.2
Typical Dimensions of Some Known DNA-Binding Drugs

Number	PDBID	Drug Name	Action	Molecular Formula	End-to-End Dimensions (in Å)	Name of Atoms	Base Pairs Spanned
1	121D	Netropsin	Antitumor, antiviral	$C_{18}H_{26}N_{10}O_3$	18.9	N1-N10	AATTT
2	3FT6	Proflavine	Anti-infectives	$C_{13}H_{11}N_3$	9.6	N15-N16	CG
3	1AU5	Cisplatin	Anticancer antibiotic	$H_6Cl_2N_2Pt$	2.9	N1-N2	GG
4	227D	Guanyl bisfuramidine	Active against P. carinii	$C_{18}H_{18}N_4O$	12.8	N1-N1′	AATT
5	1D63	Berenil	Antitrypanosomal	$C_{14}H_{15}N_7$	12.3	NA-NA′	AATT
6	182D	Nogalamycin	Antitumor	$C_{39}H_{49}NO_{16}$	12.7	O1′-C32	TG
7	202D	Menogaril	Antitumor topoisomerase II poison	$C_{28}H_{31}N_{10}$	13.07	C5M-C7M	GT
8	408D	Imidazole-pyrrole polyamide	—	$C_{31}H_{42}N_{11}O_5$	21.50	N7-C26	AGTACT

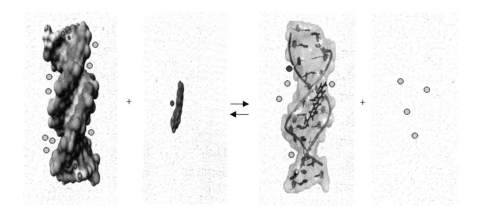

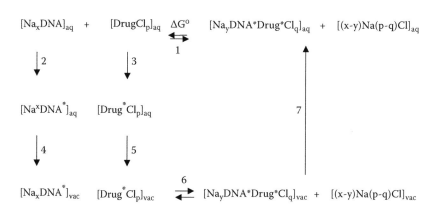

FIGURE 14.3 A thermodynamic cycle to study the energetics of DNA–drug binding. The formation of the DNA–DAPI complex (PDB id: 1D30), a typical minor-groove-binding drug–DNA complex is shown. The electrostatic potential of the face of the drug that contacts the minor groove is very positive as opposed to the remaining surface, which is mostly neutral. Note that the most negative electrostatic potential of DNA for this AT-rich sequence is in the deep and narrow minor groove rather than along the sugar–phosphate backbone. This strong negative potential originates from both the anionic sugar–phosphate backbone and electronegative O_2 of thymine and N_3 atoms of adenine. The binding position of the drug is shown. The curved surface of the drug (in blue sticks) fits tightly in the minor groove of the DNA. The color scheme used in these surface electrostatic potential maps is as follows: yellow is the most negative and green is the most positive. White is neutral. Red and blue represent negative and positive electrostatic potentials, respectively.

popular choice for binding to AT-rich minor groove regions while intercalators seem to prefer GC steps [24] (see, however, Ref. 25). One obvious choice apart from Dervan's architectures [26–29] is to bind to the minor groove in AT-rich regions and intercalate into GC regions. A rational combination of this strategy to cover the desired DNA sequences of any length is yet to emerge.

14.2 BACKGROUND AND TECHNIQUE

14.2.1 DNA Structure-Based Drug Discovery

In the absence of x-ray crystallographic or NMR structures of target DNA sequences, drug design efforts require sophisticated molecular simulation techniques to capture the subtle base sequence effects on the groove widths and the spatial disposition of the chemical functional groups. Also, very accurate computational tools for determining the electrostatic potentials of DNA with different base sequences are necessary for reliable drug design efforts. Fortunately, these have become available in the recent years.

The Ascona B-DNA consortium [30,31] was formed with the objective of developing state-of-the-art molecular dynamics simulation protocols to describe DNA accurately in aqueous media and the simulation results vis-à-vis 3D structures in the PDB/NDB databases are very encouraging. Statistical mechanical theory that can utilize the molecular simulation trajectories to determine the energetics of binding has also been worked out (see Appendix). Empirical potential functions to analyze single structures/snapshots of molecular dynamics or Monte Carlo simulations that yield results in good correlation with pertinent experimental data have also been developed [32–38].

Advances in finite difference Poisson–Boltzmann (FDPB) methodology [39–44] and the analyses of electrostatic potentials of DNA [45–48] obtained as solutions to the nonlinear PB equation [39] now make it possible to elucidate potential recognition/binding sites that result from the base sequence and/or from the phosphodiester backbone. The computed surface electrostatic potential profiles of DNA and its cationic drug partner allow one to assess the electrostatic potential complementarities from a qualitative perspective. However, the development of more quantitative electrostatic potential metrics is vital for drug design.

14.2.2 Energetics of DNA–Drug Binding

A drug molecule that competes with a regulatory protein has to generate a binding free energy in the range of −9 to −15 kcal/mol [49]. Minor groove binders have to achieve this via a snug fit and electrostatic complementarity without much structural distortion and intercalators via stacking in order to overcome the energy penalty associated with local unwinding/structural adaptation of the DNA. In order to design drugs that target DNA with enhanced-binding specificity and affinity, many groups have proposed that a better understanding of different noncovalent interactions, such as van der Waals, and electrostatics along with structural information is warranted. An energy component analysis of the binding of several known DNA-binding drugs with an empirical potential function is shown in Figure 14.4. Except for the entropies, all the energy components, electrostatics, van der Waals, and hydrophobic/cavity terms, seem to favor binding to varying degrees. Design efforts have long been focused on optimizing the number of hydrogen bonds and good steric fits in the grooves.

Since DNA and the cationic organic drugs that either intercalate into the DNA or bind to its grooves are charged molecules, it is not surprising that nonspecific and long-range electrostatic interactions have been found to be vital to the binding process. It is conceivable that in the initial binding step, the long-range and non-specific electrostatic interactions help steer these charged molecules into their initial or transient encounter complex. At this stage of the binding process, the fine atomic details of the charge distribution do not matter and electrostatics is favorable and dominated by the Coulomb attraction between the oppositely charged-binding partners [50]. At a later stage of binding that leads to the final docked bound state, short-range interactions such as van der Waals/hydrophobic contributions appear to play a large role in driving the formation of the stable complex (Figure 14.4). Also, short-range and directional electrostatic interactions such as hydrogen bonding and salt bridges play a critical role for the specificity of the final complex formation.

No experimental approach is available to directly measure or quantify the contribution of electrostatic interactions to the association reaction between biomolecules, at least with the present biophysical techniques (however, see Ref. 51). Thus, different methods have been proposed to infer information about the role of electrostatics in the binding process. Their quantitative and qualitative aspects, merits, and limitations are briefly discussed below.

A very well-established approach that is widely used to extract the electrostatic contribution to the biomolecular binding processes relies on measuring the equilibrium binding constant at varying added salt concentrations. The classic log–log plot of the

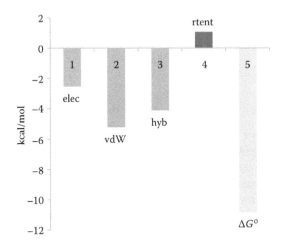

FIGURE 14.4 An energy component analysis of DNA–drug binding through PreDDICTA [32, 33], an empirical scoring function, for minor groove binders. The PreDDICTA method refers to the final state analysis in step 1 of Figure 14.3. The figure depicts a consensus view of favorable and unfavorable energy components contributing to the binding free energy (column 5) emerging from several systems (PDB ids: 127d, 264d, 109d, 1d63, 2dbe, 121, 2dnd, 227d, 298d, 289d, 1fmq, 1eel, 1fms, 1prp). Elec, electrostatics; vdW, van der Waals; hyb, hydrophobic; rtent, rotational translational entropy. Binding free energy estimate (BFEE) = ΔG°.

observed binding constant (K_{obs}) as a function of the 1:1 salt concentration [M$^+$] often portrays a linear relationship between these two quantities. This is true at least over a moderate range of 1:1 salt concentration and in the absence of competing multivalent ions for charged biomolecular complexes [52]. The slope of this linear log–log plot is referred to as SK_{obs} (= dlog K_{obs}/dlog [M$^+$]) in the literature. This linear log–log plot has become a signature of the polyelectrolyte effect for charged ligand–nucleic acid complexes [53–56]. The electrostatic contribution to the binding free energy is then obtained as $\Delta G_{el} = SK_{obs}$ ln [M$^+$] [54,56,57], as inferred from counterion condensation theory (CCT). The SK_{obs} metric is often considered to reflect the net charge on the cationic drug and the number of "condensed" counterions released from DNA upon drug binding, which is entropically favorable. A very strong support for the use of this simple electrostatic model, where the fine atomic details of the binding molecules and dielectric discontinuity effects are ignored, comes from the fact that CCT can correctly predict the experimentally observed SK_{obs} for small-charged ligands binding to polymeric DNA or RNA [54,56].

Interestingly, theoretical studies using the nonlinear Poisson–Boltzmann equation with a formal charge distribution and no dielectric discontinuity show that SK_{obs} equals the net charge of the drug in agreement with the CCT prediction [44,50,58–66]. It is important to stress that only the nonlinear solution of the PBE provides SK_{obs} values that agree with both CCT and experimental binding data [44,67–72]. However, the linear relationship $\Delta G_{el} = SK$ ln [M$^+$], which is now widely used to parse the total binding free energy into electrostatic and nonelectrostatic contributions [57,73–75], does not hold when dielectric discontinuity effects are considered. The importance of field discontinuities in correctly portraying electrostatic features of biomolecules has been shown by several PB studies [41,76]. Moreover, according to this popular relationship between SK_{obs} and ΔG_{el}, any drug having the same net charge but different shape and charge distribution will have the same ΔG_{el}. The same is not true according to the all-atom PB approach. These results highlight some issues yet to be resolved in deciphering the electrostatic contribution to the overall binding energetics from continuum or implicit solvent models. This notwithstanding, theory (CCT or PB) provides excellent predictions of SK_{obs}. This finding confirms the dominant role of long-range nonspecific electrostatics in controlling the salt sensitivity of the binding energetics. Thus, it is paradoxical and even counterintuitive that a large SK_{obs} (in terms of its magnitude) does not necessarily imply a large electrostatic contribution to binding as often assumed in the literature. In fact, many examples in the literature and from our work clearly show the opposite trend. For instance, some nucleic acid-binding proteins containing a large number of anionic residues, such as the tRNA synthetases and elongation factor Tu when bound to its nucleic acid-binding partner, have SK_{obs} absolute values that are very small or close to zero, but the electrostatic contribution to binding energy is quite large and unfavorable [41,77,78]. A strong correlation between electrostatic complementarity metrics and the electrostatic binding free energy of nucleic acid complexes is yet to be established.

Some of the issues in theoretical studies that are currently under microscopic scrutiny in diverse laboratories include more quantitative measures of electrostatic potentials, sensitivity of the PB predictions to parameters such as radii, the relative permittivities of solvent and solute, the boundary conditions, molecular surface definition, the compensatory nature of Coulomb, and desolvation interactions. A resolution of these issues is expected soon.

Molecular dynamics (MD) simulation studies on drug–DNA complexes provide additional insights into the structure, dynamics, and energetics of binding [79–88]. Results of a *post facto* molecular mechanics/generalized born surface area (MM/GBSA) analysis [36,89–92] of the MD trajectories of the unbound DNA d(CGCAAATTTGCG)$_2$, berenil, and the complex of DNA with berenil are shown in Figure 14.5. While the overall picture remains similar to the empirical energy function analyses (Figure 14.4) on drug–DNA complexes and protein–DNA complexes [93,94], MD trajectory analyses take into account structural adaptation of the interacting molecules, solvation/desolvation effects, and the role of explicit ions in binding. While some care is necessary in calling a component as favorable to binding due to the compensatory nature of several components with opposing effects contributing to binding, electrostatic complementarities, good steric fit of the drug in the grooves, as well as hydrophobic components are highlighted as important for berenil–DNA binding. Rotational, translational entropies, vibrational/configurational entropies (not computed), and adaptation are expected to be unfavorable to drug–DNA binding. The magnitudes of the components provide a good indication of the order of importance of the energy components favorable to binding and prove valuable in drug design efforts.

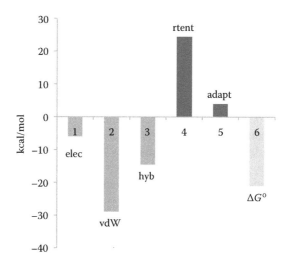

FIGURE 14.5 *Post facto* analyses of molecular dynamics trajectories on DNA, drug, and the complex using MM/GBSA theory and the thermodynamic cycle (steps 2 through 7) in Figure 14.3. An 8ns molecular dynamics simulations were performed on the DNA berenil complex (pdb id: 1d63), on free DNA, and on the drug with explicit solvent and counterions [30]. About 100 structures spaced at equal time intervals were culled from each of the three trajectories and averages of each energy component contributing to binding were computed. The net value of each energy component (DE) is calculated as $\Delta E = [E_{\text{Complex}} - E_{\text{(DNA + Drug)}}]$. Elec, electrostatics; vdW, van der Waals; hyb, hydrophobic; rtent, rotational translational entropy; adapt, adaptation energy. Binding free energy estimate (BFEE) = $\Delta G°$.

14.2.3 WEB TOOLS FOR MODELING DNA AND DRUG–DNA COMPLEXES

Theoretical methodologies are increasingly being converted into user-friendly software to facilitate drug design endeavors. Some of the web tools for analyzing DNA structure [95–107] and for assessing DNA–drug binding [108–119] are given in Tables 14.3 and 14.4.

TABLE 14.3
Web Tools for DNA Structure Generation and Analysis

Tool	Website	Description
Curves [95,96]	http://gbio-pbil.ibcp.fr/ Curves_plus/Curves + .html	Program to analyze DNA structure
NUPARM [96–102]	http://nucleix.mbu.iisc.ernet. in/nuparm/nuparm.shtml	Program to analyze sequence-dependent variations in nucleic acid (DNA and RNA) double helices
3DNA [103,104]	http://rutchem.rutgers. edu/~xiangjun/3DNA/	Program to analyze, rebuild, and visualize three-dimensional nucleic acid structures
NUCCGEN [95]	http://nucleix.mbu.iisc.ernet. in/nucgen/index.htm	Program to generate a curved or nonuniform helix
AMBER [105,106]	http://ambermd.org/	Program to generate canonical A- and B-duplex geometries of nucleic acids
DNA sequence to structure [107]	http://www.scfbio-iitd.res.in/ software/drugdesign/bdna.jsp	Program to generate canonical A and B DNA and molecular dynamics-averaged DNA structure

TABLE 14.4
DNA–Ligand and DNA–Protein Docking and Scoring Software

Docking Softwares	Website	Description
Surflex [108]	http://www.tripos.com/index. php?family = modules,SimplePage,,, &page= surflex_dock&s = 0	DNA–ligand docking software
Autodock 4.0 [109]	http://autodock.scripps.edu/	DNA–ligand docking software
DNA–ligand docking [110]	http://www.scfbio-iitd.res.in/dock/dnadock.jsp	DNA–ligand docking software
GOLD [111–117]	http://www.ccdc.cam.ac.uk/products/ life_sciences/gold/	Software for searches of databases for DNA–binding compounds and for DNA–protein docking
HADDOCK [118]	http://www.nmr.chem.uu.nl/haddock/	DNA–protein docking software
Escher NG [119]	http://users.unimi.it/~ddl/escherng/index. htm	DNA–protein docking software

14.3   PERSPECTIVES AND CONCLUSION

Atomic models and molecular modeling and simulation methodologies of drug binding to DNA have matured to a stage where it is conceivable to generate reliable *in
silico* suggestions of candidate molecules to bind to any specific base sequence of
DNA. The stage is set for DNA-targeted drug discovery.

ACKNOWLEDGMENTS

B. Jayaram acknowledges the program support to SCFBio from the Department of
Biotechnology, Government of India. Marcia O. Fenley acknowledges the support
from NIH-GM078538-01 (PI: Dr. Michael S. Chapman, OHSU, Co-PI: MOF, FSU)
and SBIR NIH 2R44GM073391-02 (PI: Dr. Alexander H. Boschitsch).

APPENDIX

A statistical mechanics theory for DNA–drug binding in aqueous salt media on the
lines of protein–DNA binding [120] and protein–ligand binding [121] is presented
here.

Let D (denoting DNA) and dr (denoting drug) be the reactants and D*dr*, the
product of binding in aqueous salt medium.

$$[D]_{aq} + [dr]_{aq} = [D*dr*]_{aq} \tag{14.1}$$

At equilibrium,

$$\mu_{D.aq} + \mu_{dr.aq} = \mu_{D*dr*.aq} \tag{14.2}$$

where $\mu_{D.aq}$ is the chemical potential of species D in the ionic solvent medium (partial
molar Gibbs free energy) and $\mu^{\circ}_{D.aq}$ is its standard chemical potential, that is, under
condition of 1 bar in gaseous state and 1 M (designated as C°) in liquid state.

$$\mu^{\circ}_{D.aq} + RT \ln\left(\gamma_D C_D / C^{\circ}\right) + \mu^{\circ}_{dr.aq} + RT \ln\left(\gamma_{dr} C_{dr} / C^{\circ}\right)$$
$$= \mu^{\circ}_{D*dr*.aq} + RT \ln\left(\gamma_{D*dr*} C_{D*dr*} / C^{\circ}\right) \tag{14.3}$$

where γ_D is the activity coefficient of species D and C_D is its concentration. The
standard molar Gibbs free energy of the reaction (standard absolute molar Gibbs free
energy of binding) is

$$\Delta G^{\circ}_{aq} = \mu^{\circ}_{D*dr*.aq} - \left(\mu^{\circ}_{D.aq} + \mu^{\circ}_{dr.aq}\right)$$
$$= -RT \ln\left[\gamma_{D*dr*} C_{D*dr*} C^{\circ} / \left(\gamma_D C_D\right)\left(\gamma_{dr} C_{dr}\right)\right] = -RT \ln K_{aq} \tag{14.4}$$

In terms of the canonical partition function (Q)

$$\Delta G^\circ_{aq} = \Delta A^\circ_{aq} + P\Delta V^\circ_{aq} = -RT \ln K_{aq}$$

$$= -RT \ln\left[\left\{Q_{D*dr*.aq}/(N_A Q_w)\right\}/\left\{\left(Q_{D \cdot aq}/(N_A Q_w)\right)\left(Q_{dr.aq}/(N_A Q_w)\right)\right\}\right] \quad (14.5)$$

$$+ P\Delta V^\circ_{aq}$$

where ΔA° is the standard Helmholtz free energy of the reaction, $P\Delta V^\circ_{aq}$ is the pressure volume correction to Helmholtz free energy in the solvent medium, $K_{eq.aq}$ is the equilibrium constant for the reaction in (1), N_A is the Avogadro number, and Q_w denotes the partition function for pure solvent (water).

Assuming that translations and rotations are separable from intrasolute degrees of freedom as well as those of the solvent,

$$\Delta G^\circ_{aq} = -RT \ln[\{Q^{tr}_{D*dr*} Q^{rot}_{D*dr*} Z^{int}_{D*dr*.aq} Q^{el}_{D*dr*} N_A Q_w\}/$$

$$\{(Q^{tr}_D Q^{rot}_D Z^{int}_{D.aq} Q^{el}_D)(Q^{tr}_{dr} Q^{rot}_{dr} Z^{int}_{dr.aq} Q^{el}_{dr})\}] + P\Delta V^\circ_{aq} \quad (14.6)$$

The superscript "int" denotes the internal contribution, and Z^{int} is the configurational partition function. It includes contributions from the intermolecular interactions and internal motions as well as solvation (hydration) effects. The translational and rotational terms have been separated out.

Z^{int} is determined via

$$Z^{int}_{D \cdot aq} = \int \cdots \int \exp\{(-E(X^N_D, X^M_w)/k_B T\}dX^N_D dX^M_w \quad (14.7)$$

$$= \langle \exp(E(X^N_D, X^M_w)/k_B T \rangle$$

X^N_D and X^M_w represent the configurational space accessible to the solute D and solvent W, respectively, in the presence of each other. $E(X^N_D, X^M_w)$ denotes the total potential energy of the system describing nonidealities.

The electronic partition function Q^{el} is assumed to be unity for noncovalent associations,

$$Q^{el}_D = Q^{el}_{dr} = Q^{el}_{D*dr*} = 1 \quad (14.8)$$

The standard free energy can be expressed as a sum of external (translational and rotational) and internal (intramolecular, intermolecular, and solvation) contributions.

$$\Delta G^\circ = -RT \ln\left[Q^{tr}_{D*dr*} N_A/(Q^{tr}_D Q^{tr}_{dr})\right] - RT \ln[Q^{rot}_{D*dr*}/(Q^{rot}_D Q^{rot}_{dr})]$$

$$- RT \ln\left[(Z^{int}_{D*dr*.aq} Q^w)/(Z^{int}_{D.aq} Z^{int}_{dr.aq})\right] + P\Delta V^\circ_{aq} \quad (14.9)$$

Equation 14.9 is an exact expression for evaluating binding free energies for noncovalent associations in aqueous medium. The first two terms on the right-hand side of

Equation 14.9 can be computed analytically. The third term is accessible to free energy molecular simulations configured in the canonical ensemble such as the perturbation method, thermodynamic integration, potential of mean force method, etc., albeit they are computationally expensive for a single ligand and not practical in a high-throughput sense even on supercomputers.

Here, some simplifications are considered to bring the binding free energy computations into the feasibility domain. The molecular translational partition function of D is

$$q^{tr}_D = V/\Lambda^3_D = V/(h^2/2\pi m_D k_B T)^{3/2} \tag{14.10}$$

where V is the volume, Λ_D is the thermal wavelength of D, h is the Planck's constant, k_B is the Boltzmann constant, T is the temperature, and m is the mass.

The molar partition function of D is

$$Q^{tr}_D = \left(q^{tr}_D\right)^{NA} \tag{14.11}$$

The volume V has been included in the translational part consistent with ideal gas statistical mechanics. This would require that the Z^{int} be divided by V to quantify nonidealities (excess free energies). The translational part of the free energy in Equation 14.9 is now given by the Sackur–Tetrode [10] equivalent as

$$\Delta G^{\circ}_{tr} = -RT \ln\left[\left(N_A/V\right)\left(\Lambda^3_D \Lambda^3_{dr}/\Lambda^3_{D*dr*}\right)\right]$$
$$= -RT \ln\left[\left(N_A/V\right)(h^2/2\pi k_B T)^{3/2}\left\{m_{D*dr*}/\left(m_D m_{dr}\right)\right\}^{3/2}\right] \tag{14.12}$$

The expression in square brackets in Equation 14.12 is dimensionless. (N_A/V) may be replaced by a concentration term ensuring that in the transfer to aqueous medium, standard free energies are recovered with the reference state with a molar concentration of unity. This expression is the same whether in gas phase or liquid phase, provided that the translational and rotational motions of the solute are unaffected by the solvent. This will be true only in a continuum, friction-less solvent influencing the position-dependent potential energy but not the velocity-dependent kinetic energy of the solute. Hence, in a transfer process (an experiment involving transfer of species D from one phase to another such as from gas phase to liquid phase, octanol to water, etc.), this term cancels out. In the binding processes however, no such cancellation occurs. Also, if D, dr, and D*dr* could be seen as a collection of nonbonded monoatomic particles, then again the translational partition function for each species could be written as a product of the individual partition functions of the constituent atoms and since the number of atoms is conserved during binding, these terms would cancel out. Again, this is not so for polyatomic species where the mass in translational partition function $m_D = \Sigma_i m_i$ is evaluated as a sum of the masses of the constituent atoms. It is thus recommended that Sackur–Tetrode equation be applied not in the aqueous medium directly where it is invalid but upon transfer to vacuum via a suitable thermodynamic cycle shown in Figure 14.3.

Similar arguments apply to the rotational partition functions. Separating the rotational part from internal motions implies working under the rigid rotor approximation.

The molecular rotational partition function of D is

$$q^{\text{rot}} = \sigma^{-1}\left(k_B T/hc\right)^{3/2}\left(\pi/I^a I^b I^c\right)^{1/2} \tag{14.13}$$

$$\Delta G^\circ{}_{\text{rot}} = -RT\ln\Big[\left(\sigma_D\sigma_{\text{dr}}/\sigma_{D*\text{dr}*}\right)(1/(8\pi^2))(h^2/2\pi k_B T)^{3/2}$$
$$*\left\{\left(I^a{}_{D*\text{dr}*}I^b{}_{D*\text{dr}*}I^c{}_{D*\text{dr}*}\right)/\left(I^a{}_D I^b{}_D I^c{}_D I^a{}_{\text{dr}}I^b{}_{\text{dr}}I^c{}_{\text{dr}}\right)\right\}^{1/2}\Big] \tag{14.14}$$

$I^a{}_D$, $I^b{}_D$, and $I^c{}_D$ are the components of moments of inertia of species D along the principal axes and s_D its symmetry number.

Considering Equations 14.12 and 14.14, Equation 14.9 may now be written as

$$\Delta G^\circ = \Delta G^\circ{}_{\text{tr}} + \Delta G^\circ{}_{\text{rot}} - RT\ln\Big[\left(Z^{\text{int}}{}_{D*\text{dr}*.\text{aq}}Q_w V\right)/\left(Z^{\text{int}}{}_{D.\text{aq}}Z^{\text{int}}{}_{\text{dr}.\text{aq}}\right)\Big] + P\Delta V^\circ{}_{\text{aq}} \tag{14.15}$$

Free energy contributions from internal motions that are coupled to solvent are best handled via molecular simulations. Separating the two will amount to an approximation.

$$Z^{\text{int}}{}_{D.\text{aq}} = Z_D{}^{\text{intra}}Z_D{}^{\text{solvn}}$$

$$Z^{\text{int}}{}_{D.\text{aq}} \approx \int\cdots\int\exp[-\{(E(X^N{}_D)+E(X_D{}^{\text{Nfixed}}, X^M{}_w))\}/k_B T\}dX^N{}_D dX^M{}_w$$
$$\approx \Big[\int\cdots\int\exp\left\{-E(X^N{}_D)k_B T\right\}dX^N{}_D\Big] \tag{14.16}$$
$$\times\Big[\int\cdots\int\exp\{(-E(X_D{}^{\text{Nfixed}}, X^M{}_w)/k_B T\}dX^M{}_w\}\Big]$$

Equations similar to Equation 14.16 can be written for D and D*dr* and converted to excess free energies. Such a separation allows

$$\Delta G^\circ = \Delta G^\circ{}_{\text{tr}} + \Delta G^\circ\text{rot} + \Delta G^\circ{}_{\text{int}} + \Delta G^\circ{}_{\text{solvn}} \tag{14.17}$$

Equation 14.17 forms the theoretical basis for the additivity assumed in free energy computations as employed in master equation methods [115,116]. The $P\Delta V^\circ{}_{\text{aq}}$ term in Equation 14.9 is often neglected in liquid-state work. If Equations 14.16 and 14.17 are employed for each structure generated according to Boltzmann distribution either via molecular dynamics or Metropolis Monte Carlo and averages are computed with a suitably calibrated dielectric continuum solvent model for solvation energy for each structure, the results are expected to correspond to Equation 14.9, which is exact (subject to the separability of translations/rotations from internal motions). In inferring free energies of drug–DNA binding from molecular simulations, care must be taken to ensure that the counterions/coions (whether they are "bound" or "released") are treated as a part of the solute, viz., the DNA or the drug or the complex and their stoichiometries maintained during binding (as in Figure 14.3). Added salt effects are yet to be incorporated in this theory.

$$\Delta G^\circ{}_{\text{int}} = \Delta H^\circ{}_{\text{int}} - T\Delta S^\circ{}_{\text{int}} \tag{14.18}$$

$$\Delta H^\circ{}_{\text{int}} = \Delta H^\circ{}_{\text{intermolecular}} + \Delta H^\circ{}_{\text{intramolecular}}$$

$$\Delta H^{\circ}_{\text{intermolecular}} = \Delta H^{\circ}_{\text{el}} + \Delta H^{\circ}_{\text{vdw}} = \langle \Delta E^{\circ}_{\text{intermolecular}} \rangle = \langle \Delta E^{\circ}_{\text{el}} + \Delta E^{\circ}_{\text{vdw}} \rangle \qquad (14.19)$$

$$\Delta H^{\circ}_{\text{intramolecular}} = \langle \Delta E^{\circ}_{\text{intramolecular}} \rangle \qquad (14.20)$$

In the above equations, $\Delta E^{\circ}_{\text{el}}$ and $\Delta E^{\circ}_{\text{vdw}}$ represent the electrostatic and van der Waals components of the intermolecular interaction energy between the DNA and the drug.

$\Delta E^{\circ}_{\text{intramolecular}}$ represents changes in both bonded and nonbonded contributions to the intramolecular energy of the DNA and the drug upon binding. All these quantities can be computed from a molecular mechanics force field either for a fixed structure (from minimization studies) or for an ensemble of structures from MD simulations.

$$\Delta S^{\circ}_{\text{int}} = \Delta S^{\circ}_{\text{vib,config}} \qquad (14.21)$$

Entropy changes can be calculated by a normal mode analysis of an energy-minimized structure ($\Delta S^{\circ}_{\text{vib}}$) or by a quasi-harmonic approximation introduced by Karplus and Kushick [122] and subsequently extended and adapted to MD simulation by Schlitter [123] and van Gunsteren [124]. Equation 14.17 is utilized to generate the energetics of drug–DNA binding from MD simulations.

REFERENCES

1. Berman, H. M., Westbrook, J., Feng, Z., Gilliland, G., Bhat, T. N., Weissig, H., Shindyalov, I. N. and Bourne, P. E. The protein data bank. *Nucleic Acids Res.* **28**, 235–242, 2000.
2. Overington, J. P., Al-Lazikani, B. and Hopkins, A. L. How many drug targets are there? *Nat. Rev. Drug Discov.* **5**, 993–996, 2006.
3. Shaikh, S. A., Jain, T., Sandhu, G., Latha, N. and Jayaram, B. From drug target to leads-sketching, a physicochemical pathway for lead molecule design in silico. *Curr. Pharm. Des.* **13**, 3454–3470, 2007.
4. Wishart, D. S., Knox, C., Guo, A. C., Shrivastava, S., Hassanali, M., Stothard, P., Chang, Z. and Woolsey, J. Drugbank: A comprehensive resource for *in silico* drug discovery and exploration. *Nucleic Acids Res.* **34**, 668–672, 2006.
5. Berman, H. M., Olson, W. K., Beveridge, D. L., Westbrook, J., Gelbin, A., Demeny, T., Hsieh, S.-H., Srinivasan, A. R. and Schneider, B. The nucleic acid database: A comprehensive relational database of three-dimensional structures of nucleic acids. *Biophys. J.* **63**, 751–759., 1992.
6. Carlton, J. M., Adams, J. H., Silva, J. C., Bidwell, S. L., Lorenzi1, H., Caler, E., Crabtree, J. et al. Comparative genomics of the neglected human malaria parasite *Plasmodium vivax*. *Nature*. **455**, 757–763, 2008.
7. Young, J. A., Johnson, J. R., Benner, C., Frank Yan, S., Chen, K., Roch, K. G. L., Zhou, Y. and Winzeler, E. A. *In silico* discovery of transcription regulatory elements in *Plasmodium falciparum*. *BMC Genomics*. **9**, 1–21, 2008.
8. Ptashne, M. A. *Genetic Switch Phage Lambda and Higher Organisms*. Cell Press and Blackwell Scientific: Cambridge, MA, 1992.
9. http://en.wikipedia.org/wiki/Zinc_finger_nuclease.
10. http://en.wikipedia.org/wiki/Zinc_finger.
11. Liu, Q., Segal, D. J., Ghiara, J. B. and Barbas, C. F. Design of polydactyl zinc-finger proteins for unique addressing within complex genomes. *Proc. Natl. Acad. Sci. USA*. **94**, 5525–5530, 1997.

12. Kim, J. S. and Pabo, C. O. Getting a handhold on DNA: Design of poly-zinc finger proteins with femtomolar dissociation constants. *Proc. Natl. Acad. Sci. USA.* **95**, 2812–2817, 1998.

13. Moore, M., Choo, Y. and Klug, A. Design of polyzinc finger peptides with structured linkers. *Proc. Natl. Acad. Sci. USA.* **98**, 1432–1436, 2001.

14. Moore, M., Klug, A. and Choo, Y. Improved DNA-binding specificity from polyzinc finger peptides by using strings of two-finger units. *Proc. Natl. Acad. Sci. USA.* **98**, 1437–1441, 2001.

15. Isalan, M., Klug, A. and Choo, Y. A rapid, generally applicable method to engineer zinc fingers illustrated by targeting the HIV-1 promoter. *Nat. Biotechnol.* **19**, 656–660, 2001.

16. http://www.genecards.org/cgi-bin/carddisp.pl?gene=CAGE1&search=cage1.

17. http://www.genecards.org/cgi-bin/carddisp.pl?gene=BRCA1&search=brca1.

18. http://www.genecards.org/cgi-bin/carddisp.pl?gene=ABCB1&search=ABCB1.

19. http://www.genecards.org/cgi-bin/carddisp.pl?gene=ABCB5&search=ABCB5.

20. http://www.genecards.org/index.php?path=/Search/keyword/ABI1.

21. Lipinski, C. A. Lead- and drug-like compounds: The rule-of-five revolution. *Drug Discov. Today: Technol.* **1**, 337–341, 2004.

22. Lipinski, C. A., Lombardo, F., Dominy, B. W. and Feeney, P. J. Experimental and computational approaches to estimate solubility and permeability in drug discovery and development settings. *Adv. Drug Delivery Rev.* **23**, 3–25, 1997.

23. Hong, M., Fitzgerald, M. X., Harper, S., Luo, C., Speicher, D. W. and Marmorstein, R. Structural basis for dimerization in DNA recognition by Gal4. *Structure.* **16**, 1019–1026, 2008.

24. Bachur N. R., Johnson, R., Yu, F., Hickey, R., Applegren, N. and Malkas, L. Antihelicase action of DNA-binding anticancer agents: Relationship to guanosine-cytidine intercalator binding. *Mol. Pharmacol.* **44**, 1064–1069, 1993.

25. Sun, J. S., Francois, J. C., Garestier, T. M., Behmoaras, T. S., Roig, V., Thuong, N. T. and Helene, C. Sequence-specific intercalating agents: Intercalation at specific sequences on duplex DNA via major groove recognition by oligonucleotide-intercalator conjugates. *Proc. Natl. Acad. Sci. USA.* **86**, 9198–9202, 1989.

26. Dervan, P. B. Molecular recognition of DNA by small molecules. *Bioorg. Med. Chem.* **9**, 2215–2235, 2001.

27. Dervan, P. B. and Edelson, B. S. Recognition of the DNA minor groove by pyrrole-imidazole polyamides. *Curr. Opin. Struct. Biol.* **13**, 284–299, 2003.

28. Pilch, D. S., Poklar, N., Baird, E. E., Dervan, P. B. and Breslauer, K. J. The thermodynamics of polyamide-DNA recognition: Hairpin polyamide binding in the minor groove of duplex DNA. *Biochemistry.* **38**, 2143–2151, 1999.

29. Kielkopf, C. L., White, S., Szewczyk, J. W., Turner, J. M., Baird, E. E., Dervan, P. B. and Rees, D. C. A Structural basis for recognition of AzT and TzA base pairs in the minor groove of B-DNA. *Science.* **282**, 111–115, 1998.

30. Lavery, R., Zakrzewska, K., Beveridge, D., Bishop, T. C., Case, D. A., Cheatham, T., Dixit, S. et al. A systematic molecular dynamics study of nearest neighbor effects on base pair and base pair step conformations and fluctuations in B-DNA. *Nucleic Acids Res.* **38**, 299–313, 2010.

31. Beveridge, D. L., Barreiro, G., Byun, K. S., Case, D. A., Cheatham III, T. E., Dixit, S. B., Giudice, E. et al. Molecular dynamics simulations of the 136 unique tetranucleotide sequences of DNA oligonucleotides. I. Research design and results on d(CpG) Steps. *Biophys. J.* **87**, 3799–3813, 2004.

32. Shaikh, S. A., Ahmed, S. R. and Jayaram, B. A molecular thermodynamic view of DNA–drug interactions: A case study of 25 minor-groove binders. *Arch. Biochem. Biophys.* **429**, 81–99, 2004.

33. Shaikh, S. A. and Jayaram, B. A swift all-atom energy-based computational protocol to predict DNA-ligand binding affinity. *J. Med. Chem.* **50**, 2240–2244, 2007.

34. Jain, T. and Jayaram, B. An all atom energy based computational protocol for predicting binding affinities of protein-ligand complexes. *FEBS Lett.* **579**, 6659–6666, 2005.

35. Arora, N. and Jayaram, B. Energetics of base pairs in B-DNA in solution: An appraisal of potential functions and dielectric treatments. *J. Phys. Chem.* B. **102**, 6139–6144, 1998.

36. Jayaram, B., Swaminathan, S., Beveridge, D. L., Sharp, K. and Honig, B. Monte Carlo simulation studies on the structure of the counterion atmosphere of B-DNA. Variations on the primitive dielectric model. *Macromolecules.* **23**, 3156–3165, 1990.

37. M. A. Young, Jayaram, B. and Beveridge, D. L. Local dielectric environment of B-DNA in solution: Results from a 14 ns molecular dynamics trajectory. *J. Phys. Chem.* B. **102**, 7666–7669, 1998.

38. Jayaram, B., DiCapua, F. M. and Beveridge, D. L. A theoretical study of polyelectrolyte effects in protein-DNA interactions: Monte Carlo free energy simulations on the ion atmosphere contribution to the thermodynamics of lambda repressor-operator complex formation. *J. Am. Chem. Soc.* **113**, 5211–5215, 1991.

39. Jayaram, B., Sharp, K. A. and Honig, B. The electrostatic potential of B-DNA. *Biopolymers.* **28**, 975–993, 1989.

40. Boschtisch, A. H., Fenley, M. O. and Zhou, H.-X. Fast boundary element method for the linear Poisson–Boltzmann equation. *Phys. Chem.* **106**, 2741–2754, 2002.

41. Srinivasan, A. R., Sauers, R. R., Fenley, M. O., Boschitsch, A. H., Matsumoto, A., Colasanti, A. V. and Olson, W. K. Properties of the nucleic-acid bases in free and Watson–Crick hydrogen-bonded states: Computational insights into the sequence-dependent features of double-helical DNA. *Biophys. Rev.* **1**, 13–20, 2009.

42. Boschtisch, A. H. and Fenley, M. O. Hybrid boundary element and finite difference method for solving the nonlinear Poisson–Boltzmann equation. *J. Comp. Chem.* **25**, 935–955, 2004.

43. Bredenberg, J., Boschtisch, A. H. and Fenley, M. O. The role of anionic protein residues on the salt dependence of the binding of aminoacyl-tRNA synthetases to tRNA: A Poisson–Boltzmann analysis. *Commun. Comput. Phys.* **3**, 1051–1070, 2007.

44. Boschtisch, A. H. and Fenley, M. O. A new outer boundary formulation and energy corrections for the nonlinear Poisson–Boltzmann equation. *J. Comp. Chem.* **28**, 909–921, 2007.

45. Fenley, M. O., Harris, R., Jayaram, B. and Boschitsch, A. H. Revisiting the association of cationic groove-binding drugs to DNA using a Poisson–Boltzmann Poisson–Boltzmann approach. *Biophys. J.* **99**, 879–886, 2010.

46. Xu, D., Landon, T., Greenbaum, N. L. and Fenley, M. O. The electrostatic characteristics of GU wobble base pairs. *Nucleic Acids Res.* **35**, 3836–3847, 2007.

47. Xu, D., Greenbaum, N. L. and Fenley, M. O. Recognition of the splicesomal branch site RNA helix on the basis of surface and electrostatic features. *Nucleic Acids Res.* **33**, 1154–1161, 2005.

48. Bredenberg, J. H., Russo, C. and Fenley, M. O. Salt-mediated electrostatics in the association of TATA binding proteins to DNA: A combined molecular mechanics/Poisson–Boltzmann study. *Biophys. J.* **94**, 4634–4645, 2008.

49. Jen-Jacobson, L. Protein–DNA recognition complexes: Conservation of structure and binding energy in the transition state, *Biopolymers.* **44**, 153, 1997.

50. Pineda De Castro, L. F. and Zacharias, M. DAPI binding to the DNA minor groove: A continuum solvent analysis. *J. Mol. Recogn.* **15**, 209–220, 2002.

51. Stafford, A. J., Ensign, D. L., and Webb, L. J. Vibrational stark effect spectroscopy at the interface of Ras and Rap1A bound to the Ras binding domain of RaIGDS reveals an electrostatic mechanism for protein-protein interactions, *J. Phys. Chem.* B. **114**, 15331–15344, 2010.

52. Bertonati, C., Honig, B. and Alexov, E. Poisson–Boltzmann Poisson–Boltzmann calculations of nonspecific salt effects on protein-protein binding free energies. *Biophys. J.* **92**, 1891–1899, 2007.

53. Anderson, C. F. and Record, M. T. J. Salt dependence of oligoion-polyion binding: A thermodynamic description based on preferential interaction coefficients. *J. Phys. Chem.* **97**, 7116–7126, 1993.

54. Manning, G. S. The molecular theory of polyelectrolyte solutions with application to the electrostatic properties of polynucleotides. *Q. Rev. Biophys.* **11**, 179–246., 1978.

55. Record, M. T. J., Anderson, C. F. and Lohman, T. M. Thermodynamic analysis of ion effects on the binding and conformational equilibria of proteins and nucleic acids: The roles of ion association or release, screening, and ion effects on water activity. *Q. Rev. Biophys.* **11**, 103–178, 1978.

56. Record, M. T. J., Zhang, W. and Anderson, C. F. Analysis of effects of salts and uncharged solutes on protein and nucleic acid equilibria and processes: A practical guide to recognizing and interpreting polyelectrolyte effects, Hogmeister effects and osmotic effects of salts. *Adv. Prot. Chem.* **51**, 281–353, 1998.

57. Chaires, J. B. Calorimetry and thermodynamics in drug design. *Annu. Rev. Biophys.* **37**, 135–151, 2008.

58. Rohs, R. and Sklenar, H. Methylene blue binding to DNA with alternating AT base sequence: Minor groove binding is favored over intercalation. *J Biomol. Struct. Dyn.* **21**, 399–711, 2004.

59. Sharp, K. A. Polyelectrolyte electrostatics: Salt dependence, entropic, and enthalpic contributions to free energy in the nonlinear Poisson–Boltzmann model. *Biopolymers.* **36**, 227–243, 1995.

60. Misra, V. K. and Honig, B. On the magnitude of the electrostatic contribution to ligand-DNA interactions. *Proc. Natl. Acad. Sci.* **92**, 4691–4695, 1995.

61. Misra, V. K., Sharp, K. A., Friedman, R. A. and Honig, B. Salt effects on ligand-DNA binding: Minor groove binding antibiotics. *J. Mol. Biol.* **238**, 245–263, 1994.

62. Sharp, K. A., Friedman, R. A., Misra, V., Hecht, J. and Honig, B. Salt effects on polyelectrolyte-ligand binding: Comparison of Poisson–Boltzmann, and limiting law/counterion binding models. *Biopolymers.* **36**, 245–262, 1995.

63. Baginski, M., Fogolari, F. and Briggs, J. M. Electrostatic and non-electrostatic contributions to the binding free energies of anthracycline antibiotics to DNA. *J. Mol. Biol.* **274**, 253–267, 1997.

64. Chen, S.-W. and Honig, B. Monovalent & divalent salt effects on electrostatic free energies defined by the nonlinear Poisson–Boltzmann equation: Application to DNA binding reactions. *J. Phys. Chem.* **101**, 9113–9118, 1997.

65. Rohs, R., Sklenar, H., Lavery, R. and Roder, B. Methylene blue binding to DNA with alternating GC base sequence: A modeling study. *J. Am. Chem. Soc.* **122**, 2860–2866, 2000.

66. Misra, V. K. and Honig, B. On the magnitude of the electrostatic contribution to ligand-DNA interactions. *Proc. Nat. Acad. Sci. USA.* **92**, 4691–4695, 1995.

67. Strekowski, L. and Wilson, B. Noncovalent interactions with DNA: An overview. *Mutat. Res. Fundam. Mol. Mech. Mutagen.* **623**, 3–13, 2007.

68. Wilson, W. D., Tanious, F. A., Mathis, A., Tevis, D., Hall, J. E. and Boykin, D. W. Antiparasitic compounds that target DNA. *Biochimie.* **90**, 999–1014, 2008.

69. Lane, A. N. and Jenkins, T. C. Thermodynamics of nucleic acids and their interactions with ligands. *Q. Rev. Biophys.* **33**, 255–306, 2000.

70. Mazur, S., Tanious, F. A., Ding, D., Kumar, A., Boykin, D. W., Simpson, I. J., Neidle, S. and Wilson, W. D. A thermodynamic and structural analysis of DNA minor-groove complex formation. *J. Mol. Biol.* **300**, 321–337, 2000.

71. Nguyen, B., Lee, M. P. H., Hamelberg, D., Joubert, A., Bailly, C., Brun, R., Neidle, S. and Wilson, W. D. Strong binding in the DNA minor groove by an aromatic diamidine with a shape that does not match the curvature of the groove. *J. Am. Chem. Soc.* **124**, 13680–13681, 2002.

72. Miao, Y., Lee, M. P. H., Parkinson,G. N., Batista-Parra, A., Ismail, M. A., Neidle, S., Boykin,D. W. and Wilson, W. D. Out-of-shape DNA minor groove binders: Induced fit interactions of heterocyclic dications with the DNA minor groove. *Biochemistry.* **44**, 14701–14708, 2005.

73. Chaires, J. B. Energetics of drug-DNA interactions. *Biopolymers.* **44**, 201–215, 1998.

74. Lin, P.-H., Yen, S.-L., Lin, M.-S., Chang, Y., Louis, S. R., Higuchi, A. and Chen, W.-Y. Microcalorimetric studies of the thermodynamics and binding mechanism between L-tyrosinamide and aptamer. *J. Phys. Chem.* **112**, 6665–6673, 2008.

75. Hossain, M. and Kumar, G. S. DNA binding of benzophenanthridine compounds sanguinarine versus ethidium: Comparative binding and thermodynamic profile of intercalation. *J. Chem. Thermodynamics.* **41**, 764–774, 2009.

76. Rohs, R., West, S. M., Sosinky, A., Liu, P., Mann, R. S. and Honig, B. The role of DNA shape in protein-DNA recognition. *Nature.* 461, 1248–1253, 2009.

77. Eargle, T. J., Black, A. A., Sethi, A., Trabuco, L. G. and Luthey-Schulten, Z. Dynamics of recognition between tRNA and elongation factor. *J. Mol. Biol.* **377**, 1382–1405, 2008.

78. Paleskava, S. A., Knevega, A. L. and Rodnina, M. V. Thermodynamic and kinetic framework of selenocysteyl-tRNASec recognition by elongation factor. *J. Biol. Chem.* **285**, 3014–3020, 2010.

79. Spiegel, K. and Magistrato, A. Modeling anticancer drug–DNA interactions via mixed QM/MM molecular dynamics simulations. *Org. Biomol. Chem.* **4**, 2507–2517, 2006.

80. Dolenc, J., Oostenbrink1, C., Koller J. and Gunsteren W. F. van. Molecular dynamics simulations and free energy calculations of netropsin and distamycin binding to an AAAAA DNA binding site. *Nucleic Acids Res.* **33**, 725–733, 2005.

81. Orzechowski, M. and Cieplak, P. Application of steered molecular dynamics (SMD) to study DNA–drug complexes and probing helical propensity of amino acids. *J. Phys. Condens. Matter.* **17**, 1627, 2005.

82. Magistrato, A., Ruggerone, P., Spiegel, K., Carloni, P. and Reedijk, J. Binding of novel azole-bridged dinuclear platinum(II) anticancer drugs to DNA: Insights from hybrid QM/MM molecular dynamics simulations. *J. Phys. Chem.* B. **110**, 3604–3613, 2006.

83. Ma, G. Z., Zheng, K. W., Jiang, Y. J., Zhao, W N., Zhuang, S. L. and Yu, Q. S. Minor groove binding between norfloxacin and DNA duplexes in solution: A molecular dynamics study. *Chinese Chem. Lett.* **16**, 1367–1370, 2005.

84. Tajmir-Riahi, H.A. AZT binding to DNA and RNA: Molecular modeling and biological significance. *J. Iranian Chem. Soc.* **2**, 78–84, 2005.

85. Spiegel, K., Rothlisberger, U. and Carloni, P. Duocarmycins binding to DNA investigated by molecular simulation. *J. Phys. Chem.* B. **110**, 3647–3660, 2006.

86. Vargiu, A. V., Ruggerone, P., Magistrato, A. and Carloni, P. Anthramycin–DNA binding explored by molecular simulations. *J. Phys. Chem.* B. **110**, 24687–24695, 2006.

87. Fang, Y. Morris, V. R., Lingani, G. M., Long, E. C. and Southerland, W. M. Genome-targeted drug design: Understanding the netropsin-DNA interaction. *Open Conf. Proc. J.* **1**, 157–163, 2010.

88. Marco, E. and Gago, F. DNA structural similarity in the 2:1 complexes of the antitumor drugs trabectedin (yondelis) and chromomycin A3 with an oligonucleotide sequence containing two adjacent TGG binding sites on opposing strands. *Mol. Pharmacol.* **68**, 1559–1567, 2005.

89. Still, W. C., Tempczyk, A., Hawley, R. C. and Hendrickson, T. J. Semianalytical treatment of solvation for molecular mechanics and dynamics. *J. Am. Chem. Soc.* **112**, 6127–6129, 1990.

90. Jayaram, B., Liu, Y. and Beveridge, D. L. A modification of the generalized Born theory for improved estimates of solvation energies and pKa shifts. *J. Chem. Phys.* **109**, 1465–1471, 1998.

91. Hawkins, G. D., Cramer, C. J. and Truhlar, D.G. Parameterized models of aqueous free energies of solvation based on pairwise descreening of solute atomic charges from a dielectric medium. *J. Phys. Chem.* **100**, 19824–19839, 1996.

92. Jayaram, B., Sprous, D. and Beveridge, D. L. Solvation free energy of biomacromolecules: Parameters for a modified generalized born model consistent with the AMBER force field. *J. Phys. Chem.* B. **102**, 9571–9576, 1998.

93. Jayaram, B., McConnell, K., Dixit, S. B., Das, A. and Beveridge, D. L. Free-Energy component analysis of 40 protein-DNA complexes: A consensus view on the thermodynamics of binding at the molecular level. *J. Comput. Chem.* **23**, 1–14, 2002.

94. Jayaram, B. and Jain, T. The role of water in protein-DNA recognition. *Annu. Rev. Biophys. Biomol. Struct.* **33**, 343–61, 2004.

95. Lavery, R. and Sklenar, H. The definition of generalized helicoidal parameters and of axis curvature for irregular nucleic acids. *J. Biomol. Struct. Dyn.* **6**, 63–91, 1988.

96. Lavery, R. and Sklenar, H. Defining the structure of irregular nucleic acids: Conventions and principles. *J. Biomol. Struct. Dyn.* **6**, 655–667, 1989.

97. Bhattacharya, D. and Bansal, M. A general procedure for generation of curved DNA molecules. *J. Biomol. Struct. Dyn.* **6**, 93–104, 1988.

98. Bhattacharya, D. and Bansal M. A self-consistent formulation for analysis and generation of non-uniform DNA structures. *J. Biomol. Struct. Dyn.* **6**, 635–653, 1989.

99. Bhattacharyya, D. and Bansal, M. Local variability and base sequence effects in DNA crystal structures. *J. Biomol. Struct. Dyn.* **8**, 539–572, 1990.

100. Bhattacharya, D. and Bansal, M. Groove width and depths of B-DNA structures depend on local variation is slide. *J. Biomol. Struct. Dyn.* **10**, 213–226, 1992.

101. Bansal, M., Bhattacharya, D. and Ravi B. NUPARM and NUCGEN: Software for analysis and generation of sequence dependent nucleic acid structures. *Comput. Appl. Biosci.* **11**, 281–287, 1995.

102. Mukherjee, S., Bansal, M. and Bhattacharyya, D. Conformational specificity of non-canonical base pairs 4 and higher order structures in nucleic acids: Crystal 5 structure database analysis. *J. Comp. Aided Mol. Des.* **20**, 629–645, 2006.

103. Lu, X. J. and Olson, W. K. 3DNA: A software package for the analysis, rebuilding and visualization of three-dimensional nucleic acid structures. *Nucl. Acids Res.* **31**, 5108–5121, 2003.

104. Lu, X. J. and Olson, W. K. 3DNA: A versatile, integrated software system for the analysis, rebuilding and visualization of three-dimensional nucleic-acid structures, *Nat. Protoc.* **3**, 1213–27, 2008.

105. Case, D. A., Cheatham, III, T.E., Darden, T., Gohlke, H., Luo, R., Merz, Jr., K.M., Onufriev, A., Simmerling, C., Wang, B. and Woods, R. The Amber biomolecular simulation programs. *J. Computat. Chem.* **26**, 1668–1688, 2005.

106. Ponder, J. W. and Case, D. A. Force fields for protein simulations. *Adv. Prot. Chem.* **66**, 27–85, 2003.

107. Arnott, S., Campbell-Smith, P.J. and Chandrasekaran, R. In Fasman, Ed. *Handbook of Biochemistry and Molecular Biology*, 3rd ed. Nucleic Acids, Vol. II, G.P. CRC Press: Cleveland, 1976. pp. 411–422.

108. *Surflex, version 2.11.* Tripos, Inc.: St. Louis, MO, 2007.

109. *Autodock, version 4.* The Scripps Research Institute: La Jolla, CA, 2007.

110. Gupta, A., Gandhimathi, A., Sharma, P. and Jayaram, B., 2007 ParDOCK: An all atom energy based Monte Carlo docking protocol for protein–ligand complexes. *Protein Pept. Lett.* **14**, 632–646, 2007.

111. Jones, G., Willett P. and Glen, R. C. Molecular recognition of receptor sites using a genetic algorithm with a description of desolvation. *J. Mol. Biol.* **245**, 43–53, 1995.

112. Jones, G., Willett, P., Glen, R. C., Leach, A. R. and Taylor, R. Development and validation of a genetic algorithm for flexible docking. *J. Mol. Biol.* **267**, 727–748, 1997.

113. Nissink, J. W. M., Murray, C., Hartshorn, M., Verdonk, M. L., Cole, J. C. and Taylor, R. A new test set for validating predictions of protein-ligand interaction. *Proteins.* **49**, 457–471, 2002.

114. Verdonk, M. L., Cole, J. C., Hartshorn, M. J., Murray, C. W. and Taylor, R. D. Improved protein-ligand docking using GOLD. *Proteins.* **52**, 609–623, 2003.

115. Cole, J. C., Nissink, J. W. M. and Taylor, R. In B. Shoichet, J. Alvarez Eds. Protein-ligand docking and virtual screening with GOLD. *Virtual Screening in Drug Discovery* Taylor & Francis: Boca Raton, Florida, USA, 2005.

116. Verdonk, M. L., Chessari, G., Cole, J. C., Hartshorn, M. J., Murray, C. W., Nissink, J. W. M., Taylor, R. D. and Taylor, R. Modeling water molecules in protein-ligand docking using GOLD *J. Med. Chem.* **48**, 6504–6515, 2005.

117. Hartshorn, M. J., Verdonk, M. L., Chessari, G., Brewerton, S. C., Mooij, W. T. M., Mortenson, P. N. and Murray, C. W. High-quality test set for the validation of protein-ligand docking performance *J. Med. Chem.* **50**, 726–741, 2007.

118. Dominguez, C., Boelens, R. and Bonvin, A. M.J.J. HADDOCK: A protein-protein docking approach based on biochemical and/or biophysical information. *J. Am. Chem. Soc.* **125**, 1731–1737, 2003.

119. Ausiello, G., Cesareni, G. and Helmer Citterich, M. ESCHER: A new docking procedure applied to the reconstruction of protein tertiary structure, *Proteins.* **28**, 556–567, 1997.

120. Jayaram, B., McConnell, K. J., Dixit, S. B. and Beveridge, D. L. Free energy analysis of protein-DNA binding: The ecoRI endonuclease—DNA complex. *J. Comput. Phy.* **151**, 333–357, 1999.

121. Kalra, P., Reddy, T.V. and Jayaram, B. Free energy component analysis for drug design: A case study of HIV-1 protease-inhibitor binding. *J. Med. Chem.* **44**, 4325–4338, 2001.

122. Karplus, M. and Kaushick, J. N. Method for estimating the configurational entropy of macromolecules. *Macromolecules.* **14**, 325–332, 1981.

123. Schlitter, J. Estimation of absolute and relative entropies of macromolecules using the Covariance matrix. *Chem. Phys. Lett.* **215**, 617–621, 1993.

124. Schafer, H., Mark, A. E. and van Gunsteren. Absolute entropies from molecular dynamics simulation trajectories. *J. Chem. Phys.* **113**, 7809–7817, 2000.

15 Computational Studies of RNA Dynamics and RNA–Ligand Interactions

Julia Romanowska, Dariusz Ekonomiuk, and Joanna Trylska

CONTENTS

In this chapter, we give an overview of common simulation techniques used to investigate the dynamics of RNA and its interactions with small molecules. We also describe the docking protocols applied to RNA targets. Applications to small RNA structures, such as ribozymes and riboswitches, as well as to large ribosomal RNA are shown as examples.

15.1 INTRODUCTION

RNAs, together with proteins and DNA, form the essential elements of the cellular machinery. While the roles of DNA and proteins have been appreciated for several decades, RNA has been recognized as a player of the same level only much later. At first, RNA was regarded mainly as a template (mRNA) for translating the message of the DNA nucleotide sequence to the protein amino acid sequence. Later, numerous classes of noncoding RNAs (e.g., tRNA, ribosomal RNA, microRNA, small nuclear RNA, siRNA, ribozymes) were discovered showing not anticipated RNA functions. RNA is involved in processes such as recognition (ribosomal RNA), cleavage (ribozymes, ribosomal RNA), regulation (riboswitches), protein translation initialization (viral RNA can mimic several protein initiation factors), or gene silencing (small interference RNA) [1–3]. Certainly, many classes of functional RNAs are still awaiting discovery.

The diverse functional roles of RNA make it an excellent drug target [4,5], especially validated for antibacterial treatment. Molecules found to target ribosomal RNA are, for example, aminoglycoside antibiotics [6], which have been successfully used in medical therapy for over 60 years to combat severe bacterial infections. The perspective of finding drugs targeting the viral RNA of human immunodeficiency virus (HIV) or hepatitis viruses (HCV, HDV) [7,8] has also emerged, showing great promise for treating viral infections.

RNA backbone differs from DNA by only one 2′OH group on the sugar ring. However, this difference has important implications for the RNA architecture [9], leading to a plethora of well-defined three-dimensional (3D) RNA folds [1]. RNA, due to numerous non-Watson–Crick base pairs (for example, by the 2′OH ribose group [10–12]), forms a variety of motifs such as hairpins, loops, kink-turns, helices, junctions, and bulges [13] (exemplary structures are presented in Figure 15.1). These motifs can form various binding pockets that constitute attractive targets for specific and selective binding of ligands.

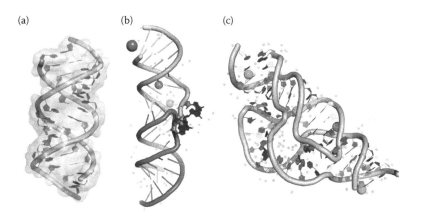

(a)   (b)   (c)

FIGURE 15.1 Examples of 3D structures adopted by RNA: (a) classical A-type RNA helix (PDB code: 3GM7); (b) the DIS complex (PDB code: 1XPF) with the flipping bases; and (c) the hammerhead ribozyme (PDB code: 2OEU). The oxygen atoms of water molecules and ions (Na^+, Mn^{2+}, Mg^{2+}) as van der Waals spheres.

Diverse functional roles of RNA arise from its flexible nature. RNA retains inherent plasticity, with base pairs often forming and breaking hydrogen bonds on a millisecond timescale [14]. For example, shortening RNA sequence by only one nucleotide can lead to a change in motions of the motif and consequently, to a change in the activity, as it was found in analogous RNA fragments of HIV-1 and HIV-2 viruses [15]. Even though RNA has some unique cavities that allow for specific ligand recognition, its structural variability makes it a promiscuous binder. The same RNA can adjust its structure when cocrystallized with different inhibitors (e.g., aminoglycoside ribosomal RNA-binding site [16] or HIV-1 TAR RNA bulge [17]). Also, RNA flexibility makes some ligands bind to many RNA targets [18,19].

The structure of RNA determines its dynamics, interactions with ligands, and biological function. Predicting how the structure of RNA is connected with its specific function is crucial in order to design drugs that could modify this function. Experiments that directly investigate RNA flexibility and RNA–ligand binding are difficult, very expensive, and time consuming, and computational tools suitable to investigate RNA–ligand recognition are often helpful. Thanks to structural biology techniques, there are about 800 (as of December 2010) high-resolution atomistic 3D structures of RNA and RNA–small molecule complexes available in the Protein Data Bank [20] and Nucleic Acid Database [21]. These data usually serve as a starting point for simulations that, if well performed and analyzed, can provide useful information complementing structural and biochemical experiments.

Computational methods using atomistic descriptions were initially developed for proteins, and simulations of nucleic acids lagged behind. There are numerous reasons why modeling of nucleic acids, and especially of RNA, is more difficult. First, RNA has a hierarchical and multiform structure, usually nonglobular with many tertiary motifs, which results in higher intrinsic RNA flexibility. A stable helix (Figure 15.1a), as is typical in the case of B-DNA, does not dominate the RNA structure. Second, the high charge of the RNA backbone requires careful treatment of electrostatics and reliable representation of long-range interactions. Third, water molecules and ions are pivotal for maintaining the correct RNA tertiary structure (e.g., Figure 15.1c) [22,23] and for interacting with small molecules (Figure 15.2) [23–28]; thus, they need to be represented reliably (often positioned explicitly and well equilibrated). Next, divalent cations, such as Mg^{2+}, are frequently tightly bound to RNA, assisting in folding and catalysis (Figure 15.1b and c) [29,30], and affecting RNA mobility. Also, these counterions as well as long-resident water molecules can be displaced upon binding of a ligand [24].

Numerous efforts were undertaken to improve the performance of techniques to study the dynamics of RNA with an intensive development of methods, parameters, and software (for comprehensive reviews, see, e.g., [26,31–35]). The studies focused on benchmarking the performance, accuracy, and predictive power of these methods. Here, we describe some methods used to study internal flexibility of RNA, the energetics of RNA–ligand interactions, and docking of small molecules to RNA with an emphasis on aminoglycoside antibiotics.

15.2 SIMULATING RNA AND ITS COMPLEXES WITH SMALL MOLECULES

15.2.1 ALL-ATOM MOLECULAR MECHANICS MODEL OF RNA

X-ray crystallography, nuclear magnetic resonance (NMR), and comparative modeling provide atomic-resolution 3D structures of RNA [20,21,33] used in computational studies. However, these "static" structures do not directly reveal the dynamics that can be characterized by three most common methods: molecular dynamics (MD) [36,37], normal mode analysis (NMA) [37], and Monte Carlo (MC) sampling techniques [38–40].

To study the dynamics, one needs to know the conformation-dependent potential energy of the molecule. The formula for this energy with a set of parameters defines a force field (FF). In most biomolecular simulations, interatomic interactions are described using classical molecular mechanics (MM) model. The potential energy function (V_{FF}) depending on atomic positions usually takes the form

$$V_{FF} = \sum_{i=1}^{N_{bonds}} \frac{1}{2} k_{bi} \, (b_i - b_{i0})^2 + \sum_{i=1}^{N_{angles}} \frac{1}{2} k_{\varTheta i} \, (\theta_i - \theta_{i0})^2$$

$$\underbrace{+ \sum_{i=1}^{N_{dihedrals}} \sum_n K_n \, (1 + \cos(n\tau_i + \delta_n)) +}_{\text{bonded interactions}}$$

$$\underbrace{+ \sum_{i>j}^{N_{pairs}} \left(\frac{A_{ij}}{r_{ij}^{12}} - \frac{B_{ij}}{r_{ij}^6} + \frac{q_i q_j}{r_{ij}} \right)}_{\text{nonbonded interactions}} \tag{15.1}$$

Equation 15.1 includes the bonded and nonbonded (van der Waals and Coulomb interaction) contributions. It allows for calculating the potential energy of the molecule and of its first (gradient) and second (Hessian) derivatives. The parameters of Equation 15.1 (such as equilibrium bond lengths b_{i0} and angles $\varTheta_{i0}$, force constants k_{bi}, $k_{\varTheta i}$, partial atomic charges q_i) are derived based on empirical data and quantum mechanical calculations. Details can be found, for example, in [37,41,42].

In recent years, several FFs for RNA have been developed. The most popular are parm99 [43] and parmbsc0 [44] from the Amber suite [45], and the CHARMM FF [46]. While modeling RNA, it is essential to have realistic descriptions of the surrounding ions and water molecules. Common models of water molecules include TIP3P [47] and SPC [48], which have been extensively tested. However, divalent cations such as Mg^{2+} are poorly characterized by classical FFs. Due to high charge accumulated in a small volume, simple parameters of standard FFs are unable to correctly capture the complex ionization effects that are observed around these ions in experiments [28,31,49,50]. The classical FFs assume fixed partial charges placed at atomic nuclei, which do not account for polarizable nature, for example, of phosphate groups or water molecules. Advancements in the FFs include the use of their polarizable versions [37,51–53] or specific parameters for the nonstandard RNA base pairs derived from quantum mechanical calculations [54].

15.2.2   Methods to Study the Dynamics of RNA

NMA predicts molecular motions at equilibrium by decomposing the intrinsic flexibility of the molecule into internal vibrational modes. Formally, NMA solves an eigenvalue problem:

$$\mathbf{A}^T \mathbf{V} \mathbf{A} = \lambda \qquad (15.2)$$

where $\mathbf{A}$ is the matrix of mode amplitudes, T denotes the transpose, λ is the diagonal matrix of eigenvalues, and $\mathbf{V}$ is the Hessian, that is, the matrix of second derivatives of the potential energy function $\mathbf{V}_{FF}$ [37]. The complete solution gives $(3N)^2$ modes, where N is the number of atoms. Only a couple of the lowest-frequency normal modes usually correspond to functional motions. When applying NMA to large structures, the most time- and memory-consuming step is the calculation of the Hessian; therefore, several approximations have been proposed to overcome this problem [26,55].

Recently, another method to specifically study RNA motions has been developed [56], which yields similar results to NMA and unveils the global motions directly from single RNA structures. The method is based on topological information and on counting-defined constraints, and is less computationally demanding than NMA.

15.2.3   Molecular Dynamics

MD is used to describe the time-dependent dynamical evolution of molecules (typically at atomistic resolution) by calculating the positions and velocities of individual atoms as a function of time (trajectory). This is achieved by computing the forces exerted on atomic nuclei (from the potential energy of each conformation defined in Equation 15.1) and by numerically solving the Newton's equations of motions in finite time steps. The details of the MD technique with the description of the most popular integration algorithms can be found in [34,37,41,42,57,58].

To execute an MD simulation, one starts from atomic coordinates (obtained from x-ray, NMR, or modeling) and usually needs to predict the positions of hydrogens. Next, the solute is immersed in a periodic box containing water molecules that fill the area of at least 10–15 Å from the molecule's surface. Counterions are added to neutralize the negative charge, coming in case of RNA, from the phosphate groups. To better reproduce the RNA natural environment, one can also create a nonzero ionic force. In practice, monovalent ions (e.g., Na^+ and Cl^-) are added at random positions to the solvent. Also, excess salt conditions (i.e., much higher ion concentration than found in natural environment) can be used in order to facilitate some conformational transitions in RNA structures [59,60]. Before the production trajectory is calculated, the system undergoes minimization, thermalization, and equilibration protocols (for descriptions of these steps, see [34,41,42] and for adjustments made to study RNA, see, e.g., [29,50,61]). Coupling to a bath keeps the temperature and/or pressure constant.

In atomistic MD, the time of one simulation step is usually 1–2 fs, that is, an order of magnitude smaller than the highest-frequency motions corresponding to a single bond stretching. Therefore, the total practical simulation time is of nanoseconds and does not typically exceed a microsecond. The time of classical MD is limited not only by the computational power and system size but also by the errors resulting from numerical integration of Newton's equations of motions and the poor performance of FFs at long timescales. The longest simulations of RNA to date are in the range of 300 ns [59], while the largest simulated system containing RNA in all-atom representation was composed of about 10^6 atoms (the ribosome [60,62] or the complete tobacco mosaic virus [63]).

The explicit treatment of water and ions in MD of large RNAs may be too time consuming to be practical. Also, for ligand–RNA docking (described further), rapid calculations of the solvation effects are required. In such cases, the solvent can be treated implicitly with one of the popular continuum solvent models, the Poisson–Boltzmann (PB) [64] or Generalized Born (GB) [65].

15.2.4 ENHANCED SAMPLING METHODS

MD sampling is not necessarily efficient because the system can cross only small barriers on the potential energy surface and does not sample the entire available phase space. Several techniques exist that can increase sampling in MD or indirectly extend the simulation time (reviewed, e.g., in [66] and [67]). For example, locally enhanced sampling (LES) [34] uses several noninteracting copies of a specified fragment of the system giving multiple trajectories of this region from a single simulation. Accelerated MD [68,69] smoothes the potential energy surface which enables overcoming even high potential energy barriers and expedites the dynamics. Also, reducing the system's size by treating a group of atoms as one pseudoatom, like in coarse-grained (CG) models [70], speeds up the simulations, although it decreases the amount of details that can be observed and, in case of nucleic acids, lacks accurate description of electrostatics. In replica exchange MD (REMD) [34,71], many copies (replicas) of the system are simulated simultaneously, but at different temperatures. By swapping two chosen replicas, the conformational changes can be more easily triggered due to a change in the environment temperature. When different end-point conformations of a molecule are known and one would like to describe a pathway between these states, a targeted MD [72] can be applied.

Another technique that has been also applied to RNA is Brownian dynamics (BD) [73], used to estimate the association rate constants of diffusion-controlled reactions and pathways of bimolecular association. In BD, the two molecules undergo random motions in a continuum solvent and their intermolecular electrostatic interactions are described with the Poisson–Boltzmann implicit solvent model [64]. The association rates are estimated based on statistics derived from multiple trajectories. BD can be also used to identify possible ion-binding sites in RNA target [74]. The description of BD algorithms and their applications to estimate the reaction rates of diffusion-controlled bimolecular reactions can be found in [42].

15.2.5 ANALYSIS OF DATA FROM MD SIMULATIONS

The sampling techniques provide multitude of data that should be further examined and carefully interpreted. The convergence of these techniques might be an issue but it will not be discussed here [35,75,76]. Typically, deviation from a reference conformation (e.g., from the starting x-ray structure) such as root mean square displacement (RMSD) and residual root mean square fluctuations (RMSF) is monitored. In general, nucleotides in RNA motifs are more mobile than in DNA [9,77] so a scheme of base pairing and stacking is examined to check their deviation from a reference structure [14,78,79]. The properties of the environment such as hydration patterns, distributions of ions, or ion residence times in the binding pocket can be calculated. For the free RNA, the location of ions and water molecules points to possible locations of the ligand's charged moieties [50,74]. The global collective motions can be analyzed by principal component analysis (PCA, also called essential dynamics) [80] that shows directions of large-magnitude movements. PCA of the MD trajectory is analogous to NMA but may give more details, since NMA is restricted to the ensemble around the energetic minimum. The clustering algorithms allow grouping of representative conformations (the Ptraj program of AmberTools [81], CHARMM [82], or GROMACS [83] packages provides clustering tools). Correlations between motions of different atoms or fragments can be derived from a dynamical correlation matrix (see, e.g., [50,84]). Distributions of various degrees of freedom in the system such as RNA end-to-end distance or specific hydrogen bond distance distribution can be measured. Naturally, the method of analysis depends on the studied problem and questions one would like to answer.

15.2.6 ENERGETICS OF LIGAND BINDING

When considering the interactions of RNA with ligands, it is often desirable to determine the energetics of binding, for example, relative binding free energies for a set of compounds [85–87]. One approximate but relatively fast way is to combine explicit solvent MD with MM and implicit solvent methods—called MM-PB/SA or MM-GB/SA, depending on the type of continuum solvation model: Poisson–Boltzmann or Generalized Born, respectively [85,88]. MD generates an ensemble of structures that are later used to calculate solvation free energies with implicit solvent models. The total average energy is given by

$$\langle \Delta G \rangle = \langle E_{MM} \rangle + \langle G_{PBSA} \rangle - T \langle S_{MM} \rangle \tag{15.3}$$

where $\langle E_{MM} \rangle$ is the average energy from MM calculations dependent on the FF, $\langle G_{PBSA} \rangle$ (or $\langle G_{GBSA} \rangle$) is the continuum solvation free energy, and $\langle S_{MM} \rangle$ is the average entropy of the solute (which is typically either ignored when comparing similar ligands or estimated with the use of the NMA technique). The above can be used to calculate the relative binding free energies:

$$\Delta\Delta G = \langle \Delta G \rangle_{complex} - \langle \Delta G \rangle_{RNA} - \langle \Delta G \rangle_{ligand} \tag{15.4}$$

The more accurate but more computationally demanding method to determine the free energy changes is the thermodynamic integration (TI) where one generates an ensemble of intermediate structures λ_i $(i = 0, 1,\ldots, N)$ between the initial and final states (e.g., before and after substituting a chemical group in a given ligand in order to modify its affinity). The parameters λ_i are changed slowly so that at each stage along the path the system is in equilibrium which ensures that the path is reversible. The free energy of the transition is calculated as the sum of the average energy of each of the intermediates [85]:

$$\Delta G = \int_{0}^{1} \left\langle \frac{dV_{(\lambda,\vec{x})}}{d\lambda} \right\rangle_{\lambda} d\lambda$$

$$\overset{TI}{\approx} \sum_{i=0}^{N} \left[\left\langle \frac{\partial V_{(\lambda,\vec{x})}}{\partial \lambda} \right\rangle_{\lambda(i)} + \left\langle \frac{\partial V_{(\lambda,\vec{x})}}{\partial \lambda} \right\rangle_{\lambda(i+1)} \right] \left[\frac{\lambda_{(i+1)} - \lambda_{(i)}}{2} \right]$$

(15.5)

Another method, free energy perturbation (FEP) is based on the fact that the binding free energy is a state function. (It depends on the current state of the system and not on the path through which the state was acquired.) Therefore, in order to calculate the free energy difference between a reference and a target state, for example, showing the effects of mutation on ligand binding (ΔG_1), one can simply introduce small perturbations to the unbound and bound states (ΔG_2 and ΔG_4) (see Scheme 15.1). In FEP, one uses Equation 15.6 to calculate the total energy difference and the differences in energy between each pair of conformations, i and j [85]. The vertical transformations (Scheme 15.1) are easier to achieve through nonquantum methods, such as MD.

$$\Delta\Delta G = \Delta G_2 - \Delta G_4 = \Delta G_1 - \Delta G_3;$$

$$\Delta G_i = -RT \ln \left\langle e^{-(E_i - E_j)/RT} \right\rangle$$

(15.6)

where R is the gas constant, T is the temperature, and the energy, E_i, is obtained from the force field.

A more recent technique that was applied specifically to RNA–ligand complexes [89,90] is the free energy perturbation molecular dynamics (FEP/MD) combined with generalized solvent boundary potential (GSBP) method [91]. It requires many

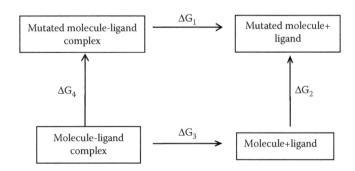

SCHEME 15.1 Exemplary free energy cycle depicting the influence of a mutation in the receptor on the ligand binding affinity.

simulations of different ligands bound to the same receptor, but it reduces the computational cost through dividing the system into two areas: explicit, located near the binding site, and implicit, further from the binding site. The GSBP method enables the two areas to be seamlessly joined together. Therefore, even such large structures as the whole 50S ribosomal subunit [89] can be investigated with this technique.

15.2.7   DOCKING SMALL MOLECULES TO RNA TARGETS

Atomic-level description of RNA–small molecule recognition is of great importance for rational drug design. The problem of computational determination of the binding mode of a small molecule (ligand) interacting with a macromolecule (receptor) is referred to as the docking problem [92]. The docking procedure involves searching various conformations of a ligand in the binding site of a receptor. This procedure generates many possible ligand–receptor complexes that have to be properly scored in order to identify the one representing the experimentally determined binding mode. The main challenges associated with the development of the docking programs are application of exhaustive search algorithms to properly account for the complexity of the conformational space of the system [66,93] and improvement of the accuracy of the scoring functions [94].

While computational drug design aimed at targeting proteins is already a well-established field, comparatively little is known on docking to RNA. Theories describing RNA–ligand recognition are similar to protein–ligand ones. The simplest concept assumes that the receptor and ligand complement each other like "a lock and a key" [95]. Obviously, this model does not take into account conformational rearrangements and may be applied to only a few cases (see, e.g., [96]). For most systems, the receptor changes its conformation upon ligand binding and a dynamic description is needed, referred to as the "induced fit" model [97]. The induced fit explains the fact that a receptor can have different conformations when crystallized with different inhibitors [17,18]. However, it does not describe the dynamics of the receptor before the binding takes place. RNA adopts various distinct conformations in solution, which provides evidence for a "conformational selection" theory [98]. In the light of this theory, a ligand binds to a complementary conformation of the receptor that is already present in the ensemble of conformations. From the thermodynamic standpoint, the latter two models are equivalent [99]. Whether the binding is driven by induced fit or conformational selection depends on the specific RNA–ligand complex [100,101].

To make docking computationally affordable, the accuracy and approximations must be properly balanced. The simplest docking programs implement the "lock and key" model [102] where the inhibitor is completely static. Currently, many programs (e.g., DOCK [103], Autodock [104,105], FlexE [106]) take into account ligand flexibility, but receptor flexibility is still subject to development [16,99,107,108]. Receptor flexibility is especially important for RNA because ligand binding can induce significant conformational changes [109–111]. One possible approach to consider the flexibility of the receptor is to use multiple conformations from NMR experiments or different x-ray structures [112,113], as well as from MD simulations.

Such an ensemble of diverse conformations of the unbound receptor provides structures that can be used as docking targets. Docking to multiple MD conformations in a method called "relaxed complex" is presented in [114–116]. However, a ligand may bind to conformations that occur infrequently in the dynamics of the receptor and may not be present in the chosen ensemble. Another idea is the "soft receptor" method [117,118], a rather computationally inexpensive approach, where the atoms of the ligand and receptor are allowed to partially overlap by using a smaller van der Waals repulsive term.

Other problems connected with docking small molecules to RNA are possible multiple-binding clefts, even in short RNA fragments, that make the scanning more time consuming. The water molecules can also be problematic because one cannot know *a priori* which bonds are better formed directly and which ones require a water bridge [26].

15.3 CASE STUDIES

15.3.1 MD Simulations of Ribozymes, Aptamers, and Viral RNA

Ribozymes are catalytic RNAs that enhance the reaction rates and can be even more specific than protein enzymes [1]. The best studied computationally are hammerhead ribozyme (Figure 15.1c), hairpin ribozyme, and hepatitis delta virus (HDV) ribozyme—all carrying out the same self-cleavage reaction but having substantially different 3D structures (for recent reviews, see, e.g., [30,119]). Also, one of the first theoretically studied chemical processes in RNA was the autocleavage of hammerhead ribozyme [29]. From MD [29,120] and BD [74] simulations, Westhof and coworkers proposed the reaction mechanism and ruled out contributions of two Mg^{2+} ions in the self-cleavage process, out of six ions identified in the experimental x-ray structure. The hairpin ribozyme was also the subject of theoretical studies, for example, Otyepka et al. [121] performed MD simulations (spanning from 50 to 150 ns) of the hairpin ribozyme that pointed out subtle conformations of bases required for its proper activity. They also found that introducing a small modification in one base (2′-methoxy group instead of 2′-OH) had a significant impact on the architecture of the active site. This small but important modification inhibits the reaction and is often used to obtain the ribozyme crystals.

Other studies focused on RNA aptamers—short RNA sequences specifically structured in order to bind metabolites such as adenine or lysine [1]. Aptamer regions are often found in riboswitches—RNA structures that naturally help control the initialization of some processes in a cell and could be used as specifically designed switches, for example, in cancer treatment. MD simulations of the SAM-II riboswitch [110] and adenine riboswitch [84,122] unveiled important differences between the free and the ligand-bound states. It appears that the pseudoknot conformation of SAM-II is stabilized by the bound ligand (S-adenosylmethionine), leading to loss of its interactions with the ribosome, and to a premature termination of translation. Some hypotheses on the mechanisms of the reactions triggered by these riboswitches were proposed. For example, Kollman and coworkers [27] performed energetic studies (MM-PB/SA and TI) of RNA aptamer–theophylline complex and concluded that the van der Waals and nonpolar interactions drive the binding of theophylline, while electrostatics opposes binding. They also showed that TI performs better when one needs quantitative results but the less computationally demanding MM-PB/SA method still provides good qualitative outcome.

Functional viral RNA is also a promising drug target [1]. An important viral RNA fragment is the dimerization initiation site (DIS)—35 nucleotides folded into a stem-loop structure. The two self-complementary regions of DIS form the so-called kissing complex (Figure 15.1b), which is transformed into an extended helix as one of the first steps in the initiation of viral replication process [123]. Classical MD and LES of different structures of this complex [124,125] showed that the bases at the interface between the two stacked helices flip in and out of the helix, maintaining the stacking interactions for the majority of time. These studies found that the Amber and CHARMM FF parameters give corresponding results, although some melting of the helix was observed with the latter FF (see also [50]).

15.3.2 SIMULATIONS OF THE RIBOSOME AND RIBOSOMAL RNA

The ribosome, composed of ribosomal RNA (rRNA) and proteins, is a workbench for protein synthesis in the cell. It is formed of two unequal subunits (Figure 15.2, left): large (50S in prokaryota, 60S in eukaryota) and small (30S in prokaryota, 40S in eukaryota). The framework for the bacterial 30S subunit is the 16S RNA (~1500 nucleotides), and for the 50S subunit—23S RNA (~2900 nucleotides) and 5S RNA (~120 nucleotides). Peptide synthesis progresses according to the nucleotide sequence of the mRNA, with tRNAs as amino acid carriers, and with the help of external translation factors. The chemical reaction is catalyzed by the RNA in the large subunit; hence, the ribosome is a ribozyme. The growing polypeptide leaves the ribosome through a tunnel in the 50S subunit whose walls are formed mainly by RNA with two protruding protein extensions.

Bacterial ribosomal RNA is an important target for antibiotics [126–128]. For example, macrolides such as erythromycin bind to 23S RNA in the entrance of the tunnel and abort the growth of the peptide by restricting its egress. The binding site of lincosamides is located in the so-called peptidyl transferase center (PTC) of the 50S subunit, where the drug blocks proper positioning of the newly attached amino acid causing premature dissociation of the peptidyl-tRNA from the ribosome. Aminoglycoside antibiotics bind to one of the tRNA-binding sites (A-site) in the 16S RNA of the 30S subunit and interfere with decoding, as well as the self-assembly of the small subunit.

Various modeling techniques were applied to investigate the flexibility of the ribosome, the details of decoding, the egress of the growing peptide, and its interactions with antibiotics. Atomistic MD simulations of the whole ribosome gave insight into the decoding process [60,62,129] and the behavior of peptides inside the tunnel [130,131]. Coarse-grained MD [132] and NMA [133–135] elucidated the global collective motions and confirmed the ratchet-like rotation of the subunits. Studies of the dynamics of peptides inside the tunnel showed partial formation of its secondary

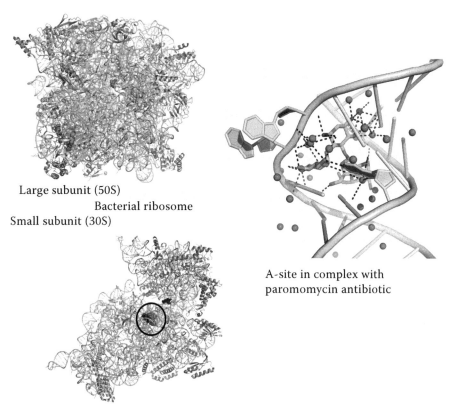

Large subunit (50S)
              Bacterial ribosome
Small subunit (30S)

A-site in complex with
paromomycin antibiotic

FIGURE 15.2   The visualization of bacterial ribosome complexed with various antibiotics: large subunit (PDB code: 1NJI) and small subunit (PDB code: 1FJG).

structures. For an overview of modeling studies of the entire ribosome and its subunits, see [136].

Because modeling of the entire ribosome at the atomic level is computationally demanding, the studies of ribosomal RNA fragments have been conducted. The flexibility of ribosomal RNA fragments such as helices H38 [137], and H40 and H68 [59] of 23S rRNA were reported. H38 is involved in translocation of the ribosome along the mRNA and in the assembling of the large subunit. Classical MD proposed a mechanism of dislocation of the ribosomal fragments due to hinge-like movements of the H38 helix, leading to sliding of the whole ribosome along the mRNA [137]. The helices H40 and H68 [59] form unusual 3D structures inside the ribosome, significantly different from the minimum energy conformation that they adopt in solution. Classical MD accompanied by LES and targeted MD simulated the transition pathways between these different structures. This study also compared the parm99 and parmbsc0 Amber FFs—over 20 simulations, from 20 ns to 300 ns, proved that both FFs yield similar results for these RNA systems. The BD technique was also employed [74] to analyze ion-binding sites in the so-called loop E of 5S ribosomal RNA. This work unveiled possible Mg^{2+}-binding regions and helped fixing the NMR data concerning the positions of the bound ions.

15.3.3 INTERACTIONS OF RIBOSOMAL RNA WITH AMINOGLYCOSIDE ANTIBIOTICS

Aminoacylated-tRNA binds in the A-site RNA located in the small subunit (Figure 15.2). A group of compounds that target the A-site are aminoglycosides, positively charged sugar derivatives [128,138]. Although subsequent generations of these drugs are more and more potent, they still are quite toxic upon long-term treatment. Therefore, a lot of research is conducted to understand the details of aminoglycoside mechanism of action in order to improve their specificity and efficacy. X-ray or NMR structures of the model A-site RNA with and without aminoglycosides are available in PDB (see, e.g., 139–141). Complexes with the entire 30S subunit can also be accessed [142–144]. The A-site comprises a short helix and a bulge formed by three nucleotides: universally conserved A1492 and A1493, and A1408 on the opposite RNA strand (which is a G in eukaryota; Figure 15.3, left). A1492 and A1493 are highly mobile both in solution [49,109,145] and in the environment of the small subunit [146], because through this movement the ribosome can recognize cognate tRNAs [6,147]. The flipped-out state is adopted upon aminoglycoside binding [148] (Figure 15.2, right).

MD simulations of systems containing the A-site RNA helix confirmed that A1492 and A1493 acquire both intra- and extrahelical states [49,50,111,149,150] (Figure 15.2, right and Figure 15.3, right). REMD studies of the A-site [149] showed that energetically A1492 and A1493 slightly prefer the flipped-in state; however, this quasi-equilibrium is drastically changed upon aminoglycoside binding, as shown by MD and MM-PB/SA analysis [111]. Both experiments [151] and simulations [25,50]

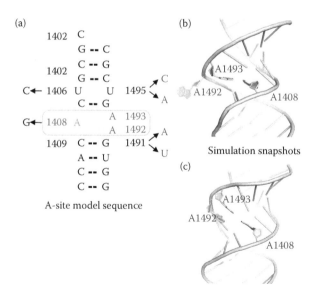

FIGURE 15.3 Aminoglycoside-binding site of the small ribosomal subunit. (a) The sequence of the A-site is highly conserved; however, some bacteria with mutations in this region developed resistance against aminoglycoside antibiotics. The mobility of A1492 and A1493 is crucial for proper recognition of cognate tRNAs during translation. These bases adopt multitude of conformations in MD simulations: both flipped out (see Figure 15.2), only A1493 inside the RNA helix (b) or both flipped in (c).

of aminoglycosides bound to the A-site model revealed that the electrostatic interactions are the most important in the final stages of forming the complex, however, not necessarily when the drug "scans" the ribosomal RNA surface in search for its binding site [152,153]. REMD simulations and potential of mean force energy landscape analysis proved that the binding occurs in a stochastic gating manner (i.e., through the selection of an appropriate conformation from an already existing one) [150], rather than an induced fit, as was suggested earlier [109].

Bacteria with specific point mutations in the A-site gain resistance against aminoglycosides [154] (e.g., Figure 15.3, left). MD simulations performed for the A-site RNA model with nucleotide substitutions helped understand the nature of this phenomenon [50,155]. Specific single- or double-nucleotide substitutions affect the electrostatic potential, shape, and volume of the bulge, the mobility of A1492 and A1493, as well as the overall flexibility of the A-site; all of these result in hindering of aminoglycoside binding. MD studies [24,25] additionally confirmed the critical role of water molecules in forming indirect bonds between the antibiotic and RNA (Figure 15.2, right). Certain water molecules exhibited residence times up to ~8 ns [25], while typically this time does not exceed ~2 ns [24,25].

From NMA and MD, the contributions of electrostatic, entropic, or nonpolar components to the total binding free energies of aminoglycoside complexes [25] were estimated. The relative energy differences between aminoglycosides were in accord with experiments. PB/SA studies showed that a change in one chemical group (e.g., from OH in kanamycin A to H in tobramycin) results in a 1 kcal/mol decrease in binding free energy [156]. Calculations showed that this is due to changes in the electrostatic and van der Waals energy [25]. Simpler PB/SA calculations maintained an overall relative order of aminoglycoside binding similar to the one obtained from experiments [156]. From BD simulations of aminoglycosides diffusing toward the model A-site RNA [153] and the 30S subunit [152], the association rate constants and pathways leading to encounter were estimated.

15.3.4   RNA DOCKING STUDIES

Several RNA docking studies (e.g., [157–162]) were performed underlying the importance of further developing RNA docking tools. Here, we describe three interesting test cases that try to incorporate the flexibility of the receptor [16,162,163].

Guilbert and James [162] presented a new flexible docking procedure MORDOR (molecular recognition with a driven dynamics optimizer) accounting for both the RNA and ligand flexibility. During docking, a ligand can change its position and explore a low-energy path at the surface of the receptor. The driving force permitting the ligand to probe the receptor's surface is obtained by conducting minimization of the potential energy supplemented by additional energy term, including root-mean-square-distance constraints [108]. Additionally, the permitted local movements of the receptor are large enough to accommodate the changes associated with the induced fit but, on the other hand, extra restraints are applied to keep the receptor close to the minimized native structure. The procedure was validated on 57 RNA complexes and the experimental binding poses were reconstructed within 2.5 Å for 74% of tested RNA–ligand complexes.

In another study, Lang et al. used the DOCK program [103] to model RNA–small molecule complexes [163]. DOCK handles flexible molecules by an "anchor-and-grow" sampling algorithm [164]. First, the largest and most rigid fragment is identified and oriented in the active site and, subsequently, the remaining moieties are reconstructed by optimizing the torsion angles. The algorithm generates many possible poses that need to be properly scored. Because the RNA backbone is charged, it is especially important to adequately score electrostatic interactions [151,165,166]. The newest DOCK version applies physics-based scoring functions. When docked conformations are rescored with the implicit solvent models (Amber PB/SA or GB/SA) in combination with explicit water molecules and sodium counterions, the success rate increases to 80% with PB/SA and for less than seven rotatable bonds, but for <13 rotatable bonds, the rates are 58% with GB/SA and 47% with PB/SA. The higher the number of rotatable bonds, the more time consuming is the calculation and the lower is the success rate [163]. The authors recommend using this approach for small compounds with less than seven rotatable bonds. Larger and more flexible compounds with more than 13 rotatable bonds were neglected in this study. This limitation does not allow considering an important class of RNA binders, aminoglycosides.

The study that focused on docking aminoglycosides to RNA [16] used autodock [104,105] with its exhaustive genetic algorithm for conformational sampling. The original autodock protocol with the default FF was not able to correctly reproduce the binding modes of 11 aminoglycosides in the A-site of 16S rRNA—the average RMSD was only 3.25 Å. To decrease the RMSD, the authors added the water molecules that could bridge the contacts between the ligand and RNA, as well as included the flexibility of the receptor through soft van der Waals potential and a single average RNA structure computed from nine experimentally determined conformations. Such treatment of the receptor flexibility improved autodock accuracy. Furthermore, a new potential that accounted for the effects related to water-mediated ligand binding was implemented. The new potential included two minima, one corresponding to the case when the water molecule is taken into account and the second one when the water molecule is removed, thus allowed considering displaceable water molecules. These efforts led to an average RMSD of 1.95 Å and a 78% success rate. The modified autodock algorithm correctly reproduced experimental data; however, the knowledge of experimentally determined structures was required *a priori*.

15.4 CONCLUSIONS

Predicting RNA dynamics and interactions with ligands from static structures is a challenge for computational biology. Simulations of RNA can complement and help explain experiments that are expensive, time consuming, and may not access all the relevant timescales. Indeed, modeling RNA flexibility and interactions with drugs has in the past decade shown a lot of progress. Today's computers are capable of simulating systems containing several millions atoms or on several hundred nanosecond timescales [63,129]. The advances in the FF development, binding free energy, and docking computations helped resolve some issues pertaining to theoretical studies of RNA such as treatment of electrostatics, divalent ions, and water molecules. For example, simulations helped elucidate the details of ribosome function [136], chemical reactions of ribozymes [30], and interactions between RNA and drugs [24,25].

15.5 PROSPECTS AND OUTLOOK

Understanding RNA flexibility and interactions with small molecules is important for designing novel drugs targeting RNA that could be useful in future therapeutic applications. Computational studies of the dynamics and energetics of RNA complexes that give atomistic details have seen recent fast development. Even large RNA-containing structures (such as the ribosome or viral RNA) can be simulated in MD for longer times. However, the methodology is as yet far from perfect and is not necessarily transferable among various RNA targets and ligands. Parameterizing FFs is challenging due to RNA-charged backbone, diverse and flexible structure, the presence of counterions, and bridging water molecules. More work is needed to check the validity of the FFs for microsecond MD timescales as well as their performance for noncanonical RNA structures such as internal loops, bulges, or hairpins that often form the sites to accommodate drugs [31]. The energy-scoring functions used in docking are typically parameterized on a small set of compounds and therefore, their transferability and predictive power are also limited. The scoring functions for RNA–ligand interactions need to be evaluated and redesigned to properly capture the balance between the charged and nonpolar interactions. Computationally efficient implementation of both the ligand and receptor flexibility is the future of the docking studies.

Typically, simulations are performed for isolated RNA complexes without taking into account molecules' natural crowded environment (what would involve simulating multiple molecules). Including crowded conditions requires coarse-grained and/or multiscale models that represent different parts of the system with a different level of detail [70]. Such multiscale approaches will need to be more commonly applied also for RNA complexes. How these crowded conditions influence the dynamics and interactions of RNA with ligands is yet to be investigated.

ACKNOWLEDGMENTS

The authors acknowledge support from ICM University of Warsaw (BST and G31-4), Polish Ministry of Science and Higher Education (N N301 245236 and N N301 033339), and the Foundation for Polish Science—Focus and Team projects (TEAM/2009-3/8 cofinanced by European Regional Development Fund operated within Innovative Economy Operational Programme).

REFERENCES

1. Gesteland, R.F., Cech, T.R., and Atkins, J.F. (eds). 2006. *The RNA World*, 3rd ed. Cold Springs Harbor Laboratory Press: New York.
2. Tor, Y. and Westhof, E. 1998. Deciphering RNA recognition: The hammerhead ribozyme. *Chem. Biol.*, **5**, R277–R283.
3. Hannon, G.J. 2002. RNA interference. *Nature*, **418**, 244–251.
4. Hermann, T. and Westhof, E. 1998. RNA as a drug target: Chemical, modelling, and evolutionary tools. *Curr. Opin. Biotech.*, **9**, 66–73.

5. Tor, Y. 2006. The ribosomal A-site as an inspiration for the design of RNA binders. *Biochimie*, **88**, 1045–1051.

6. Arya, D.P. (ed.) 2007. *Aminoglycoside Antibiotics: From Chemical Biology to Drug Discovery*, 1st ed. Wiley-Interscience: Hoboken, NJ.

7. Schroeder, R., Waldsich, C., and Wank, H. 2000. Modulation of RNA function by aminoglycoside antibiotics. *EMBO J.*, **19**, 1–9.

8. Houghton, J.L., Green, K.D., Chen, W., and Garneau-Tsodikova, S. 2010. The future of aminoglycosides: The end or renaissance? *Chembiochem*, **11**, 880–902.

9. Pan, Y. and MacKerell, A.D. Jr. 2003. Altered structural fluctuations in duplex RNA versus DNA: A conformational switch involving base pair opening. *Nucleic Acids Res.*, **31**, 7131–7140.

10. Leontis, N.B., Stombaugh, J., and Westhof, E. 2002. The non-Watson-Crick base pairs and their associated isostericity matrices. *Nucleic Acids Res.*, **30**, 3497–3531.

11. Leontis, N.B. and Westhof, E. 2001. Geometric nomenclature and classification of RNA base pairs. *RNA*, **7**, 499–512.

12. Kondo, J. and Westhof, E. 2010. Base pairs and pseudo pairs observed in RNA–ligand complexes. *J. Mol. Recognit.*, **23**, 241–252.

13. Batey, R.T., Rambo, R.P., and Doudna, J.A. 1999. Tertiary motifs in RNA structure and folding. *Angew. Chem. Int. Ed. Engl.*, **38**, 2326–2343.

14. Priyakumar, U.D. and MacKerell, A.D. Jr. 2006. Computational approaches for investigating base flipping in oligonucleotides. *Chem. Rev.*, **106**, 489–505.

15. Frank, A.T., Stelzer, A.C., Al-Hashimi, H.M., and Andricioaei, I. 2009. Constructing RNA dynamical ensembles by combining MD and motionally decoupled NMR RDCs: New insights into RNA dynamics and adaptive ligand recognition. *Nucleic Acids Res.*, **37**, 3670–3679.

16. Moitessier, N., Westhof, E., and Hanessian, S. 2006. Docking of aminoglycosides to hydrated and flexible RNA. *J. Med. Chem.*, **49**, 1023–1033.

17. Gait, M.J. and Karn, J. 1993. RNA recognition by the human immunodeficiency virus Tat and Rev proteins. *Trends Biochem. Sci.*, **18**, 255–259.

18. Vicens, Q. and Westhof, E. 2003. RNA as a drug target: The case of aminoglycosides. *Chembiochem*, **4**, 1018–1023.

19. Hermann, T. 2007. Aminoglycoside antibiotics: Old drugs and new therapeutic approaches. *Cell. Mol. Life Sci.*, **64**, 1841–1852.

20. Berman, H.M., Westbrook, J., Feng, Z., Gilliland, G., Bhat, T.N., Weissig, H., Shindyalov, I.N., and Bourne, P.E. 2000. The protein data bank. *Nucleic Acids Res.*, **28**, 235–242.

21. Berman, H.M., Olson, W.K., Beveridge, D.L., Westbrook, J., Gelbin, A., Demeny, T., Hsieh, S.-H., Srinivasan, A.R., and Schneider, B. 1992. The nucleic acid database: A comprehensive relational database of three-dimensional structures of nucleic acids. *Biophys. J.*, **63**, 751–759.

22. Auffinger, P., Bielecki, L., and Westhof, E. 2004. Anion binding to nucleic acids. *Structure*, **12**, 379–388.

23. Rázga, F., Zacharias, M., Réblová, K., Koča, J., and Šponer, J. 2006. RNA kink-turns as molecular elbows: Hydration, cation binding, and large-scale dynamics. *Structure*, **14**, 825–835.

24. Vaiana, A.C., Westhof, E., and Auffinger, P. 2006. A molecular dynamics simulation study of an aminoglycoside/A-site RNA complex: Conformational and hydration patterns. *Biochimie*, **88**, 1061–1073.

25. Chen, S.-Y. and Lin, T.-H. 2010. A molecular dynamics study on binding recognition between several 4,5 and 4,6-linked aminoglycosides with A-site RNA. *J. Mol. Recognit.*, **23**, 423–434.

26. Fulle, S. and Gohlke, H. 2010. Molecular recognition of RNA: Challenges for modelling interactions and plasticity. *J. Mol. Recognit.*, **23**, 220–231.

27. Gouda, H., Kuntz, I.D., Case, D.A., and Kollman, P.A. 2003. Free energy calculations for theophylline binding to an RNA aptamer: Comparison of MM-PBSA and thermodynamic integration methods. *Biopolymers*, **68**, 16–34.

28. Krasovska, M.V., Sefcikova, J., Réblová, K., Schneider, B., Walter, N.G., and Šponer, J. 2006. Cations and hydration in catalytic RNA: Molecular dynamics of the hepatitis delta virus ribozyme. *Biophys. J.*, **91**, 626–638.

29. Hermann, T., Auffinger, P., Scott, W.G., and Westhof, E. 1997. Evidence for a hydroxide ion bridging two magnesium ions at the active site of the hammerhead ribozyme. *Nucleic Acids Res.*, **25**, 3421–3427.

30. Wu, Q., Huang, L., and Zhang, Y. 2009. The structure and function of catalytic RNAs. *Sci. China Ser. C*, **52**, 232–244.

31. McDowell, S.E., Špackova, N., Šponer, J., and Walter, N.G. 2006. Molecular dynamics simulations of RNA: An in silico single molecule approach. *Biopolymers*, **85**, 169–184.

32. Ditzler, M.A., Otyepka, M., Šponer, J., and Walter, N.G. 2010. Molecular dynamics and quantum mechanics of RNA: Conformational and chemical change we can believe in. *Accounts Chem. Res.*, **43**, 40–47.

33. Laing, C. and Schlick, T. 2010. Computational approaches to 3D modeling of RNA. *J. Phys: Condens. Mat.*, **22**, 283101.

34. Šponer, J. and Lankaš, F. (eds). 2006. *Computational Studies of RNA and DNA*. Springer: Dordrecht, Netherlands.

35. Hashem, Y. and Auffinger, P. 2009. A short guide for molecular dynamics simulations of RNA systems. *Methods*, **47**, 187–197.

36. McCammon, J.A., Gelin, B.R., and Karplus, M. 1977. Dynamics of folded proteins. *Nature*, **267**, 585–590.

37. Becker, O.M., MacKerell, A.D. Jr., Roux, B., and Watanabe, M. (eds) 2001. *Computational Biochemistry and Biophysics*, 1st ed. Taylor & Francis: New York.

38. Lafontaine, I. and Lavery, R. 1999. Collective variable modelling of nucleic acids. *Curr. Opin. Struc. Biol.*, **9**, 170–176.

39. Åqvist, J., Wennerström, P., Nervall, M., Bjelic, S., and Brandsdal, B.O. 2004. Molecular dynamics simulations of water and biomolecules with a Monte Carlo constant pressure algorithm. *Chem. Phys. Lett.*, **384**, 288–294.

40. Hu, J., Ma, A., and Dinner, A.R. 2006. Monte Carlo simulations of biomolecules: The MC module in CHARMM. *J. Comp. Chem.*, **27**, 203–216.

41. Leach, A.R. 2001. *Molecular Modelling: Principles and Applications*. Pearson Education Limited: England.

42. Schlick, T. 2006. *Molecular Modeling and Simulation. An Interdisciplinary Guide*, 1st ed. Springer: New York.

43. Wang, J.M., Cieplak, P., and Kollman, P.A. 2000. How well does a restrained electrostatic potential (RESP) model perform in calculating conformational energies of organic and biological molecules? *J. Comp. Chem.*, **21**, 1049–1074.

44. Pérez, A., Marchan, I., Svozil, D., Šponer, J., Cheatham, T.E. III, Laughton, C.A., and Orozco, M. 2007. Refinement of the AMBER force field for nucleic acids: Improving the description of a α/γ conformers. *Biophys. J.*, **92**, 3817–3829.

45. Case, D.A., Darden, T.A., Cheatham, T.E. III, Simmerling, C.L., Wang, J., Duke, R.E., Luo, R., et al. 2010. *Amber 11*. University of California: San Francisco.

46. Foloppe, N. and MacKerell, A.D. Jr. 2000. All-atom empirical force field for nucleic acids: I. Parameter optimization based on small molecule and condensed phase macromolecular target data. *J. Comp. Chem.*, **21**, 86–104.

47. Jorgensen, W.L., Chandrasekhar, J., Madura, J.D., Klein, M.L., and Impey, R.W. 1983. Comparison of simple potential functions for simulating liquid water. *J. Chem. Phys.*, **79**, 926–935.

48. Berendsen, H.J.C., Grigera, J.R., and Straatsma, T.P. 1987. The missing term in effective pair potentials. *J. Phys. Chem.*, **91**, 6269–6271.

49. Réblová, K., Lankaš, F., Rozga, F., Krasovska, M.V., Koča, J., and Šponer, J. 2006. Structure, dynamics, and elasticity of free 16s rRNA helix 44 studied by molecular dynamics simulations. *Biopolymers*, **82**, 504–520.

50. Romanowska, J., Setny, P., and Trylska, J. 2008. Molecular dynamics study of the ribosomal A-site. *J. Phys. Chem. B*, **112**, 15227–15243.

51. Ponder, J.W., Wu, C., Ren, P., Pande, V.S., Chodera, J.D., Schnieders, M.J., Haque, I., et al. 2010. Current status of the AMOEBA polarizable force field. *J. Phys. Chem. B*, **114**, 2549–2564.

52. Warshel, A., Kato, M., and Pisliakov, A.V. 2007. Polarizable force fields: History, test cases, and prospects. *J. Chem. Theory Comput.*, **3**, 2034–2045.

53. Lopes, P.E.M., Roux, B., and Mackerell, A.D. 2009. Molecular modeling and dynamics studies with explicit inclusion of electronic polarizability. Theory and applications. *Theor. Chem. Acc.*, **124**, 11–28.

54. Mládek, A., Sharma, P., Mitra, A., Bhattacharyya, D., Šponer, J., and Šponer, J.E. 2009. Trans Hoogsteen/sugar edge base pairing in RNA. Structures, energies, and stabilities from quantum chemical calculations. *J. Phys. Chem. B*, 113, 1743–5555.

55. Skjaerven, L., Hollup, S.M., and Reuter, N. 2009. Normal mode analysis for proteins. *J. Mol. Struc-Theochem.*, **898**, 42–48.

56. Fulle, S. and Gohlke, H. 2008. Analyzing the flexibility of RNA structures by constraint counting. *Biophys. J.*, **94**, 4202–4219.

57. Karplus, M. and McCammon, J.A. 2002. Molecular dynamics simulations of biomolecules. *Nat. Struct. Biol.*, **9**, 646–652.

58. Allen, M.P. and Tildesley, D.J. 1993. *Computer Simulation of Liquids*. Oxford University Press: New York.

59. Réblová, K., Střelcová, Z., Kulhánek, P., Beššeová, I., Mathews, D.H., Nostrand, K. van Yildirim, I., Turner, D.H., and Šponer, J. 2010. An RNA molecular switch: Intrinsic flexibility of 23S rRNA helices 40 and 68 5′-UAA/5′-GAN internal loops studied by molecular dynamics methods. *J. Chem. Theory Comput.*, **6**, 910–929.

60. Sanbonmatsu, K.Y., Joseph, S., and Tung, C.-S. 2005. Simulating movement of tRNA into the ribosome during decoding. *Proc. Natl. Acad. Sci. USA*, **102**, 15854–15859.

61. Mura, C. and McCammon, J.A. 2008. Molecular dynamics of a κB DNA element: Base flipping via cross-strand intercalative stacking in a microsecond-scale simulation. *Nucleic Acids Res.*, **36**, 4941–4955.

62. Whitford, P.C., Geggier, P., Altman, R.B., Blanchard, S.C., Onuchic, J.N., and Sanbonmatsu, K.Y. 2010. Accommodation of aminoacyl-tRNA into the ribosome involves reversible excursions along multiple pathways. *RNA*, **16**, 1196–1204.

63. Freddolino, P.L., Arkhipov, A.S., Larson, S.B., McPherson, A., and Schulten, K. 2006. Molecular dynamics simulations of the complete satellite tobacco mosaic virus. *Structure*, **14**, 437–49.

64. Fogolari, F., Brigo, A., and Molinari, H. 2002. The Poisson–Boltzmann equation for biomolecular electrostatics: A tool for structural biology. *J. Mol. Recognit.*, **15**, 377–392.

65. Srinivasan, J., Cheatham, T.E., Cieplak, P., Kollman, P.A., and Case, D.A. 1998. Continuum solvent studies of the stability of DNA, RNA, and phosphoramidate–DNA helices. *J. Am. Chem. Soc.*, **120**, 9401–9409.

66. Christen, M. and van Gunsteren, W.F. 2008. On searching in, sampling of, and dynamically moving through conformational space of biomolecular systems: A review. *J. Comput. Chem.*, **29**, 157–166.

67. Elber, R. 2005. Long-timescale simulation methods. *Curr. Opin. Struc. Biol.*, **15**, 151–156.

68. Hamelberg, D., Mongan, J., and McCammon, J.A. 2004. Accelerated molecular dynamics: A promising and efficient simulation method for biomolecules. *J. Chem. Phys.*, **120**, 11919–11929.
69. Markwick, P.R.L., Bouvignies, G., and Blackledge, M. 2007. Exploring multiple timescale motions in protein GB3 using accelerated molecular dynamics and NMR spectroscopy. *J. Am. Chem. Soc.*, **129**, 4724–47230.
70. Trylska, J. 2010. Coarse-grained models to study dynamics of nanoscale biomolecules and their applications to the ribosome. *J. Phys.: Condens. Matter*, **22**, 453101
71. Hansmann, U.H.E., Okamoto, Y., and Eisenmenger, F. 1996. Molecular dynamics, Langevin and hydrid Monte Carlo simulations in a multicanonical ensemble. *Chem. Phys. Lett.*, **259**, 321–330.
72. Mccammon, J.A. and Karplus, M. 1979. Dynamics of activated processes in globular proteins. *Proc. Natl. Acad. Sci. USA*, **76**, 3585–3589.
73. Ermak, D.L. and McCammon, J.A. 1978. Brownian dynamics with hydrodynamic interactions. *J. Chem. Phys.*, **69**, 1352.
74. Hermann, T. and Westhof, E. 1998. Exploration of metal ion binding sites in RNA folds by Brownian-dynamics simulations. *Structure*, **6**, 1303–1314.
75. Smith, L.J., Daura, X., and van Gunsteren, W.F. 2002. Assessing equilibration and convergence in biomolecular simulations. *Proteins*, **48**, 487–496.
76. Lyman, E. and Zuckerman, D.M. 2006. Ensemble-based convergence analysis of biomolecular trajectories. *Biophys. J.*, **91**, 164–172.
77. Sen, S. and Nilsson, L. 2001. MD simulations of homomorphous PNA, DNA, and RNA single strands: Characterization and comparison of conformations and dynamics. *J. Am. Chem. Soc.*, **123**, 7414–74122.
78. Parisien, M., Cruz, J.A., Westhof, E., and Major, F. 2009. New metrics for comparing and assessing discrepancies between RNA 3D structures and models. *RNA*, **15**, 1875–1885.
79. Lu, X.-J. and Olson, W.K. 2008. 3DNA: A versatile, integrated software system for the analysis, rebuilding and visualization of three-dimensional nucleic-acid structures. *Nat. Protoc.*, **3**, 1213–1227.
80. Amadei, A., Linssen, A.B.M., and Berendsen, H.J.C. 1993. Essential dynamics of proteins. *Proteins*, **17**, 412–425.
81. Macke, T.J., Brown, R.A., Bomble, Y.J., Case, D.A., Cheatham, T.E., Wang, J., Cai, Q., Ye, X., Tan, C., Luo, R., et al. 2010. *AmberTools Users' Manual*. Online manual (http://ambermd.org/).
82. Brooks, B.R., Brooks III, C.L., MacKerell, A.D.J., Nilsson, L., Petrella, R.J., Roux, B., Won, Y., Archontis, G., Bartels, C., Boresch, S., et al. 2009. CHARMM: The Biomolecular Simulation Program. *J. Comp. Chem.*, **30**, 1545–1614.
83. van der Spoel, D., Lindahl, E., Hess, B., Kutzner, C., van Buuren, A.R., Apol, E., Meulenhoff, P.J., Tieleman, D.P., Sijbers, A.L.T.M., Feenstra, K.A., et al. 2005. Gromacs User Manual version 4.0.
84. Sharma, M., Bulusu, G. and Mitra, A. 2009. MD simulations of ligand-bound and ligand-free aptamer: molecular level insights into the binding and switching mechanism of the add Ariboswitch. *RNA*, 15, 1673–92.
85. Reddy, M.R. and Erion, M.D. 2001. *Free Energy Calculations in Rational Drug Design*, 1st ed. Kluwer Academic/Plenum Publishers: New York.
86. Deng, Y. and Roux, B. 2009. Computations of standard binding free energies with molecular dynamics simulations. *J. Phys. Chem. B*, **113**, 2234–2246.
87. Kollman, P.A. 1993. Free energy calculations: Applications to chemical and biochemical phenomena. *Chem. Rev.*, **93**, 2395–2417.
88. Kollman, P.A., Massova, I., Reyes, C., Kuhn, B., Huo, S., Chong, L., Lee, M., et al. 2000. Calculating structures and free energies of complex molecules: Combining molecular mechanics and continuum models. *Accounts Chem. Res.*, **33**, 889–897.

89. Ge, X. and Roux, B. 2010. Absolute binding free energy calculations of sparsomycin analogs to the bacterial ribosome. *J. Phys. Chem. B*, **114**, 9525–9539.

90. Ge, X. and Roux, B. 2010. Calculation of the standard binding free energy of sparsomycin to the ribosomal peptidyl-transferase P-site using molecular dynamics simulations with restraining potentials. *J. Mol. Recognit.*, **23**, 128–141.

91. Im, W., Bernèche, S., and Roux, B. 2001. Generalized solvent boundary potential for computer simulations. *J. Chem. Phys.*, **114**, 2924.

92. Kitchen, D.B., Decornez, H., Furr, J.R., and Bajorath, J. 2004. Docking and scoring in virtual screening for drug discovery: Methods and applications. *Nat. Rev. Drug Discov.*, **3**, 935–949.

93. Jones, G., Willett, P., Glen, R.C., Leach, A.R., and Taylor, R. 1997. Development and validation of a genetic algorithm for flexible docking. *J. Mol. Biol.*, **267**, 727–748.

94. Pfeffer, P. and Gohlke, H. 2007. DrugScoreRNA–knowledge-based scoring function to predict RNA–ligand interactions. *J. Chem. Inf. Model.*, **47**, 1868–1876.

95. Fischer, E. 1894. Einfluss der configuration auf die wirkung der enzyme. *Ber. Dtsch. Chem. Ges.*, **27**, 2985.

96. Hubbard, S.J., Campbell, S.F., and Thornton, J.M. 1991. Molecular recognition. Conformational analysis of limited proteolytic sites and serine proteinase protein inhibitors. *J. Mol. Biol.*, **220**, 507–530.

97. Koshland, D.E. 1958. Application of a theory of enzyme specificity to protein synthesis. *Proc. Natl. Acad. Sci. USA*, **44**, 98–104.

98. Henzler-Wildman, K.A., Thai, V., Lei, M., Ott, M., Wolf-Watz, M., Fenn, T., Pozharski, et al. 2007. Intrinsic motions along an enzymatic reaction trajectory. *Nature*, **450**, 838–844.

99. Teague, S.J. 2003. Implications of protein flexibility for drug discovery. *Nat. Rev. Drug Discov.*, **2**, 527–541.

100. Leulliot, N. and Varani, G. 2001. Current topics in RNA-protein recognition: Control of specificity and biological function through induced fit and conformational capture. *Biochemistry*, **40**, 7947–7956.

101. Hermann, T. 2002. Rational ligand design for RNA: The role of static structure and conformational flexibility in target recognition. *Biochimie*, **84**, 869–875.

102. Kuntz, I.D., Blaney, J.M., Oatley, S.J., Langridge, R., and Ferrin, T.E. 1982. A geometric approach to macromolecule-ligand interactions. *J. Mol. Biol.*, **161**, 269–288.

103. Moustakas, D.T., Lang, P.T., Pegg, S., Pettersen, E., Kuntz, I.D., Brooijmans, N., and Rizz, R.C. 2006. Development and validation of a modular, extensible docking program: DOCK 5. *J. Comput. Aided Mol. Des.*, **20**, 601–619.

104. Morris, G.M., Goodsell, D.S., Halliday, R.S., Huey, R., Hart, W.E., Belew, R.K., and Olson, A.J. 1998. Automated docking using a lamarckian genetic algorithm and an empirical binding free energy function. *J. Comp. Chem.*, **19**, 1639–1662.

105. Osterberg, F., Morris, G.M., Sanner, M.F., Olson, A.J., and Goodsell, D.S. 2002. Automated docking to multiple target structures: Incorporation of protein mobility and structural water heterogeneity in AutoDock. *Proteins*, **46**, 34–40.

106. Claussen, H., Buning, C., Rarey, M., and Lengauer, T. 2001. FlexE: Efficient molecular docking considering protein structure variation. *J. Mol. Biol.*, **308**, 377–395.

107. Carlson, H.A. 2002. Protein flexibility and drug design: How to hit a moving target. *Curr. Opin. Chem. Biol.*, **6**, 447–452.

108. Guilbert, C., Perahia, D., and Mouawad, L. 1995. A method to explore transition paths in macromolecules. Applications to hemoglobin and phosphoglycerate kinase. *Comput. Phys. Commun.*, **91**, 263–273.

109. Fourmy, D., Yoshizawa, S., and Puglisi, J.D. 1998. Paromomycin binding induces a local conformational change in the A-site of 16 S rRNA. *J. Mol. Biol.*, **277**, 333–345.

110. Kelley, J.M.M. and Hamelberg, D. 2010. Atomistic basis for the on-off signaling mechanism in SAM-II riboswitch. *Nucleic Acids Res.*, **38**, 1392–1400.
111. Meroueh, S.O. and Mobashery, S. 2007. Conformational transition in the aminoacyl t-RNA site of the bacterial ribosome both in the presence and absence of an aminoglycoside antibiotic. *Chem. Biol. Drug Des.*, **69**, 291–297.
112. Barril, X. and Morley, S.D. 2005. Unveiling the full potential of flexible receptor docking using multiple crystallographic structures. *J. Med. Chem.*, **48**, 4432–4443.
113. Damm, K.L. and Carlson, H.A. 2007. Exploring experimental sources of multiple protein conformations in structure-based drug design. *J. Am. Chem. Soc.*, **129**, 8225–8235.
114. Lin, J.-H., Perryman, A.L., Schames, J.R., and McCammon, J.A. 2002. Computational drug design accommodating receptor flexibility: The relaxed complex scheme. *J. Am. Chem. Soc.*, **124**, 5632–5633.
115. Lin, J.-H., Perryman, A.L., Schames, J.R., and McCammon, J.A. 2003. The relaxed complex method: Accommodating receptor flexibility for drug design with an improved scoring scheme. *Biopolymers*, **68**, 47–62.
116. Amaro, R.E., Baron, R., and McCammon, J.A. 2008. An improved relaxed complex scheme for receptor flexibility in computer-aided drug design. *J. Comput. Aided Mol. Des.*, **22**, 693–705.
117. Jiang, F. and Kim, S.H. 1991. "Soft docking": Matching of molecular surface cubes. *J. Mol. Biol.*, **219**, 79–102.
118. Ferrari, A.M., Wei, B.Q., Costantino, L. and Shoichet, B.K. 2004. Soft docking and multiple receptor conformations in virtual screening. *J. Med. Chem.*, **47**, 5076–5084.
119. Vourekas, A., Stamatopoulou, V., Toumpeki, C., Tsitlaidou, M. and Drainas, D. 2008. Insights into functional modulation of catalytic RNA activity. *IUBMB Life*, **60**, 669–683.
120. Hermann, T. and Westhof, E. 1998. Aminoglycoside binding to the hammerhead ribozyme: A general model for the interaction of cationic antibiotics with RNA. *J. Mol. Biol.*, **276**, 903–912.
121. Mlýnský, V., Banás, P., Hollas, D., Réblová, K., Walter, N.G., Šponer, J. and Otyepka, M. 2010. Extensive molecular dynamics simulations showing that canonical G8 and protonated A38H+ forms are most consistent with crystal structures of hairpin ribozyme. *J. Phys. Chem. B*, **114**, 6642–6652.
122. Priyakumar, U.D. and MacKerell, A.D. Jr. 2010. Role of the adenine ligand on the stabilization of the secondary and tertiary interactions in the adenine riboswitch. *J. Mol. Biol.*, **396**, 1422–1438.
123. Paillart, J.-C., Shehu-Xhilaga, M., Marquet, R. and Mak, J. 2004. Dimerization of retroviral RNA genomes: An inseparable pair. *Nat. Rev. Micriobiol.*, **2**, 461–472.
124. Réblová, K., Fadrna, E., Sarzyńska, J., Kuliński, T., Kulhanek, P., Ennifar, E., Koča, J. and Šponer, J. 2007. Conformations of flanking bases in HIV-1 RNA DIS kissing complexes studied by molecular dynamics. *Biophys. J.*, **93**, 3932–3949.
125. Sarzyńska, J., Réblová, K., Šponer, J. and Kuliński, T. 2008. Conformational transitions of flanking purines in HIV-1 RNA dimerization initiation site kissing complexes studied by CHARMM explicit solvent molecular dynamics. *Biopolymers*, **89**, 732–746.
126. Hermann, T. 2005. Drugs targeting the ribosome. *Curr. Opin. Struc. Biol.*, **15**, 355–366.
127. Poehlsgaard, J. and Douthwaite, S. 2005. The bacterial ribosome as a target for antibiotics. *Nat. Rev. Micriobiol.*, **3**, 870–881.
128. Walsh, C. 2003. *Antibiotics. Actions, Origins, Resistance.* 1st ed. ASM Press: Washington, USA.
129. Sanbonmatsu, K.Y. and Tung, C.-S. 2007. High performance computing in biology: Multimillion atom simulations of nanoscale systems. *J. Struct. Biol.*, **157**, 470–480.

130. Gumbart, J., Trabuco, L.G., Schreiner, E., Villa, E. and Schulten, K. 2009. Regulation of the protein-conducting channel by a bound ribosome. *Structure*, **17**, 1453–1464.

131. Trabuco, L.G., Harrison, C.B., Schreiner, E. and Schulten, K. 2010. Recognition of the regulatory nascent chain TnaC by the ribosome. *Structure*, **18**, 627–637.

132. Trylska, J., Tozzini, V. and McCammon, J.A. 2005. Exploring global motions and correlations in the ribosome. *Biophys. J.*, **89**, 1455–1463.

133. Tama, F., Valle, M., Frank, J. and Brooks, C.L. 2003. Dynamic reorganization of the functionally active ribosome explored by normal mode analysis and cryo-electron microscopy. *Proc. Natl. Acad. Sci. USA*, **100**, 9319–9323.

134. Wang, Y., Rader, A.J., Bahar, I. and Jernigan, R.L. 2004. Global ribosome motions revealed with elastic network model. *J. Struct. Biol.*, **147**, 302–314.

135. Kurkcuoglu, O., Kurkcuoglu, Z., Doruker, P. and Jernigan, R.L. 2009. Collective dynamics of the ribosomal tunnel revealed by elastic network modeling. *Proteins*, **75**, 837–845.

136. Trylska, J. 2009. Simulating activity of the bacterial ribosome. *Q. Rev. Biophys.*, **42**, 301–316.

137. Réblová, K., Rázga, F., Li,W., Gao, H., Frank, J. and Šponer, J. 2010. Dynamics of the base of ribosomal A-site finger revealed by molecular dynamics simulations and Cryo-EM. *Nucleic Acids Res.*, **38**, 1325–1340.

138. Magnet, S. and Blanchard, J.S. 2005. Molecular insights into aminoglycoside action and resistance. *Chem. Rev.*, **105**, 477–497.

139. Vicens, Q. and Westhof, E. 2003. Molecular recognition of aminoglycoside antibiotics by ribosomal RNA and resistance enzymes: An analysis of x-ray crystal structures. *Biopolymers*, **70**, 42–57.

140. François, B., Russell, R.J.M., Murray, J.B., Aboul-ela, F., Masquida, B., Vicens, Q. and Westhof, E. 2005. Crystal structures of complexes between aminoglycosides and decoding A site oligonucleotides: Role of the number of rings and positive charges in the specific binding leading to miscoding. *Nucleic Acids Res.*, **33**, 5677–5690.

141. Hermann, T. 2006. A-site model RNAs. *Biochimie*, **88**, 1021–1026.

142. Carter, A.P., Clemons, W.M., Brodersen, D.E., Morgan-Warren, R.J., Wimberly, B.T. and Ramakrishnan,V. 2000. Functional insights from the structure of the 30S ribosomal subunit and its interactions with antibiotics. *Nature*, **407**, 340–348.

143. Murray, J.B., Meroueh, S.O., Russell, R.J.M., Lentzen, G., Haddad, J. and Mobashery, S. 2006. Interactions of designer antibiotics and the bacterial ribosomal aminoacyl-tRNA site. *Chem. Biol.*, **13**, 129–138.

144. Ogle, J.M., Brodersen, D.E., Clemons, W.M., Tarry, M.J., Carter, A.P. and Ramakrishnan,V. 2001. Recognition of cognate transfer RNA by the 30S ribosomal subunit. *Science*, **292**, 897–902.

145. Shandrick, S., Zhao, Q., Han, Q., Ayida, B.K., Takahashi, M., Winters, G.C., Simonsen, K.B., Vourloumis, D. and Hermann,T. 2004. Monitoring molecular recognition of the ribosomal decoding site. *Angew. Chem. Int. Ed.*, **43**, 3177–3182.

146. Wimberly, B.T., Brodersen, D.E., Clemons, W.M., Morgan-Warren, R.J., Carter, A.P., Vonrhein, C., Hartsch, T. and Ramakrishnan, V. (2000) Structure of the 30S ribosomal subunit. *Nature*, **407**, 327–339.

147. Fan-Minogue, H. and Bedwell, D.M. 2008. Eukaryotic ribosomal RNA determinants of aminoglycoside resistance and their role in translational fidelity. *RNA*, **14**, 148–157.

148. Fourmy, D., Recht, M.I., Blanchard, S.C. and Puglisi, J.D. 1996. Structure of the A site of *Escherichia coli* 16S ribosomal RNA complexed with an aminoglycoside antibiotic. *Science*, **274**, 1367.

149. Sanbonmatsu, K.Y. 2006. Energy landscape of the ribosomal decoding center. *Biochimie*, **88**, 1053–1059.

150. Vaiana, A.C. and Sanbonmatsu, K.Y. 2009. Stochastic gating and drug-ribosome interactions. *J. Mol. Biol.*, **386**, 648–661.

151. Wang, H. and Tor, Y. 1997. Electrostatic interactions in RNA aminoglycosides binding. *J. Am. Chem. Soc.*, **119**, 8734–8735.

152. Długosz, M. and Trylska, J. 2009. Aminoglycoside association pathways with the 30S ribosomal subunit. *J. Phys. Chem. B*, **113**, 7322–7330.

153. Długosz, M., Antosiewicz, J.M. and Trylska, J. 2008. Association of aminoglycosidic antibiotics with the ribosomal A-site studied with Brownian dynamics. *J. Chem. Theory Comput.*, **4**, 549–559.

154. Hobbie, S.N., Bruell, C., Kalapala, S., Akshay, S., Schmidt, S., Pfister, P. and Bottger, E.C. 2006. A genetic model to investigate drug-target interactions at the ribosomal decoding site. *Biochimie*, **88**, 1033–1043.

155. Romanowska, J., McCammon, J.A., and Trylska, J. 2011. Understanding the origins of bacterial resistance to aminoglycosides through molecular dynamics mutational study of the ribosomal A-site. *PLoS Comput. Biol.*, 7(7), e1002099.

156. Yang, G., Trylska, J., Tor, Y. and McCammon, J.A. 2006. Binding of aminoglycosidic antibiotics to the oligonucleotide A-site model and 30S ribosomal subunit: Poisson–Boltzmann model, thermal denaturation, and fluorescence studies. *J. Med. Chem.*, **49**, 5478–5490.

157. Chen, Q., Shafer, R.H. and Kuntz, I.D. 1997. Structure-based discovery of ligands targeted to the RNA double helix. *Biochemistry*, **36**, 11402–11407.

158. Leclerc, F. and Cedergren, R. 1998. Modeling RNA–ligand interactions: The Rev-binding element RNA-aminoglycoside complex. *J. Med. Chem.*, **41**, 175–182.

159. Filikov, A.V., Mohan, V., Vickers, T.A., Griffey, R.H., Cook, P.D., Abagyan, R.A. and James, T.L. 2000. Identification of ligands for RNA targets via structure-based virtual screening: HIV-1 TAR. *J. Comput. Aided Mol. Des.*, **14**, 593–610.

160. Detering, C. and Varani, G. G. 2004. Validation of automated docking programs for docking and database screening against RNA drug targets. *J. Med. Chem.*, **47**, 4188–4201.

161. Morley, S.D. and Afshar, M. 2004. Validation of an empirical RNA–ligand scoring function for fast flexible docking using Ribodock. *J. Comput. Aided Mol. Des.*, **18**, 189–208.

162. Guilbert, C. and James, T.L. 2008. Docking to RNA via root-mean-square-deviation-driven energy minimization with flexible ligands and flexible targets. *J. Chem. Inf. Model.*, **48**, 1257–1268.

163. Lang, P.T., Brozell, S.R., Mukherjee, S., Pettersen, E.F., Meng, E.C., Thomas, V., Rizzo, R.C., Case, D.A., James, T.L. and Kuntz, I.D. 2009. DOCK 6: Combining techniques to model RNA-small molecule complexes. *RNA*, **15**, 1219–1230.

164. Ewing, T.J., Makino, S., Skillman, A.G. and Kuntz, I.D. 2001. DOCK 4.0: Search strategies for automated molecular docking of flexible molecule databases. *J. Comput. Aided Mol. Des.*, **15**, 411–428.

165. Hermann, T. and Westhof, E. 1999. Docking of Cationic antibiotics to negatively charged pockets in RNA folds. *J. Med. Chem.*, **42**, 1250–1261.

166. Lind, K.E., Du, Z., Fujinaga, K., Peterlin, B.M. and James, T.L. 2002. Structure-based computational database screening, *in vitro* assay, and NMR assessment of compounds that target TAR RNA. *Chem. Biol.*, **9**, 185–193.

Index

T - #0719 - 101024 - C16 - 234/156/19 - PB - 9781138382039 - Gloss Lamination